The Future of Cetaceans in a Changing World

Edited by
William C.G. Burns and
Alexander Gillespie

 Transnational Publishers

Published and distributed by *Transnational Publishers, Inc.*
Ardsley Park
Science and Technology Center
410 Saw Mill River Road
Ardsley, NY 10502

Phone: 914-693-5100
Fax: 914-693-4430
E-mail: info@transnationalpubs.com
Web: www.transnationalpubs.com

Library of Congress Cataloging-in-Publication Data

The future of cetaceans in a changing world / William C.B. Burns &
    Alexander Gillespie, editors.
        p. cm.
    ISBN 1-57105-262-3
        1. Cetacea.   2. Endangered species.   3. Wildlife conservation—
    International cooperation.   4. Whaling—Management—International
    cooperation.   I. Burns, William C. B.   II. Gillespie, Alexander

    QL.737.C4F88 2002
    333.95'95—dc21

                                                        2002075684

Manufactured in the United States of America

William C. G. Burns

To Tamar, my wife, who smiled through
this entire process,
and thus made me smile.

Alexander Gillespie

This book is for my brother, Garry,
who gave me my love of the ocean.

Contents

PART I: THE FUTURE OF THE INTERNATIONAL WHALING COMMISSION

Chapter 1

Japan's Position in the International Whaling Commision   3
Yasuo Iino & Dan Goodman

Chapter 2

Culture-Based Conflict in the International Whaling Commission:
The Case of Japanese Small-Type Whaling   33
Milton M.R. Freeman

PART II: THE NORTH ATLANTIC MARINE MAMAL COMMISSION AND THE WORLD COUNCIL OF WHALERS

PART III: THE THREAT TO SMALL CETACEANS AND INSTITUTIONAL RESPONSES

Chapter 8

Chapter 9

PART IV: ANTHROPOGENIC THREATS TO CETACEANS

PART V: THE ECOSYSTEM ROLE OF CETACEANS

APPENDIX

List of Abbreviations

ACCOBAMS	Agreement on the Conservation of Cetaceans of the Black Sea, Mediterranean Sea and the Contiguous Atlantic Area
ASCOBANS	Agreement on the Conservation of Small Cetaceans of the Baltic and North Seas
CCAMLR	Convention on the Conservation of Antarctic Marine Living Resources
CITES	Convention on International Trade in Endangered Species of Wild Fauna and Flora
CMS	Conservation of Migratory Species of Wild Animals
EC	European Community
ECCO	Eastern Caribbean Cetacean Commission
EEZ	exclusive economic zone
ETP	Eastern Tropical Pacific
FAO	Food and Agriculture Organization
GDP	Gross Domestic Product
GLOBEC	Southern Ocean Global Ocean Ecosystems Dynamics program
ICES	International Council for the Exploration of the Seas
ICRW	International Convention for the Regulation of Whaling
IGO	intergovernmental organization
IATTC	Inter-American Tropical Tuna Commission
IOTC	Indian Ocean Tuna Commission
IPCC	Intergovernmental Panel on Climate Change
IUCN	World Conservation Union
IWC	International Whaling Commission
JARPA	Japanese Research Program in the Antarctic
JARPN	Japanese Research Program in the North Pacific
MoU	Memorandum of Understanding
MSY	Maximum Sustainable Yield

NACCMMR	North Atlantic Committee for Coordination of Marine Mammals Research
NAMMCO	North Atlantic Marine Mammal Commission
NGO	non-government organization
NMP	New Management Procedure
PICES	Pacific International Council for the Exploration of the Sea
POP	persistent organic pollutant
RMP	Revised Management Procedure
RMS	Revised Management Scheme
SCANS	Small Cetacean Abundance in the North Sea project
SOWER	Southern Ocean Whale and Ecosystem Research
SWGEC	Standing Working Group on Environmental Concerns
UN/U.N.	United Nations
UNCLOS	United Nations Convention on the Law of the Sea
UNFCCC	United Nations Framework Convention on Climate Change
WCW	World Council of Whalers
WHO	World Health Organization
WTO	World Trade Organization\

Contributors

William Aron

William Aron is an Affiliate Professor with the College of Ocean and Fishery Science at the University of Washington. Previously, he served as Director of the Office of Oceanography and Limnology-Smithsonian (1967–1971); Director of the Office of Ecology and Environmental Conservation, NOAA (1971–1978); Director of the Office of Marine Mammals and Endangered Species, NMFS-NOAA (1978–1980); and Director of the Northwest and Alaska Fisheries Center (later just the Alaska Fisheries Science Center) (1980–1996). He also served on the IWC's Scientific Committee from 1971–1977, was the U.S. Commissioner to the IWC in 1977, and an advisor to the U.S. Commissioner from 1971–1980. His research interests include problems of biological sampling, the importance of ocean change to the abundance and distribution of biota, marine mammal biology, management of marine resources.

William T. Burke

William T. Burke has been a Professor of Law and Marine Affairs, Emeritus, at the University of Washington since 1999. He was previously a Professor of Law at the University of Washington from 1968–1999. His research has focused on coastal state fishery regulations and authority under international law, fishing in international waters, and international law as it relates to whaling and anadromous species. His principal publications include: LAW AND PUBLIC ORDER OF THE OCEANS (with M.S. McDougal) (Yale Press, 1962); THE NEW INTERNATIONAL LAW OF FISHERIES (Oxford Press, 1994); and THE INTERNATIONAL LAW OF THE SEA—DOCUMENTS AND NOTES (Lupus Publications 1999).

William C.G. Burns

Wil Burns is a Visiting Professor in the Departments of Government and Environmental Studies at Colby College in Waterville, Maine. He also serves as the Editor in Chief of the *Journal of International Wildlife Law*

& *Policy* and is Chair of the American Society of International Law's Wildlife Interest Group. His primary research interests are whaling issues, climate change, and the interface of the World Trade Organization and multilateral environmental agreements. He has authored over 40 academic articles in scientific, policy and law publications and co-edited two books.

Robin Churchill

Robin Churchill is professor of law at the Law School, Cardiff University, UK, where he has been a member of staff since 1977. He teaches EC Law and Public International Law. His main research interests are the law of the sea and international environmental law, on both of which he has published widely. His books include The Law of the Sea (with A V Lowe) (third edition, 1999) and EEC Fisheries Law (1987). He has also acted as a consultant/adviser to various environmental and fishermen's organizations.

Michael Donoghue

Mike Donoghue manages relationships with stakeholders involved in marine issues for the New Zealand Department of Conservation. He holds an M.Sc. in Oceanography from Southampton University, and is the Scientific Adviser to New Zealand's Commissioner to the International Whaling Commission. Prior to joining the D.O.C. in 1987, he spent eight years as a self-employed longline fisherman in the Hauraki Gulf, near Auckland. He has published several articles on interactions between marine mammals and fisheries in New Zealand and has provided policy advice to successive Ministers of Conservation on conservation and protection of endemic New Zealand marine mammals, particularly Hector's dolphin and New Zealand (Hooker's) sea lion. As well as leading a research program in Tonga, Mike also co-ordinates the activities of the South Pacific Humpback Whale Consortium, which brings together biologists studying humpback whales in eastern Australia, New Caledonia, Tonga, the Cook Islands, French Polynesia, Chile and Colombia.

Milton R. Freeman

Milton M.R. Freeman is Senior Research Scholar at the Canadian Circumpolar Institute, and Professor Emeritus of Anthropology at the

University of Alberta. Dr. Freeman is an ecologist whose research has focused upon understanding local-level and state-level renewable-resource management systems, and social change and adaptation in maritime societies. He has served on commissions and committees of IUCN, UNESCO, IGBP, International Social Sciences Council, and the International Arctic Science Committee. In 1979, he chaired an IWC expert panel on Aboriginal Subsistence Whaling and for a number of years was a member of the Canadian government delegation to the IWC.

Ray Gambell

Dr Ray Gambell was the first full-time Secretary of the International Whaling Commission from 1976 until his retirement in 2000. From 1963 he was actively engaged in studying the biology and life histories of the great whales, with special emphasis on age determination, reproduction and population assessments of those of commercial importance. Most of his fieldwork was carried out in the Indian Ocean and South Africa, and he was a scientific advisor and member of the UK delegation to the International Whaling Commission from 1965 to 1975. He has been deeply involved in the regulation of the whaling industry world-wide and the administration of international scientific research, with particular interests in revising management policies for both commercial and aboriginal subsistence whaling, and the development of more humane killing techniques. Since retiring, he has been engaged in lecturing and writing on the whaling issue. He has published more than 70 scientific papers and books. He was awarded the OBE (Officer of the Order of the British Empire) in the Queen's 1994 New Year's Honors for his services to the biology and conservation of whales.

Alexander Gillespie

Alexander Gillespie has a LLB & LLM (Hons) from the University of Auckland. He completed his Ph.D at Nottingham university. He has been awarded a Rotary International Ambassadorial Fellowship; A Fulbright Fellowship (which he held at the Law School at Columbia university), and a Fulbright Fellowship which he held at the Bellagio Centre in Italy. He is the author of International Environmental Law, Policy and Ethics (Oxford University Press) and The Illusion of Progress: Unsustainable Development in International Law and Policy (Earthscan). He has been a regular attendee at the International Whaling Commission since 1998. He is currently a senior lecturer at the Law School of the University of Waikato, New Zealand.

Dan Goodman

Dan Goodman is a former Senior Advisor in the Government of Canada's Department of Fisheries and Oceans and represented the Government of Canada at meetings of the IWC from 1979 to 1996. He is currently Councillor at the Institute of Cetacean Research in Tokyo and has attended IWC meetings since 1998 as part of the delegation of Japan.

Grete K. Hovelsrud-Broda

Originally from Norway, and with extensive field experience from the Svalbard Archipelago, she received her Ph.D. in Anthropology in 1997 from Brandeis University. Her doctoral dissertation: *The Seal: Integration of an East Greenlandic Economy*, focused on the socio-economy of sealing households in East Greenland. The study was prompted by the anti-sealing controversy resulting in a trade ban in sealskins. After teaching anthropology at MIT and continuing work on sealing and whaling issues, she became a Research Fellow at the Marine Policy Centre, Woods Hole Oceanographic Institution. From February 1999, she has served as the General Secretary of the North Atlantic Marine Mammal Commission [NAMMCO], at the Commissions Secretariat in Tromsø, Norway.

Yasuo Iino

Chief, Legal Research Center, Institute for Cetacean Research, Tokyo, Japan. Faculty of Law (1990) and Graduate School of Law, LL.M., Kyoto University (1992); Assistant, Graduate School of Law, Kyoto University (1993–1996), Senior Researcher, Legal Research Section (1996) and Chief thereof (1997), Institute of Cetacean Research, Tokyo.

Bruce McKay

Bruce McKay is Founding director of Greenpeace in Montreal, Quebec, Canada, where he was involved in a range of regional environmental problems such as St. Lawrence River pollution, beluga whale protection, and acid rain. He subsequently worked for Greenpeace International for 10 years on primarily marine mammal-related issues including fisheries bycatch, contaminants, and mass mortality events. He is currently Senior Researcher for the Washington, DC-based NGO, SeaWeb, an organization with a focus on marine environment education.

Kieran Mulvaney

Kieran Mulvaney is editor of the monthly SeaWeb newsletter Ocean Update. He was previously the founding director of the Whale and Dolphin Conservation Society, and an oceans campaigner for Greenpeace International in Amsterdam. He is the author of At the Ends of the Earth: A History of the Polar Regions (Island Press, 2001), wrote the main text for the Greenpeace Book of Dolphins (Sterling, 1990) and Witness: Twenty-five Years on the Environmental Front Line (Andre Deutsch, 1996), and contributed chapters to Beyond the Bars: The Zoo Dilemma (Thorson's, 1987), Conservation of Whales and Dolphins: Science and Practice (Wiley, 1996), Whale Watching (Discovery Channel/Insight Guides, 1999) and, with Bruce McKay, Seas at the Millennium: An Environmental Evaluation (Elsevier, 2000). He has written more than 200 articles for such publications as New Scientist, Manchester Guardian, BBC Wildlife, and E Magazine.

Howard Schiffman

Howard S. Schiffman is an Adjunct Professor at the New York University School of Continuing and Professional Studies where he teaches courses in international law and international dispute settlement. He holds a B.A. from Boston University, a J.D. from Suffolk University Law School in Boston and an LL.M. in International and Comparative Law from the George Washington University Law School. He is admitted to practice law before the courts of New York State and Massachusetts (retired status) as well as before the U.S. federal courts including the United States Supreme Court. His legal career began as a staff attorney in the Criminal Defense Division of the Legal Aid Society of New York. He is currently in private practice in New York City where he concentrates in international law. He is the founder and administrator of the online resource www.InternationalLawHelp.com. Schiffman has published numerous articles and contributed to several publications devoted to various aspects of international law. His research interests include international dispute settlement, the law of the sea and marine conservation including fisheries and marine mammals.

Mark Simmonds

Mark P. Simmonds is the Director of Science for the Whale and Dolphin Conservation Society (a charitable organization with its Head Office in

Bath in the UK). He is also a Visiting Research Fellow at the University of Greenwich, in London, where he previously taught, and a member of the Scientific Committee of the International Whaling Commission. He also Chairs the UK's Marine Animal Rescue Coalition. He has authored numerous papers concerning marine wildlife and particularly the problems that they face in the modern world, especially chemical pollution.

Preface

Some of the most vituperative moments in international diplomacy and in the history of multilateral environmental agreements have taken place in the context of cetacean conservation and management issues. If the first few of meetings of the International Whaling Commission at the turn of this century are any guide, it is likely that the struggle will continue apace between those who view cetaceans as a resource that can, and should, be sustainably exploited, and those who believe that commercial whaling should be consigned to the dustbin of history. This book brings together some of the most articulate proponents of both of these positions and also looks to the possible role of other management and conservation regimes beyond the International Whaling Commission.

Additionally, it is critical to explore other anthropogenic threats to cetaceans that may ultimately prove more imperiling than the harpoon. Thus, several chapters focus on the future of cetaceans in a world experiencing dramatic environmental changes.

The first section of the book focuses on some of the most salient and contentious issues faced by the International Whaling Commission. In the opening chapter, Yasuo Iino & Dan Goodman advance the proposition that "the majority of the members of the International Whaling Commission (IWC) no longer agree with the fundamental purpose of the Commission's constitutive treaty," and consequently it "has become dysfunctional and its credibility as an international resource management body is now at serious risk." To support this contention, the authors cite a host of IWC actions in recent years that they contend contravene provisions of the International Convention for the Regulation of Whaling, or are outside its purview. They also argue that if the IWC is to return to its mandated function it is critical that decision making be guided by sound scientific findings and respect for cultural differences. The chapter also outlines Japan's position on several controversial issues, including research whaling, the Revised Management Scheme, sanctuaries, and the casting of secret ballots.

In the second chapter in this section, Professor Milton Freeman advances the cultural case for Japanese small-type coastal whaling and the need for the IWC to reconsider its rejection of harvesting quotas for Japanese coastal whaling communities. Freeman outlines the important role that cetaceans have played in Japanese culture for many centuries, focusing on the contemporary role of ritual whale meat distributions by which residents in coastal communities "collectively reaffirm and validate their sense of community and collective identity by celebrating the seasonal bounty of the sea upon which their own and their community's existence has depended and continues to depend." Dr. Freeman also cautions that a focus by some members of the IWC on cash transactions associated with small-type whaling obscures the critical non-economic aspects of the fishery. To fail to take into account the interests of coastal whaling communities and scientific evidence that Japanese small-type whaling can be conducted sustainably, Freeman concludes, contravenes the International Convention for the Regulation of Whaling.

In the third chapter in this section, Ray Gambell, former Secretary of the International Whaling Commission, provides a historical overview of the evolution of the IWC, chronicling its efforts to confront a legacy of overexploitation of the very species it was entrusted with managing, and the vituperative battles that now ensue between those parties who seek to keep the moratorium on commercial whaling in place, perhaps permanently, and those who believe that the Revised Management Procedure can ensure sustainable harvesting of whales in the future. The author also outlines some of the most controversial issues facing the IWC, including research whaling, the humaneness of whale harvesting methods, whale watching, and the broader question of whether the moratorium should ever be lifted. He concludes that a number of whale stocks could now sustain carefully regulated catches. Additionally, he questions whether non-consumptive activities, such as whale watching, fall under the rubric of the ICRW's jurisdiction. Gambell contends that the very future of the IWC may now hang in the balance as it struggles to reconcile the disparate interests of its members, as well as confront anthropogenic threats to cetaceans beyond harvesting.

In the final chapter in this section, Patricia Birnie focuses on the issue of whether it would be more appropriate in the 21st Century to look to regimes and agreements other than the International Convention for the Regulation of Whaling "to conserve whales and other cetaceans as irreplaceable components of the marine ecosystem." Professor Birnie contends that the provisions of the Convention on Biological Diversity, the Law of the Sea Convention, an array of fishing agreements and the soft law provisions of the documents growing out of the United Nations Convention on Environment and Development are germane to the management and conservation of cetaceans. She suggests that States must effectively cooperate through these bodies to ensure that cetaceans remain a viable part of marine ecosystems. Moreover, she contends that the IWC must learn how to work more closely with other relevant intergovernmental organizations that address cetacean issues, as well as the growing number of organizations concerned with the protection of marine habitats from degradation.

Section II of the book focuses on the North Atlantic Marine Mammal Commission (NAMMCO), a regional mechanism for international cooperation on conservation and management of whales, seals and walruses comprised of the Faroe Islands, Greenland, Iceland and Norway. In the first chapter of this section, Grete Hovelsrud-Broda, the Secretary to NAMMCO outlines the negotiating history and key provisions of NAMMCO. The author emphasizes several rationales for the creation of NAMMCO, including the need for a mechanism to protect the right of coastal states to utilize resources within their EEZs, especially in light of the increased power of the "preservationist" bloc in the International Whaling Commission; the need for a forum for the management and conservation of small cetaceans given the perception of some nations that the IWC is neither "useful or competent" in this context; the need for an organization to management other marine mammals in the region; the need for a multi-species ecosystems approach to the management of the interaction of whales, seals and fish stocks, and the importance of establishing an body that could address the needs of coastal communities and indigenous peoples. Dr. Hovelsrud-Broda argues that to this point NAMMCO has been able to establish sustainable use programs that meet the needs of local communities while being grounded in objective scientific principles and embracing an ecosystems approach.

In the other chapter in this section, Professor Howard Schiffman assesses the legal competence of pro-consumptive international organizations to manage cetaceans, with a focus on NAMMCO and a brief discussion of the World Council of Whalers, a non-governmental organization created to provide a forum for aboriginal and non-aboriginal community-based whaling interests. Citing several provisions of the United Nations Convention on the Law of the Sea (UNCLOS), Schiffman concludes that while NAMMCO's provisions for research and knowledge building would not contravene UNCLOS, the organization's management objectives, if implemented, might ultimately be inconsistent with several living marine resources provisions of UNCLOS. Mr. Schiffman also concludes that should NAMMCO establish an alternative management scheme for cetaceans, parties to both NAMMCO and the ICRW "will need to walk a fine line to reconcile their obligations."

Section III of the book focuses on the threats faced by small cetacean species and national and international responses. In the initial chapter in this section, Kieran Mulvaney and Bruce McKay provide an overview of the most serious threats faced by small cetaceans, the species' status, and international management regimes. The authors initially outline several serious and global threats to small cetacean species, including direct harvesting, some of which is conducted illegally, incidental catch in fisheries, pollution, and environmental change. While concluding that it is difficult to ascertain the status of particular species or populations, or to determine population trends, they conclude that a large number of species appear to be in serious trouble, especially species with limited ranges. The authors also discuss the relevant international regimes for small cetacean conservation and management, including the role of the International Whaling Commission and regional regimes. Ultimately, the authors conclude, effective conservation of small cetaceans will require a combination of regional initiatives and measures to tackle the wide array of human activities that impact small cetaceans directly and indirectly, many of which take place in less developed countries where even strict national legislation has proved to be unavailing.

In the second chapter in this section, Alexander Gillespie addresses the competence of the International Whaling Commission over small cetaceans, as well as the rights of coastal states under the United Nations

Convention on the Law of the Sea. On the basis of an analysis of the language of the ICRW and associated documents, including the Schedule and the agreement's "Nomenclature of Whales," Dr. Gillespie argues that the IWC has legal competence to manage small cetaceans. Moreover, he contends that while UNCLOS imbues coastal states with broad management authority within their respective EEZs, several provisions, including Articles 64 and 65, clearly require coastal states to protect cetaceans from extinction and to cooperate with appropriate international organizations. In addressing the latter issue, Dr. Gillespie contends that under UNCLOS regional regimes, such as NAMMCO, must play a complementary and not opposing role to the International Whaling Commission.

In the final chapter in this section, Robin Churchill discusses one of the two regional cetacean regimes created under the Bonn Convention in the 1990s, the Agreement on the Conservation of Small Cetaceans of the Baltic and North Seas (ASCOBANS). After outlining the history of ASCOBANS and its primary provisions, Professor Churchill notes some of the regime's problems, including the decision to couch the conservation measures in the treaty's Conservation and Management plan in "vague, flexible and hortatory rather than prescriptive language, a demonstrated lack of resolve to address bycatch issues, and weak mechanisms for ensuring party implementation of and compliance with treaty provisions. In assessing the role of ASCOBANS vis-à-vis other relevant regimes, the author concludes that there is an independent role for the regime in addressing several threats to cetaceans, including seismic testing, military activities, whale watching and incidental catch in fishing operations. In assessing the effectiveness of ASCOBANS to date, Professor Churchill concludes that participation levels by States in the region remains low and while its institutions have functioned reasonably well, there is room for improvement, including providing more resources to the regime's Advisory Committee and improving monitoring and reporting provisions. Given the relatively brief life of the regime, it is impossible, Professor Churchill concludes, to assess what extent ASCOBANS has affected cetacean population levels in the region.

The fourth section of the book focuses on two threats that fall under the rubric of what could prove to be the gravest long-term threat to cetacean species: the impacts of environmental change caused by anthropogenic activities. In the first chapter in this section, Mark Simmonds seeks to assess the possible impacts of pollution on cetacean species. Simmonds outlines current and emerging risk assessment procedures for evaluating the impacts of pollutants that may adversely affect cetacean populations, such as organochlorines and heavy metals. Concluding that it is doubtful that we can currently propose safe thresholds of contamination for cetaceans, Dr. Simmonds advocates the adoption of a precautionary concentration threshold that would result in many cetacean species being classified as at risk.

In the other chapter in this section, William Burns examines the possible impacts of climate change on cetaceans during this century and beyond. The author concludes that diminution of ice cover associated with rising temperatures may seriously denude cetacean prey resources in the Antarctic and Arctic, imperiling the populations of cetacean species in the region, some of which are already at very low levels due to over-exploitation by the commercial whaling industry. Additionally, Professor Burns outlines the threats that climate change poses to cetaceans in other regions in the world, including in the Mediterranean. Moving to the realm of policy, Burns also chronicles the IWC's response to climate change to date, concluding that it is devoting woefully limited resources to environmental change research, and is unlikely to substantially expand its efforts in the future. Burns concludes that, ultimately, the IWC's most judicious approach would be to serve as an advocate for the interests of cetaceans in other forums that might more effectively address the threat of climate change.

In the final section, Michael Donoghue addresses the scientific validity of the claim advanced by Japan and others at the IWC in recent years that whales consume large amounts of fish that could be consumed by humans, and thus culling stocks would be conducive to promoting a marine ecosystem management approach. Mr. Donoghue questions the scientific basis of a paper by two Japanese researchers that includes the often-quoted assertion that whales consume six times as much fish as those harvested in the world's fisheries. He contends that the estimates of whale populations in the paper has been assailed by several members of the IWC's Scientific Committee as speculative or outdated. Moreover, many whale species rarely, if ever, eat fish, and in many cases the species consumed by cetaceans are not of commercial value. Finally, the author argues that laboratory-based research on ecological energetics, used to estimate rates of prey consumption, remains subject to question. Mr. Donoghue concludes that the substantial decline in fish catches over the past few years is largely attributable to anthropogenic activities, including the over-capacity of the world's fishing fleets, pollution, and climate change.

The editors of this work have more than three decades of collective experience on the issues addressed in this book. Their involvement has ranged from academic writings, to active participation on national delegations, at both general and special meetings of the International Whaling Commission, as well as participation in regional regimes. We harbor no illusions that this book will "resolve" the issues that are set forth here; however, it is our hope that it will contribute to the colloquy, and hopefully lead to more understanding and cooperation in the future.

William C.G. Burns, Waterville, Maine, USA
Alexander Gillespie, Hamilton, New Zealand

PART I

The Future of the International Whaling Commission

Chapter 1
Japan's Position in the International Whaling Commission

Yasuo Iino & Dan Goodman[1]

1. INTRODUCTION

It is Japan's view that the majority of the members of the International Whaling Commission (IWC) no longer agree with the fundamental purpose of the Commission's constitutive treaty, the 1946 International Convention for the Regulation of Whaling (ICRW)[2] and that many of the actions and decisions of this majority subvert the intent of the Convention. Consequently, the IWC has become dysfunctional and its credibility as an international resource management body is now at serious risk.

Understanding Japan's policy in the IWC requires an examination of contextual factors. These include the current situation in the IWC, the provisions of the ICRW, customary international law related to treaty interpretation, Japan's dependency on ocean resources and its food security issues, Japan's overall fisheries policies, Japan's cultural and traditional use of whales, and finally, Japan's commitment to work within the IWC to restore its proper function as mandated by the ICRW.

2. THE CONTEXT

2.1. The International Convention for the Regulation of Whaling

The purpose of the ICRW as set out in the last paragraph of its Preamble is ". . . to provide for the proper conservation of whale stocks and thus

---

[1] The Institute of Cetacean Research, Tokyo.

[2] International Convention for the Regulation of Whaling, Dec. 2, 1946, ANNUAL REPORT OF THE INTERNATIONAL WHALING COMMISSION, at 73 (1999) [hereafter ICRW].

make possible the orderly development of the whaling industry." Thus, the ICRW was one of the first resource management conventions to embody what is now referred to as the principle of sustainable use. Moreover, while it has been argued by opponents of commercial whaling that the purpose of the Convention can be separated into two parts, that is, "conservation," and "the orderly development of the whaling industry,"[3] interpretation of the Convention's purpose using the plain and ordinary meaning of the language clearly provides that it is one, indivisible objective.[4]

In order to meet its objective, Article III of the ICRW establishes the IWC and Article V(1) empowers it to adopt "regulations with respect to the conservation and utilization of whale resources." Article V(2) however, provides that these regulations:

> (a) shall be such as are necessary to carry out the objectives and purposes of this Convention and to provide for the conservation, development and optimum utilization of the whale resources; (b) shall be based on scientific findings; (c) shall not involve restrictions on the number or nationality of factory ship or land station . . . ; (d) shall take into consideration the interests of the consumers of whale products and the whaling industry.

It is in large measure the failure by the majority of IWC members to comply with this article of the Convention in good faith that has caused the current dysfunctional situation in the IWC.

2.2.   The International Whaling Commission

Aron *et al.* have also referred to the IWC as dysfunctional and conclude that "the bitter standoff [within the IWC] violates international law, fosters tensions between otherwise friendly nations, and undermines

---

3   Patricia Birnie, *Are Twentieth-Century Marine Conservation Conventions Adaptable to Twenty First Century Goals and Principles?: Part II*, 12(4) INT'L J. MARINE & COASTAL L. 488, 491 (1997).

4   William T. Burke, *The Legal Invalidity of the IWC Designation of the Southern Ocean Sanctuary*. IWC Doc. IWC/50/27, at 4 (1998).

environmental legislation"[5] and that this sets a bad precedent for the negotiation of future multilateral environmental agreements. Their characterization of the current situation in the IWC includes strong accusations, such as "alleged data-collection problems," "deliberately dragged out negotiations," "no room for good-faith negotiations" and a proclivity to ignore or trivialize scientific research.[6]

There is abundant evidence in the verbatim records of the IWC's annual meeting to support Aron *et al.*'s portrayal of the IWC as a wholly dysfunctional institution, but the best evidence to support their portrayal of the IWC is the following. The primary task of the IWC is to adopt regulations related to commercial whaling. Yet since 1981, the IWC has adopted only two such regulations, the moratorium on commercial whaling[7] and the Southern Ocean Sanctuary,[8] both of which are illegal because they fail to meet the conditions of Article V of the Convention. During the same period, the IWC adopted 137 non-binding resolutions, the majority of which are related to matters outside the mandate of the IWC as provided by the ICRW. The failure of the IWC to implement the conservative, risk-averse quota setting scheme, the Revised Management Procedure (RMP)[9] recommended by its Scientific Committee in 1992, is further evidence of the dysfunctional nature of the IWC.

In short, the IWC is an anomaly among resource management organizations. It is dysfunctional because the majority of its members no longer support the purpose of the Convention, its decisions are made on the basis of emotional and moral judgments rather than scientific evidence,[10] and it is becoming increasingly irrelevant since most whaling,

---

5   William Aron, William T. Burke & Milton M. R. Freeman, *Flouting the Convention*, 283(5) ATLANTIC MONTHLY 22, 24 (May 1999). An expanded version of this paper by the same authors is: *The Whaling Issue*, 24 MARINE POL'Y 179 (2000).

6   *Id.* at 24 & 26.

7   ICRW, *supra* note 2, at 84.

8   *Id.* at 81.

9   44 REP. INT'L WHALING COMMISSION 145.

10   Aron *et al., supra* note 5, at 24. "Conceding that no scientific reason exists to ban all whaling, the U.S. commissioner announced in 1991 that he would defend the U.S. position on ethical grounds."

even that conducted by IWC members, occurs outside of its control.[11]

The credibility of the IWC as a resource management organization has been recently challenged by other international organizations including the World Conservation Union (IUCN) and the Secretary General for the Convention on International Trade in Endangered Species of Wild Fauna and Flora (CITES).[12] In its opening statements[13] to the annual meetings of the IWC, the IUCN has urged the IWC to complete the Revised Management Scheme (RMS) and the Secretary General for CITES stated in his letter to the Chairman of the IWC that "the IWC should soon make important progress towards the adoption of a Revised Management Scheme."[14]

Japan's concerns regarding the dysfunctional nature of the IWC and its loss of credibility have been expressed in its opening statements to the annual meetings of the Commission.[15] These statements have urged the Commission to return to its primary task as outlined in the ICRW. Japan's policy towards the IWC is therefore best understood as a strategic effort to encourage the IWC to function in accordance with the objectives and provisions of the ICRW as well as relevant principles regarding conservation and management of marine living resources, such as sustainable use based on scientific evidence. In response to recent scientific evidence concerning the interaction between cetaceans and fisheries,[16] Japan is also encouraging the IWC to adopt an ecosystem approach to the management of marine resources.

---

[11]   Notwithstanding any regulations promulgated by the IWC, Norway's commercial whaling operations and Japan's research programs are legal under the ICRW: for the former under its objection to the commercial moratorium under Article V(3) and for the latter in accordance with Article VIII.

[12]   Convention on International Trade in Endangered Species of Wild Fauna and Flora, Mar. 3, 1973, 27 U.S.T. 1087, T.I.A.S. No. 8249, 993 U.N.T.S. 243, ELR Stat. 40336.

[13]   IWC Doc. IWC/52/OS/IUCN, IWC/51/OS/IUCN and IWC/50/OS/IUCN.

[14]   A letter dated July 4, 2000 from Secretary General for the CITES to Chairman of the IWC. CITES reference: WWW/MAL/pjh.

[15]   IWC Doc. IWC/52/OS Japan, IWC/51/OS Japan and IWC/50/OS Japan.

[16]   Tsutomu Tamura & Seiji Ohsumi, *Regional Assessments of Prey Consumption by Marine Cetaceans in the World*, IWC Doc. SC/52/E6 (2000).

2.3.   Whales as Food, Japanese Culture, and Food Security

From a Japanese perspective, whales are valued marine resources that should continue to be used as a source of food, providing that the stocks are robust enough to support sustainable whaling. Japan has long looked to the oceans to supply its animal protein given its limited grazing land and adherence to Buddhist teachings that prohibit eating land mammals.[17]

The Japanese have been eating whale meat and utilizing whale bones, blubber and oil for more than 2,000 years, and large cetaceans have been hunted for more than 400 years.[18] Offshore whaling activities began after the Meiji Restoration in 1868, however some whaling continued within the traditional areas where it still plays an important role in engendering community solidarity. In the years immediately following World War II (1947–1949), whale meat constituted approximately 45 percent of the total meat consumption in Japan, after which it remained at about 30 percent. Since the 1960s, the supply of whale meat has gradually declined and consumption has been reduced accordingly.[19] However, whale meat remains a significant aspect of Japanese dietary habits. More than 30 scientific papers on this subject have been presented by Japan to the IWC.[20]

Whale meat has been not only a source of protein in the every day diets of the Japanese people, but has also been treated as a special food with regional and social significance. In the areas where whaling has

---

[17]   TOMOYA AKIMICHI et al., SMALL-TYPE COASTAL WHALING IN JAPAN 1 (1988). Japan ranks second to last among OECD nations in terms of self-sufficiency of food resources. Only 40 percent of the caloric intake of the Japanese people is derived from indigenous sources, compared to 141 percent for France, 100 percent for Germany, and 78 percent for the United Kingdom, <http://www.kanbou. maff.go.jp/www/anpo/sub611.htm> [in Japanese].

[18]   ARNE KALLAND & BRIAN MOERAN, JAPANESE WHALING 66 (1992).

[19]   JAPAN WHALING ASSOCIATION, HOGEI TO NIHON KOKUMINKEIZAI TONOKANREN NI KANSURU KOSATSU [WHALING AND NATIONAL ECONOMICS OF JAPAN] 33 (1980) [in Japanese].

[20]   GOVERNMENT OF JAPAN, PAPERS ON JAPANESE SMALL-TYPE COASTAL WHALING SUBMITTED BY THE GOVERNMENT OF JAPAN TO THE INTERNATIONAL WHALING COMMISSION 1986–1996 (1997) [hereafter JAPANESE SMALL-TYPE COASTAL WHALING].

been conducted traditionally, these dietary habits have become an integral part of community life, such that all local ceremonies or festivities include the serving of some whale meat dishes.[21]

The total protection of all whales irrespective of their stock status, as promoted by some members of the IWC and some environmental and animal welfare organizations, is contradictory to Japanese cultural values where whale meat is still eaten and where whales are still revered through religious ceremonies and festivals. This is particularly so for those communities where the local people's lives have depended on whaling activities.

In December 1995, 95 states agreed to a Declaration and Plan of Action on the occasion of the International Conference on the Sustainable Contribution of Fisheries to Food Security hosted by the government of Japan with technical assistance from the Food and Agriculture Organization (FAO). In paragraph 6, the Declaration "Call[s] for an increase in the respect and understanding of social, economic and cultural differences among States and regions in the use of living resources, especially cultural diversity in dietary habits, consistent with management objectives."[22] Sustainable whaling and the consumption of whale meat in Japan are fully consistent with this Declaration. Japan's position at the IWC therefore also includes a request for the respect of cultural diversity.

3. THE BASIS OF JAPAN'S POSITION IN THE IWC

Many commentators question why the Japanese are so adamant about the right to whale when whaling does not contribute significantly to GDP and whale meat is no longer a significant part of the diet of most Japanese. The short answer is that it is a matter of principles. The most important of these principles are that marine resources should be managed according to scientific findings; that signatories to the International

---

[21]   Lenore Manderson & Haruko Akatsu, *Whale Meat in the Diet of Ayukawa Villagers*, 30 ECOLOGY FOOD & NUTRITION 207, 216 (1993).

[22]   The Kyoto Declaration and Plan of Action on the Sustainable Contribution of Fisheries to Food Security, available at <http://www.fao.org/fi/agreem/kyoto/hlf.asp> [hereafter Kyoto Declaration].

Convention for the Regulation of Whaling should carry out their treaty obligations in good faith; and that respect for cultural differences is an important aspect of international cooperation. It is the view of Japan that by ignoring these principles, the majority of the parties to the IWC have imperiled the world community's ability to responsibly manage all our marine resources and that returning the IWC to its mandated function will have benefits beyond those related to the proper management of whale resources.

3.1.   The Need for Science-Based Decision Making in Marine Resource Management

In 1999, the government of Japan announced *The Framework for Fisheries Policy Reform*.[23] The new policy accorded priority to the sustainable use of fishery resources through appropriate conservation and management measures. It also reaffirmed a commitment to scientific research as the basis for management and emphasized the importance of ecosystem conservation and management of marine resources.

It should be no surprise that the fundamental basis of Japan's fisheries policy is that management decisions should be based primarily on scientific findings. This principle is reflected in the United Nations Convention on the Law of the Sea (UNCLOS),[24] Agenda 21,[25] the Kyoto Declaration,[26] the FAO's Code of Conduct for Responsible Fisheries[27] and, as described above, in Article V of the ICRW. However, in the context of the IWC, the imposition of the moratorium on commercial whaling and establishment of the Southern Ocean Sanctuary were not based

---

[23]   GOVERNMENT OF JAPAN, THE FRAMEWORK FOR FISHERIES POLICY REFORM (1999) [in Japanese].

[24]   United Nations Convention on the Law of the Sea, Dec. 10, 1982, arts. 61 & 119, 21 I.L.M. 1261, 1281 & 1219 (hereinafter UNCLOS).

[25]   Agenda 21, June 14, 1992, para. 17.56, available at <http://www.un.org/esa/sustdev/agenda21text.htm>.

[26]   Kyoto Declaration, *supra* note 22, at ¶ 9.

[27]   Code of Conduct for Responsible Fisheries, Oct. 31 1995, arts. 6.4, 6.5, 7.1.1, 7.2.1, 7.2.2, 7.3.1, 7.4.1, 7.5.3 and 7.5.5, available at <http://www.fao.org/fi/agreem/codecond/ficonde.asp> [hereafter FAO's Code of Conduct].

on scientific findings. The IWC's Scientific Committee has never found that either the moratorium or the Southern Ocean Sanctuary were necessary to further the ICRW's conservation objectives. Thus, Japan has filed objections to the moratorium[28] and the Southern Ocean Sanctuary.[29] For the same reason, Japan is opposed to the proposed sanctuary for the south Pacific, as advocated by Australia and New Zealand at the 51st and 52nd Annual Meetings of the IWC in 1999 and 2000. The proposed sanctuary has no scientific basis under Article V of the ICRW because it would provide no additional protection to whales in the area and is not required for conservation purposes. As a result, it has not been recommended by the IWC's Scientific Committee.

Japan has been conducting whale research programs in the Antarctic and in the western North Pacific Ocean. These research programs are authorized by the government of Japan under the terms of Article VIII of the ICRW and have been making a valuable contribution to the improvement of management under the RMP (see Section 4.1).

The sustainable use of marine living resources is a broadly accepted world standard, and a principle now reflected in many international instruments, including the UNCLOS,[30] Agenda 21,[31] the Kyoto Declaration[32] and the FAO's Code of Conduct.[33] Thus, sustainable use is a goal that is shared among members of the international community and under which international organizations for fishery management are established and operated.

---

[28]   Japan had filed an objection to paragraph 10(e) but withdrew it when the U.S. threatened to deny it a fishery quota within its EEZ.

[29]   Japan filed and maintains an objection to paragraph 7(b) to the extent that it applies to the Antarctic minke whale stocks. At the 52nd annual meeting, Japan tabled a proposal (IWC Doc. IWC/52/25) to make the nature of the sanctuary valid under the ICRW.

[30]   UNCLOS, *supra* note 24, arts. 64 & 119, 21 I.L.M. 1261 & 1291.

[31]   Agenda 21, *supra* note 25, Section C. Sustainable use and conservation of marine living resources of the high seas.

[32]   Kyoto Declaration, Preamble. *supra* note 22.

[33]   FAO's Code of Conduct, *supra* note 27, Preface and art. 7.1.1.

Regrettably, the early days of the IWC was a history of failures in managing whaling and whale resources and resulted in serious depletion of stocks of large whale species. The current situation is, however, quite different. The "Save the Whale" slogan that continues to raise hundreds of millions of dollars for anti-whaling organizations is now irrelevant. With the exception of perhaps the north Atlantic right whale that is subject to mortality from ship strikes and entanglement in fishing gear, whales have been saved. Historical over-harvesting is no longer a valid argument against the resumption of whaling now that the IWC has developed the Revised Management Procedure. The RMP, which was completed by the IWC Scientific Committee in 1992 and accepted by the Commission in 1994,[34] is designed to be risk-averse by using only two types of data, abundance estimates and catch history. This is in contrast with predecessor, the New Management Procedure (NMP) which required information from biological parameters which are sometimes difficult to estimate. When used for the purposes of establishing catch quotas, the RMP will only provide quotas for abundant stocks, enabling the IWC to achieve and promote the objective of the ICRW.

3.2. The Need for Ecosystem Management of Marine Mammal Resources

Competition between marine mammals and fisheries is now a serious concern for nations dependent on fisheries, as well as for a number of global and regional fisheries management organizations, including the FAO, which have urged the development and implementation of ecosystem approaches to the management of marine resources. Agenda 21,[35]

---

[34]   In 1992 the Scientific Committee recommended formal adoption of the RMP by the Commission. However, the Commission rejected a resolution proposing it in 1993. As a result, the Chairman of the Scientific Committee, Dr. Hammond, resigned. His letter of resignation (attached to IWC circular communication RG/VJH/19814) said:

> . . . what is the point of having a Scientific Committee if its unanimous recommendations on a matter of primary importance are treated with such contempt? . . . I can longer justify to myself being the organiser of and spokesman for a Committee whose work is held in such disregard by the body to which it is responsible.

[35]   Agenda 21, *supra* note 25, para. 17.46.

the Kyoto Declaration[36] and the Rome Declaration[37] all call for the promotion of an ecosystem approach to the management of marine resources. The following instances illustrate where organizations are seeking to apply this approach: the report[38] and resolution[39] adopted by the Indian Ocean Tuna Commission (IOTC) at its fifth session, and the continuing studies of the role of marine mammals in the marine ecosystem conducted by the North Atlantic Marine Mammal Commission (NAMMCO)[40] and the PICES.[41] Others include the agreement at the February-March 2000 meeting of FAO's Committee on Fisheries that ecosystem-based fisheries management studies particularly focusing on interactions between marine mammals and fisheries should be conducted. Furthermore, the resolution should be adopted by the Third World Fisheries Congress, which endorsed further research and other initiatives in support of the development of multi-species management approaches to managing marine resources.

It has been recently estimated that cetaceans consume three to five times the amount of marine resources harvested for human consumption.[42] Further, the FAO reported in 1997 that approximately 69 percent

---

[36]   Kyoto Declaration, *supra* note 22, at ¶¶ 12, 13 & 14.

[37]   The Rome Declaration on the Implementation of the Code of Conduct for Responsible Fisheries, Mar. 11, 1999, Preamble and para. (c), available at <http://www.fao.org/fi/agreem/declar/dece.asp>.

[38]   Report of the Fifth Session of the Indian Ocean Tuna Commission. IOTC/S/05/00/R[E] 13 (2001).

[39]   *Id.* at 68.

[40]   NAMMCO ANNUAL REPORT 1999, 129 (2000).

[41]   PREDATION BY MARINE BIRDS AND MAMMALS IN THE SUBARCTIC NORTH PACIFIC OCEAN, 14 PICES SCIENTIFIC REPORT, at 26 (George L. Hunt, Jr., Hidehiro Kato & Stewart M. McKinnel eds., 2000). "The total amount of food consumption by marine mammals in the western North Pacific Ocean during summer is estimated about 13,020,000 mt. It should be noted that this estimation must be "a great under-representation and should be understood as a minimum value of the total food consumption."

[42]   Tamura & Ohsumi, *supra* note 16. Those estimates are based on 37 out of 75 species of marine cetaceans in the world and therefore would be reasonably considered to be an underestimate.

of the major world fish resources are either mature or senescent and that these resources are in urgent need of management action to arrest the increase in fishing capacity or to rehabilitate damaged resources.[43] Given this situation, FAO also called for a 30 percent reduction in the world's fishing effort. Japan responded to this call by scrapping 132 tuna long line vessels, which reduced its fleet by 20 percent and ensured that these vessels could not reflag to elude this regulation. However, given present scientific information on the consumption of fish by marine mammals, it is clear that without effective management of cetaceans, efforts to control fishing capacity might be diminished. In summary, the principles of science-based management of resources and sustainable use of resources together with respect for cultural diversity are accepted as standards worldwide and are a prerequisite to cooperation in the international management of ocean resources. The position of those opposed to whaling, irrespective of the status of stocks, and ignoring cultural diversity is contrary to these widely accepted principles that have now been enshrined in numerous international agreements, covenants and other documents.

3.3.　Respect for the ICRW and International Law

It is a fundamental obligation that states under customary international law exercise good faith vis-à-vis treaties into which they have entered (*pacta sunt servanda*). Article 31 of The Vienna Convention on the Law of Treaties[44] reflects customary international law on treaty interpretation, and is thus germane to determining the parameters of the rights and obligations of parties to the ICRW. This article, in part, reads: "A treaty shall be interpreted in good faith in accordance with the ordinary meaning to be given to the terms of the treaty in their context and in the light of its object and purpose."

The Vienna Convention recognizes the special role of the "object and purpose" as irreplaceable guidelines and standards of implementation

---

[43]　FAO, *Review of the State of World Fishery Resources: Marine Fisheries*, at 7 (1997).

[44]　Vienna Convention on the Law of Treaties, May 23, 1969, art. 26, 8 I.L.M. 679, 690 [hereafter Vienna Convention].

and application of a treaty. In addition to the provisions of Article 31, Article 60 makes a relevant reference to the object and purpose, providing that "the violation of a provision essential to the accomplishment of the object or purpose of the treaty" constitutes a material breach of the treaty and may warrant termination or suspension of the operation of a treaty. Members of the IWC therefore have an obligation to promote, and refrain from defeating, the object and purpose of the ICRW, which is "to provide for the proper conservation of whale stocks and thus make possible the orderly development of the whaling industry."[45] Action of a member state violating this object and purpose constitutes "a material breach" and may induce termination or suspension of the operation of the ICRW. When such a violation has been made by a decision of the IWC itself, the action in question is legally invalid and therefore void.

It is Japan's position that both the IWC's decision to establish the Southern Ocean Sanctuary and the continuation of the moratorium in the light of IWC's acceptance of the RMP and compelling scientific evidence of the abundance of some stocks, are both legally invalid decisions because they contravene the object and purpose of the ICRW and fail to meet the clearly specified requirements under Article V(2).

However, the majority of the IWC members, relying primarily on the views of Professor Patricia Birnie, seek to justify the decisions on the Southern Ocean Sanctuary by arguing that because of the absence of dispute settlement provisions in the ICRW "the Commission's decision on the Sanctuary, taken through use of the normal voting procedures laid down in Article V, is determinative."[46] Clearly, this is in contradiction with the principles of customary international law on interpretation of treaties described above, with the absurd inference that the IWC may decide what it likes irrespective of the objectives and purposes and the provisions of the ICRW.

Legal analyses reject these arguments. First, the principles of interpretation "confirm both that the Sanctuary decision is inconsistent with

---

45   ICRW, *supra* note 2, at Preamble.

46   Patricia Birnie, *Opinion on the Legality of the Designation of the Southern Ocean Whale Sanctuary by the International Whaling Commission*, IWC Doc. IWC/47/41 (1995).

the ICRW and that it amounts to amending the Convention by interpretation, an action which is invalid."[47] Such amendment to the Convention must be sought by revision of the ICRW text.

Second, Professor Birnie referred to "the principle of effectiveness" and "an 'effective' and an 'evolutionary' approach to interpretation"[48] of treaties in supporting the Southern Ocean Sanctuary decision. Such efforts at interpretation must aim "to promote fulfillment of a treaty's major purpose by interpreting provisions in ways that facilitate that purpose."[49] Therefore, this principle does not support "an interpretation of the ICRW that legitimizes the choice of the Southern Ocean Sanctuary," which clearly contradicts to the objectives of the ICRW, "as a means of reaching treaty objectives."[50]

The third point relates to the lack of dispute settlement procedures under the ICRW. There is no logical relation between the lack of such procedures and the view of Professor Birnie and the IWC's majority that "the Commission's decision on the Sanctuary . . . is determinative"[51] and that, therefore, "the Commission itself is the body that can reverse the decision."[52] The absence of compulsory judicial review "does not entail the consequence that *ultra vires* acts cannot be committed by international organizations or that there is no doctrine of *ultra vires* applicable to their act."[53]

In sum, the ICRW, as well as the principles of international law, do not support the view that the IWC may make decisions which contravene the objectives and purposes of the ICRW. The objective and purpose of a treaty provide the final measure to judge the legality of actions of member states as well as those of the IWC.[54] From this perspective,

---

[47]   Burke, *supra* note 4, at 11.

[48]   Birnie, *supra* note 46, at 2.

[49]   Burke, *supra* note 4, at 5.

[50]   *Id.*

[51]   Birnie, *supra* note 46, at 2.

[52]   47 REP. INT'L WHALING COMMISSION 36 (1997).

[53]   CHITTHARANJAN FELIX AMERASINGHE, PRINCIPLES OF THE INSTITUTIONAL LAW OF INTERNATIONAL ORGANIZATIONS 166 (1996).

[54]   Vienna Convention, *supra* note 44, at arts. 31.1, 33.4, 44(a)(ii), 60.3(b).

the IWC is facing critical legal problems, especially relating to its decisions on the establishment of the Southern Ocean Sanctuary and maintaining the moratorium on commercial whaling.

4.   MAJOR ASPECTS OF JAPAN'S POSITION IN THE IWC

4.1.   Japan's Whale Research Programs

Article VIII of the ICRW specifically provides for members of the IWC to issue permits for the killing of whales for research purposes. The article begins with the words "Notwithstanding anything contained in this Convention. . . ."[55] Further, both the moratorium and the Southern Ocean Sanctuary apply only to commercial whaling, and not research whaling, so that claims that Japan is violating the moratorium or the Southern Ocean Sanctuary have no legal basis.

Japan has two whale research programs, one in the Antarctic (Japanese Research Program in the Antartic—JARPA) and the other in the western North Pacific (Japanese Research Program in the North Pacific—JARPN). The JARPA was launched in 1987 in response to claims by a number of members of the IWC that the scientific information was insufficient to properly manage whale stocks in the Antarctic. This, and the IWC's Southern Ocean Whale and Ecosystem Research, known as the "SOWER," for which Japan provides researchers and vessel crews as well as the research vessels and funding, are the only long-term research programs on whales in the Antarctic. The programs are providing valuable information related to whales and the Antarctic ecosystem. The research program in the Antarctic will be carried out over a period of 16 years. Long-term research programs are required to assess trends in the changes of various population parameters since sampling for only one or two years does not tell you what is going on in a dynamic system. The objectives of the research are described in IWC Scientific Committee documents.[56] Research results were reviewed at a special meeting of the Scientific Committee held in 1997, which concluded that, at is halfway point, "many valuable results had been

---

55   ICRW, *supra* note 2, at 74.

56   48 REP. INT'L WHALING COMMISSION 96 (1998) [hereafter The 49th IWC Report].

obtained"[57] and that "the information produced by JARPA has set the stage for answering many questions about long term population changes regarding minke whales in Antarctic Areas IV and V."[58]

The JARPN research program in the western North Pacific was originally a five-year program that began in 1994. This program had two main objectives: the study of the population structure of minke whales and the study of feeding ecology of minke whales in the western North Pacific. The program was reviewed by the IWC's Scientific Committee each year and at a review meeting held in February 2000,[59] which concluded that information obtained from JARPN was relevant to the management of minke whales.

Since some scientific issues remain outstanding following the 1994–1999 program, a second phase (JARPN II) of the research began in August 2000.[60] The priority for this phase of the research is feeding ecology involving studies on prey consumption by cetaceans, prey preferences of cetaceans and ecosystem modeling. Minke, Bryde's and sperm whales are included as part of this research. Other research objectives include the study of stock structure of minke, Bryde's and sperm whales, as well as the assessment of the impact of environmental factors, such as chemical pollution, on cetaceans and the marine ecosystem. Phase II of this program is a two-year feasibility study. The research is important not only in terms of managing whales, but also in terms of fisheries management since in the waters around Japan, catches in certain fisheries are declining while at the same time the sampling from Japan's research program reveals that minke whales are eating at least ten species of fish including Japanese anchovy, Pacific saury, walleye Pollock and other commercially important species. Both the research

---

[57]   *Id.* at 104.

[58]   Report of the Intersessional Working Group to Review Data and Results from Special Permit Research on Minke Whales in the Antarctic. IWC Doc. SC/49/Rep1 (1997), 15 [hereafter JARPA Review Report].

[59]   Report of the Workshop to Review the Japanese Whale Research Programme under Special Permit for North Pacific Minke Whales (JARPN), Tokyo, Feb. 7–10, 2000. IWC Doc. SC/52/REP2 (2000), 40 [hereafter JARPN Review Report].

[60]   IWC Doc. SC/52/O1.

programs are authorized by the government of Japan in accordance with the provisions of Article VIII of the ICRW and are perfectly legal. Both involve non-lethal research including sighting surveys and biopsy sampling as well as a small take of whales for research that cannot be effectively done by non-lethal means. This includes examination of ear plugs for age determination studies, reproductive organs for examination of maturation, reproductive cycles and reproductive rates, stomachs for analysis of food consumption, blubber thickness as a measure of condition and internal tissues for pollution studies.

Scientific research is critical to the IWC since Article V of the ICRW requires that its regulations be "based on scientific findings." Further, the moratorium adopted in 1982 contains the requirement that "This provision will be kept under review, based on the best scientific advice. . . ."[61] Clearly under these circumstances and in light of the comments on the research from the IWC Scientific Committee,[62] Japan's contribution to the IWC through its research programs should be welcomed. In this regard, many fishing nations including Norway, China, Korea, Russia and Iceland, as well as many developing countries, support Japan's whale research programs since they provide for science-based resource management decisions. On the other hand, criticism of Japan's whale research program based on emotional reasons ignores international law and is a rejection of the basic principle that resources should be managed on a scientific basis.

---

[61]   ICRW, *supra* note 2, at 84.

[62]   The 49th IWC Report, *supra* note 56, at 101: "The information produced by JARPA (Japan's Antarctic Research Program) has set the stage for answering many questions about long term population changes regarding minke whales in Antarctic Areas IV and V;" *Id.* at 101, ". . . JARPA has already made a major contribution to the understanding of certain biological parameters;" *Id.* at 103, "The Committee noted that JARPA is at the half-way point and has provided substantial improvement in the understanding of stock structure;" *Id.* at 104, ". . . there was general agreement that the stock structure data were of value to management."

JARPN Review Report, *supra* note 59, at 4, ". . . its [IWC Scientific Committee's] previous advice that the effect of a small take for a short period would be negligible"; *Id.* at 16: ". . . the Workshop noted that information obtained during JARPN had been and will continue to be used in the refinement of *Implementation Simulation Trials* for the North Pacific minke whales, and consequently were relevant to their management."

The ICRW requires that the by-products of the research be processed. Japan sells whale meat derived from its research expeditions on the open market to fulfill the ICRW's requirement that resources not be wasted. These sales do not constitute a "loophole" under the ICRW, nor are they "illegal" or "commercial whaling in disguise" as some opponents of whaling contend, and the income from the sale of by-products (meat) is used to partially offset the cost of the research. The number of samples is small relative to the size of the populations being sampled so that the research programs pose no risk to the stocks. In the Antarctic, and North Pacific, the sample size of minke whales to be taken each year is the smallest number required to obtain statistically valid results. The sample size for the Bryde's and sperm whales to be taken in the North Pacific is smaller than is required for statistically significant results since the program for 2000 and 2001 is a feasibility study.

Japan submits the results from its research to the IWC Scientific Committee for review every year. Both the quality and quantity of data from Japan's research programs have been commended by the Scientific Committee. The IWC's Scientific Committee has noted that the programs have provided considerable data that has the potential to improve the management of minke whales. The Scientific Committee has also noted that non-lethal means to obtain some of this information are unlikely to be successful particularly in the Antarctic.[63]

Why then, has the IWC consistently adopted resolutions urging that the government of Japan refrain from issuing permits for its research whaling?[64] The answer is that these resolutions are part of the dysfunctional nature of the IWC described earlier in this chapter, since they are a rejection of the fundamental principle that resource decision-making should be based on scientific findings and are contrary to the views of its own Scientific Committee. However, these resolutions have been used by some to argue that there is no need to conduct the research. Others see the resolutions as justifying their view that the research is illegal. Both positions are simply wrong.

---

[63] The 49th IWC Report, *supra* note 56, at 105.

[64] The latest examples are Resolution 2000-4 & 2000-5, Chairman's Report of the 52nd Annual Meeting 66, attached to IWC Doc. IWC.CCG.152.

This dysfunctional aspect of the IWC can be illustrated by a critical examination of that part of the IWC Scientific Committee report of 1997 related to Japan's research program in the Antarctic and the relevant Resolution 1997-5 adopted by the Commission in the same year. The key sentence in the report of the Scientific Committee says "the results of the JARPA, while not required for management under the RMP, have the potential to improve management of minke whales in the Southern Hemisphere."[65] On the other hand, the key sentence in the resolution that strongly urged Japan to refrain from issuing any further permits for the take of whales says "the Scientific Committee notes (IWC/49/4) that the results of the JARPA programme are not required for management."[66] Comparison of these two sentences reveals that the latter—the sentence from the Resolution—is a mis-quote and an intentional misrepresentation of the sentence from the report of the Scientific Committee, since it omits the words "under the RMP" from the first half of the sentence and omits the second half altogether.

This point may seem trivial to those not familiar with the IWC and particularly the RMP, but it is not. To maximize sustainable use of whale stocks under the RMP, it is necessary to clarify the distribution of whale stocks. Reducing uncertainty about stock structure is a priority of Japan's whale research programs, and results from the research will contribute to improvement of the application of the RMP. The report of the Scientific Committee reflects this point by saying that "the result of analysis of JARPA data could be used . . . to increase the allowed catch of minke whales in the Southern Hemisphere, without increasing the depletion risk. . . ."[67]

Clearly then, the IWC Resolution should have quoted the entire sentence from the report of the Scientific Committee or noted simply that "the results of the JARPA . . . have the potential to improve management of minke whales in the Southern Hemisphere." The fact that the RMP does not require data other than abundance estimates and historical catch data, and that the stated objectives of JARPA do not include acquisition

---

[65]   The 49th IWC Report, *supra* note 56, at 101.

[66]   *Id.* at 47.

[67]   *Id.* at 104.

of such data misses the point, which is that the results from JARPA will improve whale management under the RMP.

On three different occasions, Japan's whale research programs have resulted in "certification" by the United States under its domestic law known as the Pelly Amendment to the Fishermen's Protective Act.[68] The first occasion was in 1987 when Japan began its Antarctic research program, the second was in 1995 when the number of whales to be taken under JARPA was increased and the third was in 2000 when the north Pacific research program was expanded to include Bryde's and sperm whales. All of these "certifications" were followed by threats of trade sanctions as provided for under the Act. The imposition of such sanctions could be challenged under international trade rules of the World Trade Organization (WTO), which prescribe that sanctions can only be imposed if they certain requirements and that they may not be implemented in a manner that constitutes arbitrary or unjustifiable discrimination.[69]

In addition to the public criticism by governments such as the United States,[70] United Kingdom,[71] New Zealand[72] and Australia,[73] Japan's whale research programs are also a focal point for protest by a number of non-government organizations (NGOs). Japan strongly believes that the anti-whaling position of NGOs is based on deceptive campaigns designed for fundraising purposes and that most of the criticism by

[68]   22 U.S.C. §1978.

[69]   Although quantitative restrictions of trade are generally prohibited under Article XI, such restrictions may be permitted to further environmental objectives, subject to conditions under Article XX of the General Agreement on Tariffs and Trade (55 U.N.T.S. 187). Relevant provisions are paragraph (b), (g) and the chapeau language of this article.

[70]   A Letter from the Secretary of Commerce of the United States to the President on the date of September 13, 2000; A Letter from the President to the Speaker of the House of Representatives and the President of the Senate of the United States on the date of December 29, 2000.

[71]   *Save the Whales, Again*, WASH. POST, Aug. 22, 2000, at A18.

[72]   For example, Media Statement by Prime Minister of New Zealand dated November 17, 2000.

[73]   For example, a press release by Minister of the Environment and Heritage of Australia dated November 14, 2000.

governments reflects the undue influence of anti-whaling NGOs in nations where whaling has no domestic constituency.

4.2. The RMS

The RMS was first proposed as a means to implement the RMP. As set out in Resolution 1992-3 adopted by the Commission in 1992,[74] the RMS includes five elements: the RMP, minimum data standards, guidelines for conducting surveys and analyzing the results, an inspection and observation scheme and arrangements to ensure that total catches over time are within the limits set under the RMS. The final step would be to incorporate these into the Schedule. All of these elements, except the observation and inspection scheme, have been completed and accepted by the Commission.

Opponents to the resumption of whaling emphasize the need to apply a precautionary approach. They cite uncertainties associated with stock numbers and other factors relevant to management, as well as past over-harvesting, as evidence of the dangers of permitting commercial whaling to resume. In fact, however, the RMP is excessively precautious, to the point that it wastes resources. Further, it already takes uncertainty into account, including uncertainty related to abundance estimates and uncertainty related to the possible impacts of environmental change.

Japan's position[75] is to support the early completion of the RMS to allow the resumption of sustainable whaling and thereby to restore the IWC's function as mandated by the ICRW. For this purpose, Japan has presented to the IWC meetings draft texts to revise Chapter V of the Schedule on the basis of several principles generally accepted in other international fishery management regimes.[76] These include:

---

[74]   43 REP. INT'L WHALING COMMISSION 40.

[75]   Report of the Intersessional Meeting of Revised Management Scheme Working Group 10, attached to IWC Doc. IWC.CCG.151.

[76]   IWC Doc. IWC/49/RMS1 (1997); IWC Doc. IWC/52/30 (2000). In this regard, it should be noted that Norway and Japan are only the two IWC members, which have tabled comprehensive draft texts to revise the current Chapter V of the schedule.

(1) Supervision and control measures must be such as are necessary and reasonable in order to achieve the objective of the . In this context, the purpose of the RMS is to ensure that the number of whales actually taken does not exceed catch limits calculated by the RMP and that sampling and the collection of information required for those components of the RMS already agreed to are carried out efficiently;

(2) Requirements must reflect the reality of whaling operations and must be practical to implement. This is particularly the case for small-type whaling operations, where the size of vessels as well as the scale of operations are small; and

(3) The IWC should refer to measures presently applied by other international fishery management organizations.

Several proposals have been made to include other items in the observer scheme. This includes a DNA registration and market monitoring system, and a proposal to require the collection of data related to animal welfare aspects of whale killing.[77] The former is outside the competence of the IWC because the control of domestic markets is a national responsibility. The latter is also outside of IWC competence and has nothing to do with conservation and management of whale stocks. Rather, it simply imposes an undue burden on whaling operations.

The IWC's work to complete the RMS has taken an inordinately long period of time, primarily due to the stalling and continual introduction of new elements into the negotiations by those members of the IWC opposed to the resumption of whaling. In addition, anti-whaling members are insisting that the costs of implementing and operating the observer and inspection scheme should be borne entirely by the whaling nations. Japan's position is that it has made substantial compromises in these negotiations, particularly with regard to the RMP tuning levels, and that it is the uncompromising position of the majority that is blocking progress to complete the RMS. The inability of the IWC to agree on an observation and inspection scheme that would allow it to do its job under the ICRW is another clear example of the body's dysfunctionality.

---

[77]   IWC Doc. IWC/52/RMS2 (2000) presented by the government of U.K.

It is also Japan's position that if the RMS is the mechanism by which the IWC chooses to regulate commercial harvesting of whales, which is its primary function, then the costs should be borne by all members. At the 52nd Annual Meeting of the IWC, Resolution 2000-3 was adopted, expressing the Commission's view that "it is important for the future of the Commission that the process of completion of the RMS proceed expeditiously."[78] This statement is the Commission's response to the comments from IUCN and the Secretary General referred to above. Unfortunately, such views have not translated into the actions necessary to complete the RMS, with very little progress made at the most recent intersessional meeting of the RMS working group in February 2000. The call for expeditious action to complete the RMS is also discordant with the next paragraph of the Resolution, which reaffirms that the RMS should include the elements agreed to in 1992, but inserts the words *"but not limited to."*[79] This is interpreted by Japan as a clear sign of further stalling and more proposed additions to come.

There is an additional serious problem related to the negotiation process for completion of both the RMP and the RMS. In order to complete the RMP and RMS, IWC members who support a resumption of whaling have already made substantial compromises. These compromises have been made on the understanding that implementation of an RMS requires the deletion of paragraph 10 of the Schedule[80] since sub-paragraphs (a), (b) and (c) of this paragraph are related to the IWC's management procedure that will be replaced by the RMS. More importantly, sub-paragraph (d), the moratorium on factory ships, and sub-paragraph (e), the moratorium, will be unnecessary given the risk-averse nature of the RMS and its overall intent: to manage whaling on a sustainable basis.

Based on the purpose of the ICRW and its articles as well as the specific wording of sub-paragraph (e) of paragraph 10 of the Schedule and the IWC's resolutions related to the RMS (with the exception of Resolution 2000-3 referred to above), this understanding is the only pos-

---

[78]   IWC Doc. IWC/52/34Rev.

[79]   The original proposal (IWC/52/34) does not contain these words.

[80]   ICRW, *supra* note 2, at 84.

sible good faith interpretation of events related to the development of the RMS. The Resolution is a disturbing exception because its final paragraph "Confirms that this Resolution does not prejudge the positions of Contracting Governments with respect to the status of paragraphs 10(d) and 10(e) of the Schedule." In other words, this resolution confirms the view expressed by some IWC members including the U.K., New Zealand and the U.S. that they are opposed to the resumption of commercial whaling under any circumstances.

On this basis, it is not unreasonable to suggest that these IWC members have not been negotiating in good faith. This view is supported by the fact that Australia, which is also opposed to the resumption of commercial whaling under any circumstances, announced at the 49th Annual Meeting that it would no longer participate in the Commission's discussions on the RMS. It declared in its opening statement that that it would vote against any proposal to adopt the RMS and the RMP because this would facilitate the resumption of commercial whaling.[81]

On this basis, it is also not unreasonable to suggest that these IWC members, and others who share similar views, should withdraw from the IWC since they are clearly opposed to the overarching purpose of the ICRW.

4.3. Sanctuaries

Paragraph 7 of the current Schedule of the ICRW includes two sanctuaries: the Indian Ocean Sanctuary adopted in 1979 and the Southern Ocean Sanctuary adopted in 1994. Japan's position, supported by the legal analysis of Burke,[82] is that while Article V(1) of the ICRW provides for the establishment of sanctuaries, the conditions of Article V(2) must be met, and that neither the Indian Ocean Sanctuary nor the

---

[81]   IWC Doc. IWC/49/OS Australia, Add. 2.

[82]   William T. Burke, *Memorandum of Opinion on the Legality of the Designation of the Southern Ocean Sanctuary by the International Whaling Commission*, IWC Doc. IWC/47/41 (1995); *Legal Aspects of the IWC Decision on the Southern Ocean Sanctuary*, IWC Doc. IWC/48/33 (1996); *Memorandum on the IWC decision adopting the Southern Ocean Sanctuary*, IWC Doc. IWC/49/24 (1997); *The Legal Invalidity, supra* note 4.

Southern Ocean Sanctuary meets these conditions. In other words, the establishment of these sanctuaries was beyond the Commission's authority under the ICRW and they are therefore illegal. Further, the designation of the Southern Ocean Sanctuary explicitly contradicts the objective of the ICRW by prohibiting the taking of whales from the abundant stocks in the Southern Ocean. The Southern Ocean Sanctuary is by its own words anti-science and contrary to Article V(2)(b) of the ICRW, which requires that regulations "shall be based on scientific findings." This position is illustrated by its assertion that "This prohibition applies irrespective of the conservation status of baleen and toothed whale stocks. . . ."[83]

In accordance with the provisions of the ICRW, the government of Japan filed an objection to the Southern Ocean Sanctuary to the extent that it applies to the Antarctic minke whale stocks. Records of the IWC,[84] including Japan's proposals to amend paragraph 7(b) of the Schedule to make it consistent with the ICRW,[85] clearly document Japan's view that the establishment of sanctuaries without scientific basis is illegal. In this regard, it is important to note that Japan's failure to file an objection to the Indian Ocean Sanctuary and the fact that Japan's objection to the Southern Ocean Sanctuary only applies to minke whales does not imply that Japan accepts the legality of these sanctuaries. Rather it indicates Japan's willingness to cooperate with the IWC to the fullest extent possible.

The establishment of sanctuaries has become part of the strategy of those opposed to the resumption of whaling particularly in light of the fact that the IWC has accepted the RMP and is, at least in theory, working to complete the RMS. For example, Greenpeace is promoting a worldwide sanctuary and the Irish Proposal[86] would make all the oceans

---

83   ICRW, *supra* note 2, at 81.

84   Japan's opening statements, *supra* note 13 and articles by Professor Burke, *supra* note 82.

85   IWC Doc. IWC/52/25.

86   At the 1997 annual meeting, the Commissioner from Ireland expressed the view that there was a risk of the IWC breaking up. He proposed a "compromise package" that has since been referred to as the "Irish Proposal." The proposal con-

except the coastal areas of a few countries a sanctuary. At the 51st Annual Meeting of the IWC, Brazil expressed its intention to propose a sanctuary in the south Atlantic in the future[87] and the governments of Australia and New Zealand proposed a South Pacific Sanctuary.

The proposal for the South Pacific Sanctuary[88] was reviewed by the IWC's Scientific Committee at its meeting in 2000, but the Committee did not reach a consensus view. The Committee's report simply summarized arguments in favor and arguments against. But the Scientific Committee did provide one of the most compelling arguments against the sanctuary proposal:

> Sanctuaries provide no extra protection for the most vulnerable depleted stocks from actual threats that they face such as habitat destruction, pollution, shipping, fisheries interactions etc, and do not distinguish between areas of critical habitat and those of little importance. Such stocks are already protected under existing IWC management measures.[89]

In his comments on the proposed sanctuary,[90] Dr. Peter Best averred, "One can only conclude that there is little scientific basis (or urgent conservation need) for such a sanctuary." He further notes that:

> Sanctuaries can and do play an important role in marine conservation, but only if they are realistic in concept and practical in terms of enforcement. If their use in cetacean conservation is not

---

tained five elements: adoption of RMS into the Schedule; restriction of quotas to coastal areas and to nations now whaling; limiting whaling to catches for local consumption with no international trade; phase out of lethal scientific whaling and development of IWC regulations for whale watching. With the exception of the first, these elements of the proposed compromise contradict the provisions of the ICRW or are outside the mandate of the IWC. The proposal has not been voted on by the IWC.

[87]   ANNUAL REPORT OF THE INTERNATIONAL WHALING COMMISSION 1999, at 10 (2000).

[88]   IWC Doc. IWC/51/21.

[89]   IWC Doc. IWC/52/4, 81.

[90]   IWC Doc. SC/52/WP7.

to be devalued, it is important that proposals for their creation should be credible.

Japan is opposed to the proposed South Pacific Sanctuary for a number of reasons:

(1) It disregards science, ecosystem considerations and sustainable management practices (see Section 3.1);
(2) It could have significant negative impacts on fisheries resources;
(3) It is contrary to the ICRW since it does not meet the requirements of Article V(2)(b) particularly since it does not have a scientific basis;
(4) It is not needed for conservation reasons since the current moratorium on commercial whaling will only be lifted when the RMS, which will only provide sustainable quotas for abundant stocks, is implemented;
(5) It is not needed to promote research; and
(6) It is not needed to promote tourism.

Relevant research activities have not been encouraged by the establishment of the Indian Ocean Sanctuary and it has led to the situation where there is no scientific data available to manage whale stocks in the Indian Ocean. The Scientific Committee noted in its report[91] that "the Sanctuary has not provided the stimulus for research originally envisaged" and "the Committee believes that the Indian Ocean Sanctuary provides no special advantage for the scientific study of whales."

At the 52nd Annual Meeting of the IWC in 2000, the proposed South Pacific Sanctuary was defeated by a vote of 18 in favor, 11 against and four abstentions (adoption requires a three-quarter majority under IWC rules of procedure). Following its review of the proposal for a South Pacific Sanctuary in 2000, the Scientific Committee requested advice from the Commission on how to review sanctuary proposals in the future. This is a significant issue since, according to the provisions of paragraph 7 of the Schedule, the Indian Ocean Sanctuary is subject to review at the annual meeting in 2002, followed by a review of the provision for the Southern Ocean Sanctuary in 2004.[92] The Commission

---

[91]   40 REP. INT'L WHALING COMMISSION 73.

[92]   ICRW, *supra* note 2, at 81.

will consider how it intends to conduct these reviews at its meeting in 2001.

Finally, it is necessary to point out that the establishment of a sanctuary in the South Pacific would undermine almost a decade of work by the IWC to develop its RMS. This, once again, raises the question whether participation in this work has been in good faith.

4.4. Interim Quota for Small Scale Coastal Whaling

Since 1986 Japan has presented documentation on the localized and small-scale nature of community-based whaling in Japan, and the socio-economic importance of whale meat production, distribution and consumption in four small coastal whaling communities (see Section 2.3). The IWC has acknowledged the socio-economic and cultural needs of these communities and the distress to these communities that resulted from the cessation of minke whaling. In 1993, the IWC resolved to work expeditiously to alleviate this distress.[93] This commitment was reaffirmed in Resolution 2000-1[94] adopted at the 52nd annual meeting in 2000.

It is Japan's position that the failure to satisfactorily address this matter by providing a small interim quota from an abundant stock (the Scientific Committee has estimated there are 25,000 minke whales in the western north Pacific[95]) in spite of the demonstrated need and in spite of these resolutions is another demonstration of the dysfunctional nature of the IWC. This position is strengthened by the fact that Japan has documented the many similarities between the cultural and traditional aspects of whaling in the affected communities and the aboriginal communities that receive quotas under the IWC's provisions for "aboriginal/subsistence whaling."[96]

---

[93]   IWC Doc. IWC/45/51.

[94]   Chairman's Report of the 52nd Annual Meeting *supra* note 64, 64.

[95]   42 Rep. Int'l Whaling Commission 66.

[96]   Japanese Small-type Coastal Whaling, *supra* note 20.

4.5.   Administrative Matters

4.5.1   Secret Ballot

Since 1997 Japan has proposed that the IWC institute a provision for the use of secret ballots as part of its voting system. These proposals, with the exception of a variation of the proposal to provide for the use of secret ballots on matters related to the election of the Chair, Vice-Chair, the appointment of the Secretary of the Commission, and the selection of the annual meeting venue which was adopted in 1998, have been rejected by the Commission.

Japan's proposals have been made in an attempt to protect the rights of contracting parties to express their views freely, without fear of coercion or reprisals[97] and in accordance with democratic procedures. Japan is of the view that the Commission's current voting not only subject the rights of contracting parties to undue pressure from other member nations and NGOs but that the current procedures unduly influence the Commission's decisions on matters of fundamental importance. For these reasons, the use of secret ballots for deciding substantive matters is provided for in many international commissions and organizations including the Commission for the Conservation of Antarctic Marine Living Resources,[98] the Convention on the Conservation of Migratory Species of Wild Animals,[99] the Convention on International Trade in Endangered Species of Wild Fauna and Flora[100] and the International

---

[97]   At the 51st Annual Meeting of the IWC some small island developing states officially expressed their concern about "the harassment and intimidation directed at developing countries by certain NGOs and associates of the developed countries." ANNUAL REPORT OF THE INTERNATIONAL WHALING COMMISSION 1998, at 39 (1999). Evidencing their concern, an NGO stated at the same meeting that "trade sanctions and boycotts are recognised as legitimate tools of enforcing compliance with international obligations" (IWC Doc. IWC/50/OS DCA, at 2).

[98]   Rule 5, Rules of Procedure of the Commission, available at <http://www.ccamlr.org/English/e_basic_docs/e_pt3_p1.htm>.

[99]   Rule 14, paragraph 4, of the Rules of Procedure of the CMS Conference of Parties, *Proceedings of the Sixth Meeting of the Conference of the Parties Convention on the Conservation of Migratory Species of Wild Animals*, Vol. I, Cape Town, South Africa, NOV. 10–16, 1999, 126, at 130.

[100]   Rule 25, paragraph 2, of the Rules of Procedure, CITES doc. Doc. 11.1 (Rev. 2), at 11.

Commission for the Conservation of Atlantic Tunas.[101] Japan further argues that the use of secret ballots does not conflict with the need for transparency. Transparency does not mean that the individual votes of all members must, on every occasion, be made public. A secret ballot would not prevent those members of the IWC who would wish to disclose how they voted on any issue from doing so.

4.5.2. Commission Membership

The IWC is currently working to revise its system of determining membership contributions in order to allow for greater participation by developing countries. Japan supports the position that an expanded membership, particularly if it includes countries dependent on marine resources, will help to normalize the functioning of the IWC.

5.   CONCLUSION

While Japan views the current situation in the IWC as dysfunctional, it is committed to work to normalize the situation in order to restore the IWC's credibility as a resource management organization and work towards the resumption of its responsibilities under the ICRW. This includes, in particular, a commitment to enhancing scientific activities under Article VIII, the early implementation of the RMS and the recognition of the importance of ecosystem approaches that account for the consumption of marine resources by cetaceans. Japan is optimistic that its commitment and perseverance in restoring normality to the IWC will ultimately succeed. At the same time, Japan will continue consultations with neighboring states in the North Western Pacific initiated in 1998 with a view to establishing a regional regime for cooperation in research and management of cetaceans. This regime could ultimately assume the functions of the IWC if the IWC continues to reject a resumption of properly managed catches of whales from abundant stocks.[102]

---

[101]   Rule 9, paragraph 5 and 7, of the Rules of Procedure, available at <http://www2.rediris.es/iccat/>.

[102]   Informal consultations have been held four times since 1998 among Japan, China, South Korea and Russia. The press release issued following the fourth consultation held in Tokyo on January 15–16, 2001 is available at <http://www.jfa.maff.go.jp/rerys/13.01.18.1.html>.

While Japan's current research whaling activities have received some public criticism from a small number of governments, as well as some negative media commentary generated largely by anti-whaling non-government organizations as part of their fundraising activities, it is important to note that worldwide public opinion and the views of the world community of states are not opposed to whaling. Public opinion surveys in the United States, Australia, France and the United Kingdom[103] have shown that the level of public knowledge about whales is very low but that when people were informed, there was strong public support (71 percent in the United States) for regulated whaling for food. Other examples include the past two conferences of the parties to the CITES (COP 10 and 11) held in 1997 and 2000, where the majority of parties supported a proposal to resume limited trade in whale products, the 1992 UN Conference on Environment and Development, which rejected efforts by anti-whaling governments and NGOs to exclude whales from the general applicability of the principle of sustainable use, and the October 2000 IUCN World Conservation Congress where over 100 resolutions on conservation issues were adopted but none on the subject of whaling.

In fact, "sustainable use" is the widely accepted world standard supported by many international documents and instruments including the United Nations' Rio Declaration[104] and the Kyoto Declaration and Plan of Action related to the role of fisheries in sustainable food supplies.[105] The IWC should fulfill its responsibilities to ensure the integrity of the treaty and further the world community's support for the sustainable use of resources.

---

[103]   RESPONSIVE MANAGEMENT, KNOWLEDGE OF WHALES AND WHALING AND OPINIONS OF MINKE HARVEST AMONG RESIDENTS OF AUSTRALIA, FRANCE, THE UNITED KINGDOM AND THE UNITED STATES 20 (1998); AMERICAN'S OPINIONS OF MINKE WHALES HARVEST 12 (1997).

[104]   Rio Declaration on Environment and Development, June 14, 1992, available at <http://www.unep.org/unep/rio.htm>.

[105]   Kyoto Declaration, Preamble and paras. 11, 12 and 14, *supra* note 22.

Chapter 2
Culture-Based Conflict in the International Whaling Commission: The Case of Japanese Small-type Whaling

Milton M.R. Freeman[1]

1. INTRODUCTION

Most whaling activity today is community-based and conducted within the territorial waters of a small number of nations; as such, more than 90 percent of all whaling carried out today falls under national jurisdiction and is not subject to International Whaling Commission (IWC) oversight. Thus, in the absence of any large-scale industrial demand for whale products (which, in the past, resulted in the serious over-exploitation of whale stocks), current whale fisheries are not, in objective terms, a pressing global fisheries or conservation concern.

Whaling today is undertaken to meet domestic need for food rather than supply an international oil market, although a small number of fisheries wish to trade whale products surplus to their domestic needs.[2] It is likely that whale management will assume a greater degree of international attention in the future, one reason being that all fishery regulatory bodies will ultimately adopt an ecosystem approach to management. This approach will compel them to address the predatory impacts caused by growing whale populations that, in some cases, are known to be increasing at rates greater than 10 percent annually.[3]

---

[1] Senior Research Scholar, Canadian Circumpolar Institute, University of Alberta, Edmonton AB, T6G 0H1, Canada. milton.freeman@ualberta.ca.

[2] Although Norwegian whalers supply a domestic market for red meat, Norwegians do not eat blubber. However, a market for blubber exists in both Japan and Iceland.

[3] William Aron, William Burke & Milton M.R. Freeman, *The Whaling Issue*, 24 MAR. POL'Y 179, 184 (2000).

With regard to the management of whaling, the UN Convention on the Law of the Sea (UNCLOS) does not mandate any particular whale management agency, nor does it favor a global over a regional approach to whale management. UNCLOS Article 65 requires that states work through "appropriate international organizations" for managing whaling (also referenced in Article 64). States appear to be left to determine what is an "appropriate" organization. UNCLOS Article 61, in reference to establishing catch levels for living marine resources, refers to "competent international organizations, whether subregional, regional or global, where appropriate." Similar wording also appears in UNCLOS Article 63.

The question of regulatory competence has particular salience when considering the International Whaling Commission (IWC). Several commentators have noted the longstanding stalemate[4] occurring in the IWC and commented on the reasons for this on-going impasse.[5] The inability of the IWC to resolve the political deadlock that has existed for more than a decade is illustrative of the limited capability of this "global" whaling regime[6] to meet its treaty obligations under international law

---

[4]   Steinar Andresen, *NAMMCO, IWC and the Nordic Countries, in* WHALING IN THE NORTH ATLANTIC: ECONOMIC AND POLITICAL PERSPECTIVES 75, 79, 84 (Guðrun Pétursdóttir ed., 1997); Robert L. Friedheim, *Fostering a Negotiated Outcome in the IWC, in id.* at 135, 145, 146, 148, 151; Kate Sanderson, *The North Atlantic Marine Mammal Commission—in Principle and Practice, in id.* at 67, 68; John A. Knauss, *The International Whaling Commission—Its Past and Possible Future* 28 OCEAN DEV. & INT'L L. 79, 82–83 (1997); Aron, *supra* note 3, at 179, 181; IUCN, *Statement to the 52nd Meeting of the International Whaling Commission*, IWC Doc. IWC/52/OS/IUCN (2000).

[5]   One student of the IWC states the basis of the stalemate thus: "The international regime for regulating whaling is a mess . . . the regime's main legal instrument calls for sustainable management of the whaling to 'make possible the orderly development of the whaling industry.' Yet most members favor extinction of the industry." David G. Victor, *Whale sausage. Why the whaling regime does not need to be fixed, in* TOWARDS A SUSTAINABLE WHALING REGIME 292, 296–98 (Robert L. Friedheim ed., 2001). *See also,* Milton M.R. Freeman, *Is money the root of the problem? Cultural conflict at the IWC, in id.* at 123, 135–38; Elizabeth DeSombre, *Distorting global governance: membership, voting, and the IWC, in id.* at 183, 183–194; Christopher D. Stone, *Summing up: whaling and its critics, in id.* at 269, 278–80.

[6]   Although some claim the IWC is a global regime, it consists (in 2001) of

as set down in its founding charter, the 1946 International Convention for the Regulation of Whaling (ICRW).[7]

There are many reasons for the current malaise affecting the IWC, and these will not be fully explored here.[8] However, in brief, it appears that there are profound cultural disagreements about the appropriateness of the commercial consumptive use of whales. These cultural conflicts[9] exist because, *inter alia*, on the one hand are those who hold that the sustainable use of natural resources allows for the consumptive use of any non-endangered species; and on the other hand are those who believe that whales are a privileged group of animals that should not ordinarily (or under any circumstance) be subject to consumptive use.

Although the IWC regulates less than 7 percent of whaling taking place in the world today,[10] nevertheless the ongoing cultural conflicts

---

about 33 voting members, about half of whom have adopted an anti-whaling position. Given that binding decisions require a three-fourths majority of those voting, significant decisions in respect to commercial whaling are now impossible to achieve within the Commission.

[7]   The International Convention for the Regulation of Whaling, Dec. 2, 1946, 62 Stat. 1716, 161 U.N.T.S. 72.

[8]   For example: Arne Kalland, *Management by Totemization: Whale Symbolism and the Anti-whaling Campaign*, 46 ARCTIC, 124, 125–33 (1993); Friedheim, *supra* note 4, at 135–156; TOWARDS A SUSTAINABLE WHALING REGIME (Robert L. Friedheim ed., 2001); Oran R. Young, Milton M.R. Freeman, Gail Osherenko *et al.*, *Subsistence, Sustainability, and Sea Mammals: Reconstructing the International Whaling Regime* 23 OCEAN & COASTAL DEV. 117, 119 (1994).

[9]   The cultural nature of these conflicts are recognized by a number of commentators; *e.g.*, Aron *et al.*, *supra* note 3, at 181; Knauss, *supra* note 4, at 82. At the 1995 IWC meeting held in Dublin, the Irish Minister of Culture and Tourism stated: "I believe it would be wrong and in the nature of cultural imperialism to attempt to impose our cultural values on those nations whose populations have depended on whales for generations." Elsewhere, I have discussed the culturally based aversions held by western anti-whaling advocates, namely toward blood and animal slaughter, and toward commerciality; *see, respectively*, Milton M.R. Freeman, *Issues Affecting Subsistence Security in Arctic Societies*, 34 ARCTIC ANTHROPOLOGY 1–17 (1997), Milton M.R. Freeman, *Is Money the Root of the Problem?*, in Friedheim, *supra* note 5, at 122, 128–31.

[10]   Aron, *supra* note 3 at 188, states "a current estimate indicates that the IWC regulates less than five percent of the whales and other cetaceans hunted each year

create persistent problems for those societies that chose to maintain a tradition of consuming whale products for food. According to the IWC Scientific Committee, there are no objective, science-based reasons for continuing to object to the resumption of a regulated, sustainable minke whale fishery on several stocks of this species. However, the various non-objective reasons for opposing whaling remain after more than a decade and a half of often acrimonious debate and have led to inaction by the IWC.[11]

Several whaling nations have established regional science-based management regimes; for example, the North Atlantic Marine Mammal Commission (NAMMCO),[12] the Canada-Greenland Joint Commission on the Conservation and Management of Narwhal and Beluga,[13] and the Eastern Caribbean Cetacean Commission (ECCO).[14] There are also discussions taking place in the North Pacific region to explore the benefits

---

on a global basis." Excluding conservative estimates of the directed take of dolphins and porpoises, the figure rises to about 7 percent. The IWC exerts management authority over baleen whales and sperm whales, but not over long-finned and short-finned pilot whales, pygmy and false killer whales, melon-headed whales, Baird's and Cuvier's beaked whales, beluga, and narwhal which are the main whale species taken by various nations. In addition, although the IWC has placed a ban on non-aboriginal hunting of minke whales, a large majority of minke whales taken each year are hunted legally by non-aboriginal whalers under art. V.3 and art. VIII of the ICRW.

[11]   The unwillingness of a core number of anti-whaling nations (including Australia, Austria, Brazil, Finland, France, Germany, India, Netherlands, New Zealand, Spain, Sweden, U.K. and U.S) to enter into good-faith negotiations makes the deadlock in the IWC appear intractable in the immediate to mid-term future. Since entering into discussions about its small-type coastal whale fishery, Japan has reduced its initial request for a quota from 210 minke whales per year to 50 minke whales. Japan has also offered to prevent commercial profit from the minke whale fishery. However, opponents of Japanese whaling appear to be uninterested in reaching a negotiated outcome. *See* Government of Japan, *Summary of the JSTCW Discussion*, IWC/48/SEST1 at 295–96 (1996); Aron, *supra* note 3, at 190; William T. Burke, *Whaling and International Law*, in Pétursdóttir, *supra* note 4, at 117.

[12]   Sanderson, *supra* note 4, at 67.

[13]   Dan Goodman, *Land Claim Agreements and the Management of Whaling in the Canadian Arctic*, in PROCEEDINGS OF THE 11TH INTERNATIONAL ABASHIRI SYMPOSIUM 39, 44, 46 (1997).

[14]   Anon., *The Eastern Caribbean Cetacean Commission (ECCO)*, 21 INWR DIGEST 1–2 (2001).

of creating similar regional management regimes.[15] It appears that these regional bodies, operating on the basis of consensus and shared conservation objectives, benefit from the greater degree of cultural understanding that a regional regime allows. In addition, as these existing and proposed bodies are of recent origin, members enjoy a shared understanding of contemporary resource management approaches.[16] In contrast, the diverse (and now largely non-whaling) membership of the IWC lacks cultural understanding of whaling societies and includes a bloc of anti-whaling members who find the objectives and requirements of the 1946 treaty to be culturally and politically unacceptable.[17]

This chapter illustrates, by reference to research on Japanese small-type coastal whaling presented at the IWC by the government of Japan, the apparent difficulty exhibited by Euro-American IWC members in

---

[15] Anon., *NW Pacific Whaling Commissioners Meet*, 17 INWR DIGEST 3 (1999).

[16] Aron, *supra* note 3, at 181 states "Replacement of IWC by regional bodies provides adequate and more cost effective management as well as respecting sensitivity to local cultural and socio-economic differences that the IWC demonstrably lacks. The regional approach would also conform to the developing trend in respect to managing resource issues." Similar sentiments were expressed in Brian Moeran, Tomoya Akimichi, Richard Caulfield *et al. Similarities and Diversity in Coastal Whaling Operations: A Comparison of Small-scale Whaling Activities in Greenland, Iceland, Japan and Norway*, IWC/44/SEST6 at 243 (1992), where the serious disconnect between IWC and current management practices were noted, and it was concluded that "it is imperative and urgent that the IWC address and resolve the issue of equitability and sustainability in resource management, since if it fails to do so, its critics will almost certainly seek alternative means of satisfying those reasonable human needs that present IWC practice is unable to accommodate." *See also* Steinar Andresen, *The Effectiveness of the International Whaling Commission*, 46 ARCTIC 108, 115 (1993); Alf Håkon Hoel, *Regionalization of Whale Management: The Case of the North Atlantic Marine Mammal Commission*, 46 ARCTIC 116, 118, 123 (1993).

[17] Several current IWC members (including Australia, Austria, Italy, New Zealand and the U.S.) have announced that they will never support a resumption of commercial whaling, an activity the IWC was expressly created to manage; *see* Aron, *supra* note 4, at 181. The objectives of IWC are set down in the Preamble of the ICRW and include, *inter alia*, to "provide for the proper conservation of whale stocks and thus make possible the orderly development of the whaling industry" and "provide for the conservation, development and optimum utilization of whale resources" by "tak[ing] into consideration the interests of the consumers of whale products and the whaling industry" (in art. V.2).

understanding the cultural significance of whaling in a non-western whaling nation and treaty partner.

Japan introduced a series of reports detailing the socio-economic circumstances and cultural needs in its four traditional whaling communities for discussion at the annual IWC meetings, beginning in 1986.[18] These communities were impacted by the 1988 commercial whaling moratorium that reduced the resident coastal whalers' annual quota of North Pacific/Okhotsk Sea minke whales to zero from an average annual quota of about 320 whales during the preceding several decades.

Despite more than a decade of sustained effort to enter into good-faith negotiations to obtain an interim quota of 50 minkes (pending the lifting of the IWC whaling moratorium), Japan has made no progress in seeking a negotiated resolution to this problem, and consequently the zero quota remains in effect. Minke are a widely occurring and abundant whale species.[19] A quota of around 200 whales could be safely taken from the North Pacific/Okhotsk Sea stock (that numbers about 25,000) by Japanese small-type whalers, without adversely impacting the minke whale population,[20] according to the IWC's Scientific Committee.

---

[18]   Japan has adopted a methodological approach successfully used by the U.S. since 1980 to justify an increasing take of the endangered bowhead whale by Alaskan whalers. Since 1988, Japan has relied heavily on non-Japanese scientists (including two U.S. scientists who developed the US methodology referred to above) to prepare a series of more than 36 research reports relating to its traditional community-based coastal whale fishery. The research group has included members from the following countries (numbers of researchers in parentheses): Australia (1), Canada (4), Iceland (2), Israel (2), Japan (7), Norway (1), U.K. (1), U.S.A. (9). For a partial list of researchers, *see* Tomoya Akimichi, *et al., Small-Type Coastal Whaling in Japan: Report of an International Workshop* 109–111; Moeran, *supra* note 14, at 244; The Government of Japan, *List of Documents Related to Japanese Small-type Coastal Whaling and the Socio-economic Implications of a Zero Catch,* IWC/45/SEST2 at 256 (1993).

[19]   The IWC classified some minke stocks as so-called Protection Stock in 1975, and reaffirmed this classification in 1991. However, it is important to note that the reference to "protection" refers only to catch-maximization considerations, and has nothing to do with any likelihood of extinction. D.S. Butterworth, *Science and sentimentality,* 357 NATURE 532 (1992); Sidney Holt, *The debate on whaling,* 359 NATURE 9 (1992).

[20]   The IWC Scientific Committee's 1991 Comprehensive Assessment of North Pacific minke whales reports annual sustainable yields in excess of 200 animals for

As part of its efforts to facilitate informed discussion, Japan has presented its research in various IWC Technical Committee Working Groups and Sub-Committees between 1986 and 1996.[21] In 1997 an intercessional IWC meeting was held in Japan to consider ways to resolve the small-type whaling impasse. At this meeting, additional reports, summarizing the research of the previous decade, were provided.[22]

---

the Okhotsk Sea/West Pacific stock. This is based on estimates of productivity for this resource which a number of members of the committee considered overly conservative. (Anon., *Report of the Scientific Committee. Annex F. Report of the Sub-Committee on North Pacific Minke Whales*, 156, 162, 177 *in* 42nd REP. INT'L WHALING COMMISSION (1992)). There have been suggestions that there are two stocks rather than one in this area, with the Okhotsk Sea stock from which Japanese catches would be taken, the smaller of the two and hence with a lesser sustainable yield than indicated above. However, extensive research in recent years has indicated that a West Pacific stock (if it exists at all) is scarcely sufficiently abundant to Japan to impact the 200 figure as an estimate of the sustainable yield (H. Hatanaka, pers. comm).

[21]   In March 1997, the government of Japan brought most of these papers together into a book: THE INSTITUTE OF CETACEAN RESEARCH, PAPERS ON JAPANESE SMALL-TYPE COASTAL WHALING SUBMITTED BY THE GOVERNMENT OF JAPAN TO THE INTERNATIONAL WHALING COMMISSION, 1986–1996 (1997). Page numbers referred to in the IWC reports cited in this chapter are (with a single exception) from this 1997 book. The single exception is to Akimichi, *supra* note 18, tabled at the IWC as IWC/40/23, and which was not included in the 1997 book (but is generally available).

[22]   These 1997 reports were: James R. McGoodwin, *The Importance of Maintaining Diverse Economic Alternatives in Communities of Small-scale Fishers*, TC/M97/CBW/1, 1–20; Junichi Takahashi, *Reviewing the IWC Discourses on Small-type Whaling: Content Analysis of Chairman's Reports and Other Documents, 1987–1996*; TC/M97/CBW/2 at 1–10; John Beatty & Junichi Takahashi, *Formal and Substantive Economic Framework of Small-type Whaling in Japan*, TC/M97/CBW/3 at 1–7; John Beatty, *Baubles, Blubber and Beads: Trade, Economics and Commercialism*, TC/M97/CBW/4, 1–7; Milton M.R. Freeman, *A Review of Documents on Small-type Whaling, Submitted to the International Whaling Commission by the Government of Japan, 1986–95*, TC/M97/CBW/5 at 1–29; Government of Japan, *An Overview of the IWC Discussion on the Japanese Small-type Coastal Whaling (JSTCW) from 1986–1996*, TC/M97/CBW/6 at 1–11.

2.   JAPANESE SMALL-TYPE WHALING

2.1.   History

The people of Japan have relied on marine resources to sustain them since prehistoric times. This is partly the result of living in a mountainous and densely forested island nation; partly the result of the early introduction of Buddhism (with its proscriptions again killing four-legged animals); and partly due to the high biodiversity and productivity of its coastal waters.[23]

For many centuries, whales have been included among the food resources used by Japanese coastal communities. Indeed, Japan differs from most other former whaling nations by making full and varied use of all soft-tissue and cartilaginous parts of the whale for food.[24] In earliest historic times, whale hunting was conducted using hand-held harpoons (the *tsukitori-ho* method). In the 17th century, a new form of whaling, using nets, harpoons, and a variety of hunt and support boats (the *amitori* method) developed in Taiji (in present-day Wakayama Prefecture) from where it spread to many other parts of Japan.[25] This highly organized form of whaling required a large, diversified work-force to capture whales and process the carcasses. The shore-based factories manufactured various food products, whale oil-based insecticide

[23]   C.W. Nicol, *Observations on Taiji Attitudes and Diet as Reflected by their Whaling History*, IWC/41/21 at 56–57 (1989).

[24]   The Government of Japan, *Small-type Whaling in Japan's Coastal Seas*, TC/38/AS2 at 2 (1986); Akimichi, *supra* note 18, at 92–95 (1988); The Government of Japan, *Report to the Working Group on Socio-Economic Implications of a Zero Catch Quota*, IWC/41/21 at 56–58 (1989); Junichi Takahashi, Arne Kalland, Brian Moeran *et al., Japanese Whaling Culture: Continuities and Diversities*, TC/41/STW2 at 131, 134 (1989); Masami Iwasaki-Goodman & Milton M.R. Freeman, *Social and Cultural Significance of Whaling in Contemporary Japan: A Case Study of Small-type Coastal Whaling*, IWC/46/SEST3 at 270 (1994).

[25]   The Government of Japan (1986) *supra* note 24, at 1–2; Akimichi, *supra* note 18, at 10–17; Stephen R. Braund, Milton M.R. Freeman, & Masami Iwasaki, *Contemporary Sociocultural Characteristics of Japanese Small-type Coastal Whaling*, TC/41/STW1 at 107–108 (1989); Takahashi, *supra* note 24, at 123–35; Arne Kalland, The Spread of *Whaling Culture in Japan*, TC/41/STW3 at 137–50 (1989).

for use on rice fields, fertilizer and other useful items. In this way the whole whale carcass was used.[26] In addition, artisans were employed at the shore stations to make and repair the boats, ropes, nets, barrels, and tools required for whaling and whale processing. Wherever this form of large-scale shore-based whaling became established, it dominated the local economy, diet and culture; indeed, these enterprises were the largest industries in medieval Japan. These communities became widely recognized as "whaling towns," a name that most continue to bear to this day.[27]

During the 19th century, Japanese whaling grounds were discovered by American and European whalers, and within a few years the foreign whalers quickly decimated the slow-swimming baleen whale stocks, and particularly the right whales upon which *amitori*-whaling depended. Japanese whalers now had to hunt faster-swimming rorquals (*e.g.*, blue, fin, and Bryde's whales), which required replacing the hand-rowed whaling boats with motorized vessels.[28]

Throughout this period of change, small-type whaling continued in nearshore waters, based on catching and processing small-whale species (*e.g.*, minke, beaked, and pilot whales). Production from the small-type whaling operations was limited by the size and small numbers of whales taken, so that most of the meat landed in the whaling communities was consumed locally.[29] Small-type whaling boats (locally known as "minke boats"), almost always returned to port each night after a hunt averaging

---

[26]   Takahashi, *supra* note 24, at 125.

[27]   The Government of Japan (1989), *supra* note 24, at 62, 68; Theodore C. Bestor, *Socio-economic Implications of a Zero Catch Limit on Distribution Channels and Related Activities in Hokkaido and Miyagi Prefectures, Japan*, IWC/41/SE1 at 77 (1989); Braund, *supra* note 25, at 113.

[28]   The Government of Japan (1986), *supra* note 24, at 2; The Government of Japan (1989), *supra* note 24, at 57; Takahashi, *supra* note 24, at 125; The Government of Japan, *Distinguishing Between Japanese STCW and LTCW in Relation to Coastal Whale-Fishery Management*, TC/42/SEST3 at 159–60 (1990); Michael Ashkenazi & Jeanne Jacob, *Summary of Whale Meat as a Component of the Changing Japanese Diet in Hokkaido*, IWC/44/SEST2 at 214 (1992).

[29]   Akimichi, *supra* note 18, at 19–25, 32–51, 86–91; Takahashi, *supra* note 24, at 133; The Government of Japan (1989), *supra* note 24, at 160–67.

about 13 hours. Hunting required full daylight, as no electronic whale-finding equipment was used;[30] about one-third of trips succeeded in taking a whale. Minke whales were taken on average about 20 miles offshore, pilot whales and beaked whales usually closer inshore.[31]

Prior to WWII, there were never more than 20 small-type whale boats licensed to operate in Japan. However, following the loss of large-type catcher boats during the Pacific War, there was a rapid increase in small-type whaling immediately following the end of the war, when many small fishing boats were converted into small-type whaling vessels to help alleviate severe food shortages. With post-war reconstruction underway, the central authorities rationalized the whale fishery by replacing some of the small-type whaling licences (that had reached 83 in 1947) with a smaller numbers of large-type whaling licences.[32] Thus, by around 1960, only nine small-type whaling licenses remained, the same number of licenses and boats that exist today.[33]

2.2.   The Scale of the Japanese Small-type Whale Fishery

Contemporary small-type whaling is carried out from three towns on the Pacific coast of the main Japanese island of Honshu, namely Ayukawa, Taiji, and Wada, each with between 2,000 to 4,000 residents; and from the Okhotsk Sea fishing port of Abashiri (population ca. 43,000) on the northern island of Hokkaido. Ayukawa has three small-type whaling boats, and each of the other towns has two boats. Minke boats range in size from 15 to 49 tons (the maximum size allowed under Japanese law) with a five to eight-man crew.[34] When minke whaling was suspended due to the IWC whaling moratorium, the industry supported

---

[30]   Braund, *supra* note 25, at 112; Takahashi, *supra* note 24, at 129.

[31]   The Government of Japan (1989), *supra* note 24, at 163–65.

[32]   The Government of Japan (1986), *supra* note 24, at 3; Akimichi, *supra* note 18, at 15.

[33]   The Government of Japan (1986), *supra* note 24, at 3; Akimichi, *supra* note 18, at 15, 18; Braund, *supra* note 25, at 109; The Government of Japan, *Action Plan for Japanese Community-based Whaling (CBW): Revised*, IWC/47/46 at 292 (1995).

[34]   The Government of Japan (1986), *supra* note 24, at 4; Akimichi, *supra* note 18, at 18–20.

63 whalers working on the nine minke boats.[35] In addition, onshore flensing stations employed one or two flensers each, assisted by volunteer helpers (often retired whalers and other local people).[36] The total workforce engaged in Japanese small-type whaling at sea and on land in 1987 numbered about 100 people, of which 75 were full-time employees. This workforce often included members of the boat owners' immediate families.[37]

Seven of the nine small-type whaling boats are owner-operated; one other boat is operated by a Fisheries Cooperative Association, and one is operated by a former employee of a now-disbanded large-type whaling company, a resident of the whaling town.[38] Many whalers, flensers and processors belong to whaling families, reflecting the family-based nature of small-type whaling operations and the cultural importance in Japan that is placed on following family traditions in order to respect ancestors.[39] Indeed, in the Japanese whaling towns, whaling is considered by many as "a profession granted by heaven . . . an honourable way of life worthy of perpetuation."[40]

The members of each minke boat crew constitute a recognized social group and they continue socializing during the non-whaling season.[41] Although the coastal whaling season is only six months long, the boat owner continues to pay his crew a partial wage in the non-whaling season in order to maintain crew social solidarity throughout the year.[42]

---

[35] The Government of Japan (1986), *supra* note 24, at 4, 9.

[36] Akimichi, *supra* note 18, at 29–30.

[37] The Government of Japan (1989), *supra* note 24 at 23, Tables 9–12; The Government of Japan, *Socio-economic Countermeasures in the Four Japanese STCW Communities*, TC/42/SEST2 at 154–55 (1990).

[38] Akimichi, *supra* note 18, at 18–20; The Government of Japan, *How Different are Small-type and Large-type Whaling?* IWC/45/SEST1 at 248 (1993).

[39] The Government of Japan (1989), *supra* note 24, at 30, 35; Bestor, *supra* note 27, at 73; Lenore Manderson & Helen Hardacre, *Small-type Coastal Whaling in Ayukawa: Draft Report of Research, December 1988–January 1989*, IWC/41/SE3 at 96–97 (1989).

[40] Manderson, *id.* at 102.

[41] Takahashi, *supra* note 24, at 132–33.

[42] The Government of Japan (1989), *supra* note 24, at 25; The Government of

2.3. Culturally Important Aspects of Whaling

Whaling remains a culturally important activity in all present-day whaling towns, as well as a number of former whaling communities. In these whaling towns, for example, the symbolic association of whale meat with health, longevity, and vitality insures the continuing place of whales in the local food and ceremonial culture.[43]

In the whaling towns there exists an extensive system of customary gift-based ritual exchange that occurs prior to, and throughout, the whaling season. Whale meat continues to be gifted and received within the community throughout the entire year.[44]

In anticipation of gifts of whale meat being received during the whaling season, townspeople bring gifts *(omiki)* to boat owners, and to the boat and her crew. These gifts are ceremonially presented to the boat owner, who may receive several hundred such gifts each year. In 1987 about 1,500 bottles of *sake*[45] were purchased in Ayukawa for distribution during the minke whale First Catch *(hatsuryo)* ceremonies.[46] The same situation occurs in other whaling towns such as Abashiri and Wada.[47] In each

---

Japan, *Japan's answers to questions on Japanese small-type coastal whaling,* TC/42SEST9 at 193 (1990).

[43]   Akimichi, *supra* note 18, at 66–74; Manderson, *supra* note 39, at 91; Braund, *supra* note 25, at 113, 115; Stephen R. Braund, Junichi Takahashi, John A. Kruse *et al., Quantification of Local Need for Minke Whale Meat for the Ayukawa-based Minke Whale Fishery,* TC/42/SEST8 at 185–89 (1990); The Government of Japan, *The Cultural Significance of Everyday Food Use,* TC/43/SEST1 at 197–200 (1991); Ashkenazi, *supra* note 28, at 214–17; The Government of Japan, *The Importance of Everyday Food Use,* IWC/44/SEST4 at 223–27 (1992).

[44]   Akimichi, *supra* note 18, at 41–45; Bestor, *supra* note 27, at 74; Manderson, *supra* note 39, at 85–90; The Government of Japan, *supra* note 38, at 249; Iwasaki-Goodman, *supra* note 24, at 273–74.

[45]   *Sake,* a rice-based wine, is the most important item of commensality and social exchange in Japanese culture, Emiko Ohnuki-Tierney, RICE AS SELF: JAPANESE IDENTITIES THROUGH TIME 97 (1993).

[46]   Akimichi, *supra* note 18, at 43–45.

[47]   *Id.* at 43–44.

case the boat owner or crew member makes a return gift of whale meat for the *omiki* earlier received.[48]

2.4.   The Social Significance of Whale Meat Distribution

The one or two professional flensers working at the small-type whaling shore stations receive whale meat as well as a cash payment for their work.[49] Local people assisting the flenser(s) receive their payment only in whale meat[50] and relatives of whalers also receive free whale meat at the flensing station.[51] In addition, all those involved in catching and processing the whale receive a bonus *(buai-kyu)* paid in whale meat.[52] Consequently, a considerable amount of whale meat enters into non-commercial distribution throughout the whaling towns as people receive whale meat either as payments-in-kind or as gifts. These distributions occur every time a whale is caught and flensed.[53] Such produce is freely shared with neighbors, associates, relatives and friends, for everyone appreciates receiving such gifts and similar gifts are continually received from others. This ready availability of "free" whale meat in the whaling towns has led to the common local sentiment: "fish is to buy, whale is to be received as a gift."[54]

In addition to whale meat distributions and redistributions occurring throughout the whaling season, boat owners make periodic gifts of whale

---

[48]   The Government of Japan (1990), *supra* note 28, at 166.

[49]   Manderson, *supra* note 39, at 89.

[50]   Akimichi, *supra* note 18, at 43.

[51]   Bestor, *supra* note 27, at 74.

[52]   Akimichi, *supra* note 18, at 29; The Government of Japan (1989), *supra* note 24, at 55.

[53]   Bestor, *supra* note 27, at 74; Manderson, *supra* note 39, at 89–90; Takahashi, *supra* note 24, at 133; Braund, *supra* note 25, at 114; The Government of Japan (1990), *supra* note 28, at 166; The Government of Japan, *supra* note 38, at 249.

[54]   Akimichi, *supra* note 18, at 46; The Government of Japan (1989), *supra* note 24, at 30–31, 62; Manderson, *supra* note 39, at 88, 89; Braund, *supra* note 25, at 114; The Government of Japan, *Commercial Distribution of Whale Meat: An Overview*, IWC/44/SEST at 222 (1992).

meat to community institutions, such as the hospital, old peoples' home, volunteer fire brigade, school, and to temples and shrines.[55] Whale meat is a preferred choice during traditional seasonal gift giving events, such as *chugen* (the mid-summer gift) and *seibo* (the year's end gift).[56]

In winter, people may need to purchase whale meat to give away as gifts, because no new supplies from whalers will enter the community distribution after the end of the whaling season. However, the purpose of such purchases is to obtain meat required to give to another as a culturally appropriate gift. The use of money to pay for items needed for subsistence purposes is common in all societies today where money is a universal currency that serves to maintain customary practices that originated in earlier (pre-cash) times.[57]

---

[55]   Akimichi, *supra* note 18, at 46–48.

[56]   The Government of Japan (1989), *supra* note 24, at 31–32; Manderson, *supra* note 39, at 85; Braund, *supra* note 25, at 114–115; Iwasaki-Goodman, *supra* note 24, at 273.

[57]   Moeran, *supra* note 16, at 238–40; The Government of Japan, *'Commercial' vs. 'Subsistence,' 'Aboriginal' vs. 'Non-aboriginal,' and the Concept of Sustainable Development in the Context of Japanese Coastal Fisheries Management*, IWC/46/SEST1 at 263 (1994). In regard to the sale and purchase of whale products in Greenland, a Danish IWC report stated: "The role of money in the distribution channels does not justify the labelling of aboriginal subsistence hunting as commercial in the sense used in the IWC." Denmark has made widespread reference in its submissions to the IWC, to the sale of whale in the Greenlandic aboriginal subsistence hunt: R. Petersen, E. Lemche & F.O. Kapel, *Subsistence Whaling in Greenland, in* THE ANTHROPOLOGY OF COMMUNITY-BASED WHALING IN GREENLAND: A COLLECTION OF PAPERS SUBMITTED TO THE INTERNATIONAL WHALING COMMISSION 29, 34 (Marc G. Stevenson, Andrew Madsen & Elaine L. Maloney eds., 1997); P. Helms, O. Hertz & F.O. Kapel, *The Greenland Aboriginal Whale Hunt*, in *id.* at 57, 80; R. Petersen, *Communal Aspects of Preparing for Whaling, the Hunt Itself, and the Ensuing Products, in id.* at 95, 96, 98; Jens Dahl, *Hunting and Subsistence in Greenland in Light of Socioeconomic Relations, in id.* at 168, 171, 174–76; Robert Petersen, *Traditional and Present Distribution Channels in Subsistence Hunting in Greenland, in id.* at 179, 183–188; Svend E. Larsen and Klaus G. Hansen, *Inuit and Whales at Sarfaq (Greenland), in id.* at 191, 213; Erling Josefsen, *Cutter Hunting of Minke Whale in Qaqortoq (Greenland), in id.* at 223, 232–33; Richard A. Caulfield, *Greenland Inuit Whaling in Qeqertarsuaq Kommune, in id.* at 241, 252–53; Richard A. Caulfield, *Whaling and Sustainability in Greenland, in id.* at 263, 270–71.

2.5.   Commercial and Non-Commercial Whale Meat Distribution

During IWC discussions about the meat distribution associated with Japanese small-type whaling, attention always focused upon the relative proportions of the catch entering into "non-commercial" (sharing and gifting) and "commercial" (for cash) exchange.[58] In response, it was always explained that attempts to compare two (or more) wholly incomparable systems of exchange is fraught with difficulties and will only provide misleading and counterproductive conclusions.[59]

However, those who argue against Japanese small-type whaling appear to believe that it is indeed possible and useful to place monetary value on a ritual, a social bond, and a shared identity.[60] They insist that such a monetary conversion be undertaken, despite being advised that such a result would be invalid, and subject to endless dispute, as well as contributing nothing useful to understanding the social significance of what is actually being achieved in society.[61]

In the context of Japanese small-type whaling, none of these community-wide ritual whale meat distributions should be regarded as primarily economic transactions. The proportion of the total catch involved in traditional gift exchange may not be large in relation to the total landed catch (although it may be large in relation to the proportion of

---

[58]   The Government of Japan (1990), *supra* note 42, at 192–94; The Government of Japan (1992), *supra* note 54, at 221–22; The Government of Japan (1992), *supra* note 43, at 224.

[59]   Denmark has also argued at the IWC that imposing this distinction in small-scale whaling is difficult, if not impossible to sustain on objective grounds. At the 1989 IWC meeting, a Greenland Home Rule Government report stated: "One thing remains certain: the distinction between subsistence and commercial harvests is artificial. . . . Maintaining a distinction between the two only serves to undermine local control, while strengthening neo-colonial control." Jens Dahl, *Hunting and Subsistence in Greenland in Light of Socioeconomic Relations*, 168, 178, *in* Stevenson, *supra* note 57.

[60]   The Government of Japan (1990), *supra* note 42, at 192–193; The Government of Japan (1992), *supra* note 54, at 222.

[61]   The Government of Japan (1990), *supra* note 28, at 160; The Government of Japan (1990), *supra* note 42, at 194.

the first caught whale).[62] However, the true value of such circulating prestations are not commercial, but are social and cultural: they are important because they involve repeatedly almost every person in the community and they are derived from local enterprise whose several benefits are widely shared and highly valued. Local residents collectively reaffirm and validate their sense of community and collective identity by celebrating the seasonal bounty of the sea upon which their own and their community's existence has depended and continues to depend.[63] As one Wada resident explained, "through this meat gift-giving custom, we have a complex social networking arrangement. This area has a system that depends on forty whales at the moment, but if that quota disappears, so does our social system."[64]

2.6.   Religious Observances Associated with Whaling

Since the earliest times, whales have provided an important source of food and employment in Japan. Consequently, whales are equated with a widespread sense of security and prosperity, particularly in whaling towns. The whale, as an ample provider of community needs, remains symbolically important in those small coastal whaling communities that were often relatively remote until recent times. Within such communities, a number of ceremonies, both secular and religious, are practiced to show gratitude and court divine favor to ensure the benefits of whaling continue to be enjoyed. These religious obligations to whales *(kujira kuyo)* and celebrations of whaling *(kujira matsuri)* do not end when a village stops whaling, and in many communities, ceremonies to insure the peaceful repose of the souls of whales taken by villagers in the past are faithfully conducted several generations after whaling has ceased in the area.[65]

Local shrines are important in defining a community's geographic

---

62   Akimichi, *supra* note 18, at 45–46; Bestor, *supra* note 27, at 74.

63   The Government of Japan, *supra* note 38, at 249.

64   Akimichi, *supra* note 18, at 46; The Government of Japan (1989), *supra* note 24, at 30–32; Braund, *supra* note 25, at 113; Takahashi, *supra* note 24, at 132–33; Moeran, *supra* note 16, at 238.

65   Akimichi, *supra* note 18, at 54–56.

and social boundaries.[66] Throughout the whaling season, female members of whalers' families visit their local shrines to pray for whalers' safety, for a good catch, and for the souls of whales.[67] Before every coastal whaling season starts, the small-type whalers assemble in Ayukawa and visit a Shinto shrine, Kinkazan, to participate in solemn services.[68] Shinto priests also officiate at purification ceremonies conducted on board the whaling boats.[69]

In addition to these Shinto ceremonies, members of the whaling communities participate in Buddhist ceremonies, two of which are particularly important. The first involves memorial services for the souls of whales killed, and the second is for the souls of whalers, who, by taking life, seek forgiveness and spiritual compensation for the loss of karmic merit that results from the taking of life.[70] In some whaling towns, virtually the whole community participates in these services.[71] Religious services held to propitiate the souls of whales that have been killed are held early in the New Year. These ceremonies take place in Buddhist temples, where death registers *(kako-cho)* containing the names and details of dead whales are kept.[72] The whale boat spirits *(funadama)*, responsible for protecting the whale boat, are also renewed at the New Year.[73] Each boat has a small Shinto altar on board, at which sacramental offerings to the appropriate deity are always in place.[74]

Buddhist and Shinto ceremonies are often constructed around a communal meal *(naorai)* that is shared between parishioners and deities. The New Year period is when a number of religious observances take

---

[66] Takahashi, *supra* note 24, at 133.

[67] Braund, *supra* note 25, at 116; Takahashi, *supra* note 24, at 133.

[68] Takahashi, *supra* note 24, at 133.

[69] Akimichi, *supra* note 18, at 59–64; Moeran, *supra* note 16, at 241.

[70] *Id.* at 57.

[71] Takahashi, *supra* note 24, at 133.

[72] Akimichi, *supra* note 18, at 57.

[73] Manderson, *supra* note 39, at 91.

[74] Akimichi, *supra* note 18, at 64; Braund, *supra* note 25, at 116.

place, and in whaling towns the centrality of whaling to the community's existence and well-being is reflected in the prominence of whale meat in these ceremonies.[75] Thus whale meat is offered to the community's tutelary deities as well as to the domiciliary gods in peoples' home shrines over a one- or two-week period at the New Year. The importance of whale meat at these religious events is reflected by the prominence it has in the varied food offerings. It is significant that "normally it would be unthinkable for red meat to be offered on an altar to Shinto divinities, but this is precisely what happens in Ayukawa."[76]

3.   WHALES, FOOD, AND IDENTITY

3.1.   Diet and Cultural Identity

According to one recognized authority on Japanese dietary traditions:

> Eating whale meat is undeniably a part of Japanese culture, just as eating pork is a part of European culture. . . . Not eating pork defines a segment of mankind . . . just as not eating whales defines another segment. Thus eating whale meat for some Japanese contraposes "Japanese" with "non-Japanese" in a deeply felt, emotional, way. . . . Food culture, as a whole, plays a large part in daily Japanese life . . . losing access to fresh whale meat . . . means that many residents are losing access to an important part of their natal culture: a part that is also particularly intimate and has strong emotional appeal.[77]

Research undertaken to inform IWC discussions on Japanese small-type whaling demonstrated the existence of a series of localized cuisines reflecting historical practices. Thus, the cuisine in Abashiri and Ayukawa is based principally upon minke whale, that of Wada upon Baird's beaked whale, and of Taiji upon pilot whales and dolphins.[78]

---

[75]   Manderson, *supra* note 39, at 90–91, 94; Takahashi, *supra* note 24, at 133.

[76]   Manderson, *supra* note 39, at 90.

[77]   Ashkenazi, *supra* note 28, at 217.

[78]   Akimichi, *supra* note 18, at 30–31; Manderson, *supra* note 39, at 92–94; Braund, *supra* note 25, at 113–16; Takahashi, *supra* note 24, at 133.

Whaling-town residents' strong attachment to their local cuisine results from a series of cultural and personal valuations associated with hunting, processing, distributing, consuming, and celebrating whales. As a customary food, whale, invokes a variety of very positive associations in peoples' minds, such that to contemplate a future without such foods constitutes a worrisome, indeed depressing, thought.[79]

To assist IWC delegates in understanding the magnitude of the social and cultural problems resulting from the extended pause in minke whaling, the social science literature detailing the importance of maintaining peoples' culturally appropriate diets was brought to delegates' attention.[80] These presentations drew attention, *inter alia*, to the relationship between customary dietary patterns, socialization and social relationships (including intra-family and gender relations), the importance of customary diet in defining social groups and cultural identity, and the manner in which gastronomy relates to developing moral and aesthetic norms and precepts characteristic of particular cultures and social groups.

Japanese food culture reflects a remarkably complex, elaborate, and diversified food system derived from the variety of available foods. These distinctive local cuisines remain in existence despite recent outside influences. In Japanese whale-based cuisine, every non-ossified or keratinized part of the whale is eaten, so the cuisine may be highly diversified, even if based upon a single cetacean species favored at a particular locality.[81] The dietary preference for whale meat in the whaling towns transcends mere preference; rather, it is based on a number of historical, local, familial, symbolic and aesthetic valuations that "in their totality ensure a sense of well-being for the people concerned, and in their absence represent a profound and damaging loss."[82]

---

[79]   The Government of Japan (1989), *supra* note 24, at 30–31; Manderson, *supra* note 39, at 92–94; Braund, *supra* note 25, at 113–16.

[80]   The Government of Japan (1991), *supra* note 43, at 195–201; The Government of Japan (1992), *supra* note 43, at 223–28.

[81]   Akimichi, *supra* note 18, at 67–70, 90–95; The Government of Japan (1989), *supra* note 24, at 38, 63; Manderson, *supra* note 39, at 92–93; Iwasaki-Goodman, *supra* note 24, at 272–73.

[82]   The Government of Japan (1991), *supra* note 43, at 201.

Research undertaken in one whaling town (Ayukawa), elicited 30 occasions when whale meat was the culturally appropriate centerpiece of the *naorai* (celebratory meal). Such occasions included New Year's Eve, New Year's Day, Girl's Day, Boy's Day, Initiation Day, *Obon* (All Soul's Day), Birth of a Child, Child's First Birthday, Child's First Visit to Shrine, auspicious Birthday Anniversaries (viz. 60, 77, and 80 year anniversaries), Memorial Services for Ancestors, Weddings, Building or Restoring a Home or Store, *Toshi-iwai* (prayer service for safety and health in a hazardous year), Old Peoples' Day, First Day of School, Graduation, Recovery from Illness, Receiving a Special Award or Recognition, Return of Family Members to Home after Absence Elsewhere.[83]

3.2. The Social and Cultural Importance of Restaurants and Caterers

During IWC meetings, whaling opponents regard the availability of whale meat in restaurants and inns as constituting a serious impediment to the resumption of whaling. In answer, Japan has tried to explain the reason and importance of such eating establishments to cultural life and social relations in the whaling towns.

The importance of food and of eating together is commonly recognized in all societies. Expressions such as *"we are what we eat,"* and the practice of sharing a meal when important family, business or professional interactions take place attests to the fundamental role that food plays in defining, creating, and confirming social relationships. In Japan, group activities occur far more commonly than in western societies. However, Japanese homes are often small, and in small rural communities the large number of relatives, friends, and work associates taking part in such group activities requires that the meal takes place outside private homes, commonly in a restaurant or inn. Thus, in the whaling towns, providing whale meat to inns, restaurants, and other establishments that cater to large family or group meals is essential for any number of socially- and culturally-important occasions. It also follows that meals served on these significant social occasions in the whaling towns must include whale dishes if the meal is to fulfil its intended purpose.[84]

---

[83] Braund, *supra* note 43, at 186–88.

[84] The Government of Japan (1992), *supra* note 54, at 222.

There is another important reason that commercial eating establishments in whaling towns need to serve whale meat and that is to encourage tourists to visit these towns. Abashiri, Ayukawa, and Taiji are all promoting tourism to partially compensate for the problems of their declining fisheries (including whale fisheries) and the limited local opportunities for other forms of economic diversification.[85] Internal tourism in Japan is partly driven by the expectation of regional food specialities,[86] and consequently most Japanese tourists visiting a whaling town would find the visit disappointing and unmemorable if whale dishes were not available.[87] Between 1986 and 1987, servings of whale meat meals more than doubled at one Taiji eating establishment due to the concern of tourists that supplies of whale meat would be exhausted when the whaling ban came into effect in 1988.[88] When IWC discussions focus upon mitigating the socio-economic impacts of the whaling moratorium on whaling towns, anti-whaling delegates, ignorant of the expectations of Japanese tourists, persist in advocating increased tourism (in the absence of whaling) as the solution to these coastal communities' economic decline. This is despite Japanese tourist expectations being frequently reported at the IWC.

4.   COMMERCIAL AND NON-MONETIZED EXCHANGES

4.1.   Economic Aspects of Japanese Small-Type Whaling

Japanese small-type whaling involves both commercial and subsistence elements. The term "subsistence" is used in a selective and inconsistent manner by whaling opponents during IWC discussions, and the term "commercial" has never been defined by the IWC.

---

[85]   The Government of Japan (1989), *supra* note 24, at 24–26; The Government of Japan (1990), *supra* note 37, at 155–56; Nelson H.H. Graburn, *Whaling Towns and Tourism: Possibilities for Development of Tourism at the Former [sic] Whaling Towns—Taiji, Wada, and Ayukawa*, 1–6. Information Paper submitted by the Government of Japan to the 42nd IWC Meeting, Noordwijk, The Netherlands (1990).

[86]   Akimichi, *supra* note 18, at 101.

[87]   Bestor, *supra* note 27, at 78.

[88]   *Asahi Shimbun*, January 13, 1989, quoted in The Government of Japan (1989), *supra* note 24, at 35.

There is an accepted use of cash in all subsistence societies together with frequent market sales of food in the course of subsistence pursuits. It has been noted that "all people whose lives articulate in any way with the modern world require to use money as a generalized currency. However, this does not make subsistence activities any less subsistence activities."[89] The IWC accepts that aboriginal-subsistence whaling (in the case of Greenland and Russia) involves the selling and buying of meat,[90] provided the meat is used "locally" (a term which is also not defined by the IWC). So-called aboriginal-subsistence whaling is exempt from the IWC moratorium *not* because of an absence of monetary exchange or its exclusive consumption in the producing community, but rather because of a recognized dependence on whaling and whale product use by a culturally defined group of people who need to produce and consume whale meat for a variety of demonstrated social, cultural, religious, ceremonial, nutritional, and economic reasons.[91]

In this regard, Japanese small-type whaling shares the characteristics that allow North American, Greenlandic and Russian whalers to be exempt from the whaling moratorium.[92] However, this exemption is dis-

---

[89]   Akimichi, *supra* note 18, at 6; The Government of Japan (1992), *supra* note 54, at 222; The Government of Japan, *supra* note 57, at 263–64; *see also* Thomas D. Lonner, *Subsistence as an Economic System in Alaska: Theoretical Observations and Management Implications*, in CONTEMPORARY ALASKAN NATIVE ECONOMIES, 15, 21 (Steve J. Langdon ed., 1986).

[90]   The International Convention for the Regulation of Whaling does not explicitly refer to aboriginal-subsistence whaling; however, the rights of aboriginal tribes were first recognized in the 1946 Schedule to the Convention. Alexander Gillespie, *Aboriginal Subsistence Whaling: A Critique of the Inter-Relationship Between International Law and the International Whaling Commission*, 12 COLO. J. INT'L ENVTL. L. & POL'Y 77, 79 (2001); Milton M.R. Freeman, *The International Whaling Commission, Small-type Whaling, and Coming to Terms with Subsistence*, 52 HUMAN ORG. 243–48 (1993).

[91]   The Government of Japan, *A Critical Evaluation of the Relationship Between Cash Economies and Subsistence Activities*, IWC/44/SEST5 at 230–33 (1992); Moeran, *supra* note 16, at 238–42.

[92]   Some opponents of whaling assert that Japan claims its community-based whaling is "aboriginal-subsistence" in nature, and therefore should be exempt from the 1982 whaling moratorium. This is not so; Japan has repeatedly pointed out the many cultural and socio-economic similarities between aboriginal-subsistence whal-

allowed by the majority of IWC member states in the case of Japanese small-type coastal whaling. This inconsistency occurs in part because IWC has not defined what constitutes a "recognized dependence on whaling and whale products," a concept that consequently is left entirely to individual interpretation. Clearly, "dependence" (understood by the anti-whaling majority to refer to economic and/or nutritional dependence) upon hunting wild animals cannot be so easily associated with a people whose tidy communities, well-stocked food shops, and general appearance is the antithesis of the seeming poverty and neglect that is stereotypically associated in the public mind with aboriginal communities in Alaska, Siberia, and elsewhere.

In cultural terms, however, Japanese small-type whaling, in common with aboriginal-subsistence whaling in the Arctic, insures that coastal communities maintain an effective and unbroken cultural connection to their traditional past. Such connections are abundantly represented in the many Buddhist and Shinto religious rituals and beliefs, the whaling-related festivals, the ritualized whale meat sharing practices, the traditional payments-in-meat made to boat crews, flensers, and local helpers, as well as the techniques of whaling, flensing, and whale product processing. All such practices continue to be passed on by today's Japanese small-type whaling community members, much as they received them from their forebears.

The use of cash for payment of wages, fees, and taxes has existed since feudal times,[93] and is as much a respected traditional part of Japanese coastal whaling, as are the ways in which the meat is flensed, processed and eaten, or the whales celebrated in rituals and festivals.[94]

Marketing practices in the small-type whaling town of Wada illustrates well the customary nature of the whaling trade. In Wada, following

---

ing and its own community-based whaling, and contrasted both these categories to large-scale industrial whaling of the past. Akimichi, *supra* note 18, at 79–84; Moeran, *supra* note 16, at 237–44.

[93]   Akimichi, *supra* note 18, at 12; Takahashi, *supra* note 24, at 123, 131, 134; Kalland, *supra* note 25, at 138, 144; The Government of Japan (1992), *supra* note 54, at 222.

[94]   Akimichi, *supra* note 18, at 43–65; Manderson, *supra* note 39, at 85–94; Ashkenazi, *supra* note 28, at 214–17; Iwasaki-Goodman, *supra* note 24, at 270–78.

strict tradition, the price decided for whale meat is announced by the seller at the very beginning of the season and is not changed there-after—even if subsequent shortages due to poor hunting would allow a higher price to be obtained.[95] This practice indicates that profit-maxi-mization (a feature of capitalism abhorrent to many opponents of whal-ing) does not occur in the Wada whale meat market.[96]

This aspect of market relations illustrates the continuing importance in Japan of building and maintaining business relationships on a foun-dation of social amity and trust. Thus, in regard to the method of setting the price of whale meat in the small-type whaling town of Abashiri, it is reported that "intimate ties and long-term business relations between wholesalers and this [minke boat] owner enable the latter to set the most appropriate price without putting existing social relations in jeopardy."[97]

4.2. Attempts to Normalize Market Changes Affecting Japanese Small-Type Whaling

Until recently, the sale of whale meat at the fish markets in the small-type whaling towns was only to buyers supplying local or regional (rather than national) markets. This limited distribution occurred because local and neighboring markets could absorb the relatively small quantities of fresh produce available each time a whale was landed. In addition, local distributors had business connections with buyers repre-senting local retail shops, restaurants, and inns, and after supplying these customers' needs, the quantity of surplus meat was too small and uncer-tain to be of interest to larger-scale buyers supplying distant markets.[98]

---

95   Akimichi, *supra* note 18, at 89.

96   For examples of non-profit-maximizing commercial behavior in other Japanese small-type whaling communities, *see* Akimichi, *supra* note 18, at 25, 35, 39; The Government of Japan (1989), *supra* note 24, at 24; The Government of Japan (1990), *supra* note 28, at 166–67. Similar commercial behavior occurs in some Greenland markets, which enjoy the aboriginal-subsistence exemption from the whaling moratorium: Erling Josefsen, *Cutter Hunting of Minke Whale in Qaqortoq (Greenland)* in Stevenson, *supra* note 57, at 223, 232.

97   Akimichi, *supra* note 18, at 35.

98   Akimichi, *supra* note 18, at 32–40; The Government of Japan (1989), *supra*

This customary market situation began to change in the late 1980s as a direct result of IWC-imposed restrictions. Thus, with the imposition of the moratorium and the consequent dismantling of all Japanese large-scale whaling companies in 1987, non-local whale meat buyers needed new sources of supply to meet their trading partners' continuing needs. Such new sources included fish markets in Ayukawa and in two or three designated landing ports in Hokkaido where minke meat from small-type whaling operations was being sold. Thus, part of the commercial sales of coastal minke and beaked whale products moved from the local and regional markets into a wider commercial distribution, a situation which persists to this day in regard to beaked and pilot whale products.[99]

The government of Japan has made recent[100] proposals to the IWC to re-establish the traditional distribution system whereby minke whale products are exclusively distributed within the food-culture area traditionally supplied by these community-based whaling operations. These proposals have been consistently rejected at IWC meetings, as have all efforts to negotiate acceptable conditions for a partial reinstatement of community-based minke whaling in these traditional whaling towns.

During the several years of discussion of Japanese small-type whaling at IWC meetings, it appears that most discussants have insisted on evaluating the economic behavior of Japanese whaling communities in purely western monetary terms.[101] Unfortunately, such ethnocentrism

---

note 24, at 34; Bestor, *supra* note 27, Table 4; Braund, *supra* note 25, at 112; The Government of Japan (1990), *supra* note 28, at 167; The Government of Japan (1992), *supra* note 54, at 221–22; The Government of Japan, *supra* note 38, at 248–49.

[99]   The Government of Japan (1989), *supra* note 24, at 33–34; The Government of Japan (1990), *supra* note 28, at 167.

[100]   The Government of Japan, *Action Plan for Japanese Community-based Whaling (CBW)*, IWC/45/SEST3 (1993); The Government of Japan, *Action Plan for Japanese Community-based Whaling (CBW): Distribution and Consumption of Whaling Products*, IWC/46/31 Rev. 2 (1994); The Government of Japan, *Action Plan for Japanese Community-based Whaling (CBW)*, IWC/47/SEST1 (1995); The Government of Japan, *Action Plan for Community-based Whaling (CWB): Revised*, IWC/47/46 (1995).

[101]   The Government of Japan (1990), *supra* note 28, at 159.

has prevented many IWC delegates from understanding the nature and multi-dimensioned significance of Japanese small-type whaling, despite lengthy explanations accompanying research reports tabled at several successive IWC meetings. At the 40th IWC Meeting, an expert panel advised the Technical Committee:

> to social scientists, the simple notion than an economy is either "commercial" or "subsistence" (or non-commercial) is decidedly unhelpful. It is generally recognized that a monetized economy can be non-commercial, for though the term "commercial" implies an interest in market forces, it does not imply a dependence upon market forces. A thoroughly commercialized economic transaction would require all the inputs to be paid for in cash, and all the benefits likewise to be in cash received from sales of the commodity. Clearly in the small-type coastal whaling communities, as in the aboriginal/subsistence whaling communities, many of the inputs and outputs do not involve cash transfers. Indeed, a large degree of the benefits (social, psychic, nutritional, and cultural) derived from such whaling activities occur wholly in non-monetized sectors of everyday life.[102]

Again, at the 42nd IWC Meeting, delegates were reminded of the need to appreciate that Japanese society is, in cultural terms, quite different from their own. Certainly, it would assist in gaining better understanding of events being described if more IWC delegates could begin to understand the need to consider Japanese society and economic behavior in relation to Japanese cultural realities, rather than believing that Euro-American behavior is universally applicable.[103]

4.3.   The Ancient Roots of Cash Use in Japanese Whaling

The use of cash in Japan has roots in profoundly pre-modern practices.[104] For example coastal communities have, since ancient times,

---

[102]   Akimichi, *supra* note 18, at 6.

[103]   The Government of Japan (1989), *supra* note 24, at 35; Bestor, *supra* note 27, at 73; The Government of Japan (1990), *supra* note 28, at 159–60.

[104]   Akimichi, *supra* note 18, at 12; The Government of Japan (1992), *supra* note

controlled access to marine resources. In such circumstances, whaling entrepreneurs wishing to set up whaling operations in a new location had to establish an appropriate relationship with the local resource managers. In the 17th century, the *amitori* (net-whaling) enterprises made large cash payments to the feudal authorities in order to signal both goodwill and commitment to the local community. In addition, cash compensation *(ura-gane)* was paid to local fishing villages for the inconvenience caused by whaling, lump-sum tax payments *(unjo-kin)* were made for each whale landed, and ground rent *(tatami-gane)* was paid for placing look-out stations and other shore facilities on community land.[105] This use of cash in whaling operations since medieval times demonstrates the culturally appropriate means by which Japanese gain the social acceptance that allows enterprises to function in new social settings with a minimum of social friction.

Four centuries later, the use of cash in community-based whaling enterprises in Japan has not significantly changed. Thus, in 1990, when small-type whalers considered building a shore facility to flense pilot whales in southern Japan, *ura-gane* (annual compensation) of one million yen to be paid to the local community was discussed, as well as *unjo-kin* of 4.3 percent of the value of each whale landed. In the IWC context, such cash payments are viewed as strictly commercial transactions. However, as these exchanges are not occurring in a western society, it makes little sense to interpret their purpose or significance in such narrowly Eurocentric terms. On the contrary,

> In these former and current whaling districts, demonstrating conformity with historically sanctioned practices remains important. In social terms, these financial arrangements facilitate necessary social intercourse between individual insiders (villagers) and outsiders (whalers from other communities) through culturally appropriate ways and means.[106]

54, at 222. *See also* Edwin O. Reischauer & Albert M. Craig, JAPAN: TRADITION AND TRANSFORMATION 63 (1978).

[105]   Takahashi, *supra* note 24, at 124, 134; Kalland, *supra* note 25, at 159–60.

[106]   The Government of Japan (1990), *supra* note 28, at 160.

4.4.   Cultural Persistence and Change: Opposing World Views

As mentioned earlier, the basis of objections directed at Japanese small-type whaling year after year at the IWC have focused upon the use of cash in this small-scale community-based fishery, giving the impression that the questioners believe the financial returns from the fishery was the only significant event taking place. Of course, without a cash return, no fishery can operate in a modern society. In the Japanese case, coastal whale fisheries have operated with cash during the preceding four centuries, for even in pre-modern times, as discussed above, supplies, rent, wages, taxes, goodwill payments, and loans had to be paid for with cash.

However, the point was repeatedly made by Japan at the IWC[107] that in addition to these financial transactions, small-type whaling serves a number of critically important non-economic local needs relating to the continued existence and integrity of these whaling towns. These non-economic functions include meeting various social, cultural, and religious institutional needs, and sustaining local dietary practices that are important in maintaining these communities' core institutions as well as community members' sense of individual and collective identity.

In the whaling towns, whalers' families enjoyed considerable social standing due to the whalers' ability to distribute gifts of whale meat to relatives, neighbors and friends. Suddenly, with the end of whaling, these social obligations could no longer be discharged, and whalers' felt shame at their lost ability to continue these culturally valued community practices, and their inability to pass on the traditions associated with their profession.[108]

---

[107]   The Government of Japan (1986), *supra* note 24; Akimichi, *supra* note 18; Manderson, *supra* note 39; Braund, *supra* note 25; Braund, *supra* note 43; The Government of Japan (1991), *supra* note 43; Ashkenazi, *supra* note 28; The Government of Japan (1992), *supra* note 43; The Government of Japan (1993), *supra* note 38; Iwasaki-Goodman, *supra* note 24.

[108]   The Government of Japan (1989), *supra* note 24, at 27, 30–32, 35, 39; Manderson, *supra* note 39, at 102–103; Braund, *supra* note 25, at 113; The Government of Japan, *Socio-economic Implications of Zero Catch Limit: Some Examples of Small-type Whaling*, TC/43/SEST2 at 203–204 (1991).

During discussion of these matters at the IWC, some delegates would point to the inevitability of such events occurring in a changing world. They gave as examples, the loss of employment occurring in English or Dutch fishing and mining towns when fisheries collapsed or mines closed.[109] However, the circumstances in which European herring fishermen and coal miners lost their jobs has almost nothing in common with the reason Japanese community-based whalers were losing their livelihood and traditions. The North Sea herring fisheries and Dutch mines were no longer economically viable due to overfishing and the severe reduction in demand for local coal. Clearly, neither circumstance applies in the case of Japanese small-type whaling: minke whales remain abundant and a robust demand exists for the highly valued foods that whalers can provide.[110]

A further difference between the European and Japanese situations derives from Confucian obligations that exist in the Japanese situation. These obligations are more than purely economic, and include, for example, "the notion of preserving the family enterprise . . . [one's] *seigyo* [or] calling . . . a powerful cultural imperative, implying a high degree of moral obligation."[111] This issue was explained by a distinguished expert in Japanese culture in these terms:

The traditional Japanese family . . . has always been organized around the principle of family enterprise as an enduring social and economic unit, which ideally exists through generations. It is a conception of family and household that is extremely different from the notions of family that underpin domestic organization in many industrial societies, where nuclear family households are presumed to be formed and to disappear as social units in the natural cycle of personal lives.

Within the [Japanese] family system, traditional notions of obligation and of filial piety place the highest moral value on the

---

[109]   *E.g.*, The Government of Japan (1991), *id.* at 204.

[110]   The Government of Japan (1989), *supra* note 24, at 65; Manderson, *supra* note 39, at 89–94; The Government of Japan (1991), *supra* note 108, at 203–204; Ashkenazi, *supra* note 28, at 214–17.

[111]   Bestor, *supra* note 27, at 73.

responsibility of the present generation to repay the debts to the members of previous generations by nurturing the family's enterprise and property through careful stewardship so that it may be bequeathed to the members of future generations.

Thus for the whaling operator, the prospect of being unable to continue his family's enterprise looms as *a devastating personal, moral failure.*[112]

5.   CONCLUSIONS

The preamble and the articles of the International Convention for the Regulation of Whaling (ICRW) provide the IWC with its objectives and operating instructions. Article V.2(b) states that actions taken "shall be based on scientific findings" and Article V.2(d) that IWC "shall take into consideration the interests of the consumers of whale products and the whaling industry."

The government of Japan, pursuant to Article V.2(b), has tabled more than 36 scientific reports detailing the research conducted by established social scientists from universities and research institutes in eight countries.

In an effort to communicate the concerns of whale producers and consumers (as stipulated under Article V.2(d)), the chair of the Japanese Small-Type Whaling Association and the mayor of one of the small-type whaling towns regularly attend, and have relayed the views of consumers and producers during IWC meetings. In addition, the IWC convened an intercessional workshop in 1997 to consider in detail and resolve the issue of Japanese small-type whaling. Workshop participants visited the whaling town of Ayukawa, where many town residents made oral presentations and presented a petition with 4,000 signatures calling for a reinstatement of coastal minke whaling. Townspeople spoke about the various negative consequences of the minke whaling ban that continues to impact their families and community.[113] These consequences have

---

[112]   *id*. at 73 (emphasis added).

[113]   Many of these impacts are socio-economic in nature, and some also relate to negative impacts on local (non-whale) fisheries. These impacts have been

been concisely captured in the 1989 IWC chairman's report on the 41st Annual Meeting as follows:

> In Japan, the zero-catch limit has affected individuals economically, socially, culturally and in respect to health. The effects include disruption and failure of small businesses, job loss and employment at less valued positions and/or limited work in temporary or seasonal positions. Because of the nature of small-type whaling the zero-catch limit affects individuals in small villages more than in the industrial centres. The small size of the local economy has required physical moves for individuals and families in order to find employment. High levels of unemployment for former whalers result from the highly specialized nature of their work and barriers to re-employment due to age and the particularities of Japanese employment and fisheries practices. As whalers enjoyed prestige, their job loss is especially stressful. Within the family, interpersonal stress, disruption of rigid gender-related division of labor and stress on children occurs. Local businesses depending on whale products have been severely affected and the loss of revenue threatens the survival of such institutions as fishery cooperative associations. Tourism is highly dependent upon the availability of whale meat which also plays an important role in religious observances and community celebrations. These impacts pose a serious threat to the continued survival of these traditional small communities.[114]

Finally, at the 45th Meeting in 1993, Resolution IWC/45/51 was passed, committing IWC "to work expeditiously to alleviate the distress to these communities which has resulted from the cessation of minke whaling." However, since that 1993 Resolution passed, the IWC remains deadlocked in unproductive debate, whilst a small number of member nations (specifically Australia, Austria, Italy and New Zealand) have

---

addressed in various IWC reports, including Akimichi, *supra* note 18, at 104–108; The Government of Japan (1989), *supra* note 24, at 17–69; Bestor, *supra* note 27, at 71–84; Manderson, *supra* note 39, at 91–103; The Government of Japan, *supra* note 108, at 203–204; Freeman, *supra* note 22.

[114]   Chairman's Report of the 41st IWC Meeting at 4, in The Government of Japan (1990), *supra* note 37, at 151.

openly announced that their purpose at the IWC is to work toward *ending* all commercial whaling. These goals are proclaimed notwithstanding these nations' continuing obligations under the ICRW whose purpose is, *inter alia*, "to make possible the orderly development of the whaling industry."

This deadlocked situation continues today, despite the efforts of the Chairman of the IWC between 1997 and 1999 to initiate what he termed a "compromise solution" to the longstanding impasse. Compromise appears to be highly unlikely while a small number of nations insist that they will not allow any commercial whaling to resume.[115] This resolve to oppose commercial whaling occurred after the IWC Scientific Committee concluded that a number of whale stocks can sustain a regulated annual take.[116] The whaling issue remains unresolved at the IWC, and while some nations persist in preventing the commission from discharging its management responsibilities, the question of whether the present-day IWC is an "appropriate" whaling management authority would appear to be abundantly clear.[117]

---

[115]   Kristen Fletcher, *The 49th Annual Meeting of the International Whaling Commission: Prelude to the Next Fifty Years*, 1 J. INT'L WILDLIFE L. & POL'Y 6 (1998); Aron, *supra* note 3, at 181.

[116]   Aron, *supra* note 3, at 182–83.

[117]   For others, including the CITES Secretariat and IUCN, the credibility of IWC remains seriously at risk, if not already compromised. *See* Anon., *IUCN Expresses Concern about IWC, and US Scholars Urge Completion of the RMS*, 19 INWR DIGEST 2 (2000); Anon., *IWC Receives More Warnings*, and *IWC Woes Continue*, 20 INWR DIGEST 1 (2000). The long-time Secretary of the IWC, Dr. Ray Gambell, told BBC News Online on June 11, 2000, that a failure to signal the forthcoming end of the whaling ban would mean "a real danger that the [IWC] will lose its credibility totally." BBC Homepage/World Service June 11, 2000.

Chapter 3
I Am Here, Where Should I Be?[1]

Ray Gambell[2]

1. INTRODUCTION

For more than a decade, the International Whaling Commission (IWC) has struggled to reconcile the interests of member states who wish to lift the moratorium on commercial whaling imposed in the 1980s, those who believe that it is not judicious to do so at this point, and those parties who advocate a permanent ban on commercial whaling. At the 49th Annual Meeting of the IWC, held in 1997, Michael Canny, the Commissioner for Ireland, advanced a number of ideas to break the deadlock in the IWC. This so-called "Irish Proposal" recognized that some whale populations are recovering or are abundant, and that substantial progress has been made in developing a Revised Management Scheme (RMS) to regulate commercial whaling should the moratorium be lifted. Nevertheless, the IWC does not now control or regulate all whaling in the world. In particular, it does not exert complete control over the harvesting practices of all member nations. For example, Norway and Japan, under objection and special permits for scientific research, harvested 383 whales in 1992 and increased the catch to 1,043 in 1997. These levels have remained approximately the same in the ensuing years.

The range of opinions in the IWC extends from that of some governments who believe that whales are special animals that should not be killed at all, to others for whom the taking of whales is a part of their culture. In the center ground are those who view whales as a resource that may be exploited, provided this is done in a controlled and sustainable manner. In advancing its proposal, Ireland believed it was premature to agree to a resumption of commercial whaling, but it did believe that there was a basis to reach a consensus among the factions at the

---

[1] A telegram sent by G.K. Chesterton to his wife on arriving by train at Market Harborough.

[2] Former Secretary to the International Whaling Commission.

65

IWC to limit whaling in the medium term. It therefore suggested that:

- The RMS should be completed and adopted. The scheme should be conservative and provide for inspection and observation procedures that will engender public confidence;
- Where quotas are justified under the RMS, these should be restricted to coastal areas of nations who are now involved in whaling operations. This would result in a *de facto* sanctuary over most of the oceans of the world;
- Whales harvested under the RMS only would be utilized for local consumption. No international trade of whale meat or whale products would be permitted;
- Killing of whales for scientific purposes should be phased out;
- Regulations for whale watching should be prepared to minimize the impacts of disturbance on whale populations.

Many delegations expressed their appreciation for this initiative by Ireland, gave support to the proposal in varying degrees, and indicated their willingness to consider it in more detail. Some delegations did express reservations, chiefly over the question of allowing commercial whaling to resume, and issues arising under the Law of the Sea and coastal states' rights and responsibilities. However, there was a consensus to continue the dialogue, and Ireland has continued its informal discussions with other delegations in the following years.[3]

2.   THE EVOLUTION OF THE INTERNATIONAL WHALING COMMISSION

The whaling industry has failed throughout history to conduct sustainable operations. Evidence of this failure has been demonstrated since the beginning of commercial whaling in the English Channel and Bay of Biscay more than 1,000 years ago, through the expansion to the Arctic seas in the 17th century, worldwide sperm whale hunting a century later, to the final phase of Antarctic whaling in the 20th century. Each episode has been part of a constantly repeated pattern of over-

---

[3]   International Whaling Commission, *Chairman's Report of the Forty-Ninth Annual Meeting*. 48 REP. INT'L WHALING COMMISSION 17, 35 (1998).

exploitation followed by the decline of targeted species.[4] Despite efforts to limit catches in the last century by both the industry itself, in an effort to maintain the price of the primary product, oil, and governments through international agreements,[5] over-exploitation continued through the first half of the 20th century. A significant reason for unexpected declines in stocks was the initial lack of an adequate scientific basis for determining safe catch levels. Coupled with pressure from the whaling companies to set their quotas as high as possible, catch limits continued to exceed the capacity of the stocks to support them. Eventually the declining catches, particularly in the Antarctic, the primary locus of commercial whaling operations in this century, led to a call at the 1972 United Nations Conference on the Human Environment in Stockholm for a ten-year moratorium on commercial whaling. The Stockholm proposal also called for strengthening the IWC (which was initiated by the appointment of a permanent Secretariat in 1976) and an increase in scientific research.

The IWC initially resisted the proposal for a moratorium, in part because it had finally taken note of the assessment and recommendations of a special Committee of Three experts that it had appointed in 1961 to assess the population dynamics of baleen whale stocks. This committee was charged with the tasks of carrying out an independent analysis of the status of baleen whale stocks in the Antarctic and making appropriate recommendations to the IWC. The work of these experts provided the impetus for the introduction of mathematical methods and techniques that had been recently developed for fisheries assessments and management.[6] After protracted and contentious deliberations, the

---

4   RICHARD ELLIS, MEN & WHALES 38 (1991); GORDEON JACKSON, THE BRITISH WHALING TRADE 3 (1978); A.A. BERZIN, THE SPERM WHALE (Translation of KASHALOT, 1971) 312 (1972). J.N. TØNNESSEN & A.O. JOHNSEN, THE HISTORY OF MODERN WHALING (A shortened translation of DEN MODERNE HVALFANGSTS HISTORIE: OPPRINNELSE OG UTVIKLING) 16 (1982); S Risting, *The Development of Modern Whaling*, II INTERNATIONAL WHALING STATISTICS 4 (1931).

5   Ray Gambell, *International Management of Whales and Whaling: An Historical Review of the Regulation of Commercial and Aboriginal Subsistence Whaling*, 46 ARCTIC 98 (1993).

6   D.G. Chapman, K.R. Allen & S.J. Holt, *Special Committee of Three Scientists Final Report*, 40 REP. INT'L WHALING COMMISSION 14 (1990).

member nations of the IWC ultimately agreed that the catch limits would be reduced to below the estimated sustainable yield for targeted stocks. It was hoped that this would help these stocks to recover from over-exploitation.[7] Thus, the IWC was not in favor of a total ban on whaling. Rather, it adopted a compromise amendment proposed by Australia for a new management regime,[8] often referred to as the New Management Procedure (NMP).

The NMP was designed to provide for the regulation of the catches of whale stocks on an individual basis to ultimately attain the Maximum Sustainable Yield (MSY). Since whales tend to adjust their birth rates depending on population density, at particular stock levels there is a certain surplus of recruitment over natural mortality, which reaches a maximum at some 50–60 percent of the original abundance. This yield represents a harvest that can be taken in perpetuity without further depleting the stock.[9] Survivors in populations reduced by hunting experience improved feeding conditions. This results in increased growth rates and ovulation frequencies, leading to enhanced rates of recruitment through a reduction in the age of sexual maturity and an increase in pregnancy and survival rates.[10]

The NMP was regarded as a major step forward in the management of whaling. Unfortunately, it turned out that our knowledge of the biology and recruitment characteristics of whale species, which was based on available information on the age structures, mortality rates, variations in the ages of maturity and reproduction, and the population estimates needed to calculate the sustainable yields, was too imprecise to facilitate accurate calculations of MSY. The data available to the IWC yielded a wide range of values for possible harvesting quotas. This led

---

[7]   International Whaling Commission, *Chairman's Report of the Seventeenth Meeting*, 17 REP. INT'L WHALING COMMISSION 17, 20 (1967).

[8]   International Whaling Commission, *Chairman's Report of the Twenty-Sixth Meeting*, 26 REP. INT'L WHALING COMMISSION 24, 25 (1976).

[9]   Nina M. Young, *The History of the International Whaling Commission in* GUIDE TO THE REVIEW OF THE MANAGEMENT OF WHALING 1, 7 (Sidney Holt & Nina M. Young eds., 1990).

[10]   Ray Gambell, *Some Effects of Exploitation on Reproduction in Whales*, 19 J. REPRODUCTION & FERTILITY (Supplement) 534 (1973).

to protracted debates between industry representatives lobbying for higher hunting quotas and environmental advocates seeking substantially lower quotas in the face of stock abundance uncertainties and extremely limited knowledge of the critical biological parameters of targeted species.

3.   THE COMMERCIAL MORATORIUM ON WHALING

Support for the imposition of a commercial moratorium grew substantially in the 1970s and early 1980s. The primary impetus came from a number of IWC parties, notably Australia, New Zealand, the United Kingdom and the United States, as well as several international non-governmental organizations. Although the underlying principle for the development of the NMP looked very attractive, full implementation was difficult. By the early 1980s the Scientific Committee found it almost impossible to reach agreement on the classification of catch limits for stocks subject to commercial whaling, other than for those needing complete protection. However, the anti-whaling lobby had been thwarted in its efforts throughout the 1970s to secure the three-quarters majority vote necessary to amend the Schedule to the International Convention for the Regulation of Whaling (ICRW) to impose a halt to whaling.

Nonetheless, these interest groups pursued a policy in the IWC of methodical restrictions on commercial whaling activities. They secured gradual reductions in the level of catches, and through the adoption of sanctuaries, achieved limitations on the areas where whaling was permitted. Whaling by factory ships, except for minke whales, was ended in 1980 at the instigation of Panama.[11] The government of Seychelles took the lead in the creation of an Indian Ocean Sanctuary in 1979.[12] The Southern Ocean was also declared a sanctuary on the proposal of France in 1994.[13] The designation of areas of the world's oceans off-limits to commercial whaling was seen as an effective method of both

---

[11]   International Whaling Commission, *Chairman's Report of the Thirty-First Annual Meeting*, 30 REP. INT'L WHALING COMMISSION 25, 26 (1980).

[12]   *Id.* at 27.

[13]   International Whaling Commission, *Chairman's Report of the Forty-Sixth Annual Meeting*, 45 REP. INT'L WHALING COMMISSION 15, 28 (1995).

limiting the geographical boundaries for whaling and preventing any large-scale resumption of operations even if the commercial moratorium was lifted.

Non-governmental organizations opposed to whaling also encouraged sympathetic non-member IWC states to join the IWC so as to attain the requisite three-quarters majority to secure a moratorium on commercial whaling.

As is shown in Table 1, from 1976 to 1982, 24 governments adhered to the International Convention for the Regulation of Whaling (ICRW) and thus became members of the IWC. Of these, 18 voted for the introduction of zero catch limits, two were opposed and four abstained. Subsequently, the Caribbean states have changed their representation to government ministers or officials, but only two of the other countries that adhered to the ICRW during this period have continued to play an active part in IWC affairs. The moratorium was finally adopted at the 1982 Annual Meeting of the IWC, with 25 votes in favor, seven against, and with five abstentions. Commercial whaling was to effectively cease at the conclusion of the 1985/86 pelagic and the 1986 coastal seasons.

4.   THE COMPREHENSIVE ASSESSMENT

Coupled with the imposition of a moratorium on commercial whaling, the IWC agreed:

> that the provision would be kept under review and by 1990 at the latest the Commission will undertake a comprehensive assessment of the effects of this decision on whale stocks and consider modification of this provision and the establishment of other catch limits.[14]

While the intent of the Commission in developing this provision was unclear, and its interpretation ambiguous, following this decision, the Scientific Committee of the IWC embarked on the process that came to be known as the Comprehensive Assessment of whale stocks. The scientists framed this program as an in-depth evaluation of the status and trends of all whale stocks so as to facilitate research-based management

---

14   International Whaling Commission, *Chairman's Report of the Thirty-Fourth Annual Meeting*, 33 REP. INT'L WHALING COMMISSION 20, 21 (1983).

Table 1 Vote on the Moratorium by IWC Member Nations, 1982									
Original	*1974*	*1975*	*1976*	*1977*	*1978*	*1979*	*1980*	*1981*	*1982*
Argentina	Brazil		New Zealand	Nether-lands	R.o. Korea	Chile	P.R. China	Costa Rica	Antigua & Barbuda
Australia						Peru	Oman	Egypt	Belize
Denmark						Seychelles	Switzer-land	India	Germany
France						Spain		Kenya	Monaco
Iceland						Sweden		Philip-pines	Senegal
Japan								St Lucia	
Mexico								St Vincent & Grena-dines	
Norway								Uruguay	
South Africa									
USSR									
UK									
USA									

Yes = 25   No = 7   Abstain = 5

objectives and procedures.[15] The Committee completed detailed assessments of the species and stocks subject to aboriginal subsistence whaling, and of those likely to be targeted if commercial whaling was resumed. The species and stocks included the gray whales in the North Pacific, bowhead whales off Alaska, minke whales in the Southern Hemisphere, North Atlantic and western North Pacific, and fin whales of the North Atlantic.

An important aspect of this assessment program has been the development of rigorous sighting surveys and photo-identification techniques

---

[15]   G.P. Donovan, *Preface* in *The Comprehensive Assessment of Whale Stocks: the early years*, 11 REP. INT'L WHALING COMMISSION (Special Issue) (G.P. Donovan ed.,1989).

that have been coupled with robust methods of analysis of resulting data. These efforts have enabled the scientists to develop statistically reliable abundance estimates of various whale stocks.[16] The IWC has decided that only surveys and analyses of data that fulfil the guidelines formulated by the Scientific Committee can be used to determine the abundance estimates pertinent to establishing quotas under the Revised Management Procedure.[17]

5.   REVISED MANAGEMENT PROCEDURE

Another part of the comprehensive assessment program was the development of a Revised Management Procedure for setting catch quotas should the moratorium be lifted. Five different procedures were developed and tested in a series of computer simulation trials, but it was not possible to complete this work by the original deadline of 1990.[18] However, at the 1991 Annual Meeting, the Scientific Committee identified one of the procedures as suitable to replace the discredited NMP and recommended its adoption by the IWC.[19]

The IWC formally adopted this Revised Management Procedure (RMP) with some modifications at its 1994 Annual Meeting.[20] The objective of the RMP is to provide an acceptable balance between conservation and exploitation of baleen whales and to provide a simple and convenient method for determining catch limits with minimal requirements for data. It seeks to ensure that depleted stocks are replenished, and that no whaling is permitted on stocks that are below 54 percent of their initial population levels. The goal is to obtain the highest possible continuing yield, with stable catch limits, ultimately bringing all stocks

---

[16]   A.R. Hiby & P.S. Hammond, *Survey Techniques for Estimating Abundance of Cetaceans, id.* at 47. S.T. Buckland & E.I. Duff, *Analysis of the Southern Hemisphere Minke Whale Mark-Recovery Data, id.* at 121.

[17]   International Whaling Commission, *supra* note 14, at 26.

[18]   G.P. Kirkwood, *Background to the Development of Revised Management Procedures*, 42 REP. INT'L WHALING COMMISSION 236 (1992).

[19]   International Whaling Commission, *Report of the Scientific Committee*, 42 REP. INT'L WHALING COMMISSION 51, 55 (1992).

[20]   International Whaling Commission, *supra* note 14, at 26.

to the 72 percent of their initial levels. Both of these percentage levels relate to stock classifications based on the concept of the Maximum Sustainable Yield (MSY), which formed the basis of the NMP.

The only data required for the revised procedure are a current population estimate and the known catch history. The hypothetical sustainable stock size is estimated using a simplified production model that includes no biological parameters. Initially, a fixed MSY rate is assumed, but as more data accumulate the procedure gradually evolves to one based on the best estimate of the MSY rate obtained by fitting into the production model. The procedure has been tested in simulation trials for over 100 years and is adaptable to a range of potentially confounding factors, including underestimation of historic catches by up to 50 percent, variations over time in carrying capacity and recruitment, environmental degradations and a wide range of uncertainty, including stock units and differing population dynamics.

6.   REVISED MANAGEMENT SCHEME

Although the scientific component of this comprehensive management plan has been completed and accepted, the IWC has determined that it should not be implemented until all aspects of a Revised Management Scheme (RMS), including an effective inspection and observation scheme that fully addresses *inter alia* the issues of under-reporting and mis-reporting of catches, are incorporated into the Schedule.[21]

It is now well known that catch data submitted by the former Soviet Union were substantially falsified to conceal large-scale violations of the international regulations and did not constitute accurate records of the actual numbers and species taken by its pelagic fleets.[22] The official records from many operations that measured the body lengths of whales caught were also unreliable, particularly for whales close to the minimum size limits.[23] Such evidence has reinforced the need for the

---

[21]   *Id.* at 44.

[22]   A.V. Yablokov, *Validity of whaling data*, 367 NATURE 108 (1994); V.A. Zemsky *et al*, *Soviet Antarctic Pelagic Whaling After WWII: Review of Actual Catch Data*, 45 REP. INT'L WHALING COMMISSION 131 (1995).

[23]   Peter B. Best, *Some Comments on the BIWS Catch Record Data Base*, 39

establishment of credible inspection and international observation schemes for any future whaling activities. An International Observer Scheme was implemented by the IWC in 1972 after more than a decade of negotiations. Observers appointed by, and directly reporting to, the IWC were stationed at the sites of whaling operations of member states to confirm their compliance with the regulations.[24] This resulted in an improvement of the quality of Soviet recordkeeping.

Evidence of such misconduct reinforced the reluctance of individuals and nations to see a resumption of commercial whaling unless there is adherence to very strict controls under international supervision. In addition, in consideration of the inconsistencies in data collection and observance of international standards, opponents sought the establishment of large sanctuary areas to provide a geographical area in which whales would be safe from hunting.

A Working Group has tried since 1994 to resolve numerous issues, including protocols for DNA testing of market samples of whale products and satellite monitoring of whaling vessels. The ongoing activities of the Working Group include efforts to reconcile the legal basis for such activities, national sovereignty and the legal competence of the IWC under the 1946 Convention, as well as issues attendant to inspection methods, independent verification, reporting requirements, and cost and transparency issues.[25]

Some governments would like international inspection and controls to be extended beyond the initial surveys and analyses on which whale stock estimates are based, to actual catching and processing operations at sea or on shore, as well as provisions for monitoring the marketing and distribution of the products. This position has been deemed a necessary procedure due to allegations of the sale of whale products on the Japanese and Korean markets that could not have been obtained or

---

REP. INT'L WHALING COMMISSION 363, 365 (1989).

[24]   International Whaling Commission, *Chairman's Report of the Twenty-Fourth Meeting*, 24 REP. INT'L WHALING COMMISSION 27 (1974).

[25]   International Whaling Commission, *Chairman's Report of the Fiftieth Annual Meeting*, 1998 ANN. REP. INT'L. WHALING. COMMISSION 3, 23.

derived from species and whaling operations permitted by the IWC.[26]

Other IWC signatories believe that a less comprehensive approach to ensuring sustainable whaling operations would prove more effective and create less friction among the parties. In particular, some nations believe that trade in whale products should be the responsibility of the Convention on International Trade in Endangered Species of Wild Fauna and Flora. At its Annual Meeting in 2000, the IWC committed itself to serious consideration of a draft text for proposed amendments to the Schedule, which at least would establish the legal foundation for implementing the RMS.[27]

A separate, but related, issue is the establishment of accepted methods of data collection and analysis under the RMS, as well as guidelines for conducting surveys of whale populations and methods to analyze the resultant data. The Scientific Committee has developed protocols for both of these issues.[28]

7.   WHALE KILLING METHODS

Arguments used to support an end to whaling included concerns over the numbers of whales caught, the ability of the stocks to sustain these losses and the scientific uncertainties involved in the estimation of the size of stocks. However, as the scientific data and results of the analyses became more reliable and demonstrated the viability of sustainable harvesting, the focus of the justification advanced for the cessation of whaling shifted. More emphasis was placed on the methods of killing whales and the length of time that elapsed before a whale died from har-

---

[26]   C.S. Baker, F. Cipriano & S.R. Palumbi, *Molecular Genetic Identification of Whale and Dolphin Products from Commercial Markets in Korea and Japan*, 5 MOLECULAR ECOLOGY 671 (1996).

[27]   International Whaling Commission, *Chairman's Report of the Fifty-Second Annual Meeting*, 2000 ANN. REP. INT'L WHALING COMMISSION (in press).

[28]   International Whaling Commission, *Report of the Scientific Committee*, 45 REP. INT'L WHALING COMMISSION 53, 68 (1995); International Whaling Commission, *Report of the Scientific Committee*, 47 REP. INT'L WHALING COMMISSION 59, 67 (1997).

pooning. The ethics of the pain and suffering inflicted on hunted whales also became an issue for consideration in conservation policy.

The principal hunting technique now used in commercial whaling dates back to the 1860s, when a harpoon fitted with a grenade was fired from a power boat, enabling the whalers to have access to even the fastest species. When the detonation of the grenade killed the whale, the proximity of the vessel to the floating whale facilitated easy access to the fresh carcass. The efficiency of this practice has been significantly improved in recent years by both Japan and Norway through the use of penthrite, a more powerful explosive substance. In addition, better triggering mechanisms ensure that grenades explode inside the body of the whale. The IWC, through a series of specialist Workshops and a Working Group on the topic, has encouraged these developments as a means to address ethical concerns. These new tools minimize the pain and suffering and hasten the death of these sentient marine mammals. The new equipment and technologies instituted in commercial whaling operations are now also being modified and used in aboriginal subsistence hunts.[29]

A particularly salutary change to engender more humane hunting was the replacement of the electric lance as a secondary killing method in Japanese whaling operations. The electric lance was used when the first harpoon fired into a whale failed to kill the animal. In such a situation the injured whale was hauled close to the catching vessel, electric lances were placed on either side of the head and a lethal current was administered. There has been much debate in the IWC associated with the use of the lance and many delegations in the IWC encourage the use of rifles as a secondary method of killing, since they are already employed successfully by Norway. Because of domestic legislation prohibiting the private ownership rifles, Japan did not have extensive experience with this weaponry, but it has stated its intention to carry out research on the use of rifles. At the 1997 IWC Annual Meeting, in the light of further experiments and discussion, there was general consensus that the rifles appear to be more efficient and humane than the electric lance as a secondary killing technique, although Japan still considered the lance to be

---

[29]   International Whaling Commission, *Chairman's Report of the Fifty-First Annual Meeting*, 1999 ANN. REP. INT'L WHALING COMMISSION 7, 11.

equally as effective. Nonetheless, Japan announced that it intended to use rifles as the principal secondary killing method in the future. This development must be seen as a further advance in the harvesting technology employed by the whaling industry.[30]

8. COASTAL WHALING

One of the major inconsistencies in the IWC's refusal to allow commercial whaling is the close similarities that exist between the coastal whaling operations carried out by Norway at present, and by Japan in the recent past, with some of the aboriginal subsistence whaling activities that the IWC does permit.

Modern minke whaling began in Norway in the 1920s, but small cetaceans have been hunted for thousands of years. Whaling operations are conducted by households that own the vessels and provide the crews. This helps to preserve the traditions and way of life of the remote northern communities. There is considerable resentment against the prohibition of the catching activities imposed by other members of the IWC.[31]

Norway has set its own quotas since 1993. This is permissible under the ICRW because Norway lodged timely objections to both the setting of zero catch limits on commercial whaling by the IWC and the classification of the Northeast Atlantic stock of minke whales as a Protection Stock under the NMP. The catches are calculated under the RMP developed by the IWC and inspected and monitored under Norwegian national regulations. Norway has also established a DNA register of harvested whales that is equivalent to the inspection and control procedures that the IWC might ultimately adopt if the moratorium is ultimately lifted.

Since the ban on commercial whaling came into effect in 1982, Japan has repeatedly argued for recognition of the deleterious effect that the moratorium has caused in its small-type coastal whaling communities, comprised of four coastal enclaves harvesting minke whales within 30 miles of the shore. Japan contends that the whale meat obtained from

---

30 International Whaling Commission, *supra* note 3, at 20.

31 International Whaling Commission, *Chairman's Report of the Forty-Fourth Annual Meeting*, 43 REP. INT'L WHALING COMMISSION 11, 16 (1993).

these catches serves an important role in the cultural and social cohesion of these communities.[32] In seeking interim relief for these communities, it has presented considerable social, scientific and anthropological research supporting the conclusion that Japanese small-type whaling has a character distinct from other forms of industrial whaling and shares some of the features of aboriginal subsistence whaling.

The IWC has consistently refused requests for an interim relief allocation of 50 minke whales for the Japanese coastal operations, even though these operations:

(1)   would be regulated and controlled through application of the RMP; and

(2)   would be subject to an Action Plan approved by the IWC that would mandate strictly non-commercial distribution system of the products.[33]

The nature of Japan's whaling operations, as well as distribution and consumption of whale products, is similar in many respects to the aboriginal subsistence hunts. The IWC purports to recognize the special socio-economic and cultural roles of whaling in the lives of the aboriginal communities.

The difficulty for some governments in accepting these claims emanates from the commercial aspects of small-type whaling operations in Norway and Japan. However, the catches of minke and fin whales taken by the small coastal vessels under the aboriginal subsistence whaling arrangements in Greenland, for example, do include a commercial element. Similarly, the catch of humpback whales on St. Vincent and the Grenadines is sold commercially and the Alaskan bowhead hunt and Siberian operations for gray whales also involve monetary transactions to cover the costs of equipment and supplies.

---

[32]   ARNE KALLAND & BRIAN MOERAN, ENDANGERED CULTURE, JAPANESE WHALING IN CULTURAL PERSPECTIVE 1 (1990).

[33]   International Whaling Commission, *Chairman's Report of the Forty-Seventh Annual Meeting*, 46 REP. INT'L WHALING COMMISSION 15, 18 (1996).

9. SPECIAL PERMIT CATCHES

Although Japan has reluctantly accepted the commercial moratorium, Japan conducts research whaling in the Antarctic and the North Pacific. As a Contracting Government of the ICRW, under Article VIII, Japan is accorded the right to grant itself a special permit for the taking of whales for scientific research. Following two years of feasibility studies starting in the 1987/88 season to resolve problems in collecting representative samples, Japan embarked on a 12-year research program in the Antarctic that included an annual catch of 300 (and 400 from 1995/96) ± 10 percent minke whales. This program permitted Japan to establish biological parameters that could be used for improved management, particularly data on natural mortality, stock identity, and the elucidation of the role of whales in the Antarctic ecosystem.[34]

Japan has also carried out a program since 1994 to clarify the stock structure and mixing rates of minke whales along its coasts, which could help the Scientific Committee to resolve problems associated with assessing the Northwestern Pacific stocks. The program has included catches of 100 minke each year, and Japan has now established an expanded program to include catches of Bryde's and sperm whales to further investigate the ecosystem role of whales in this area.

Norway also carried out a research program under Article VIII from 1988 to 1995 that included the taking of 289 minke whales to study and monitor this species in the Northeast Atlantic, including investigations on feeding ecology, age determination and aspects related to energetics. This was part of a broader ecological program designed to provide information germane to future multi-species management in the Barents Sea.[35]

Prior to its withdrawal from the IWC, Iceland carried out a four-year program of research on the biology, genetics and ecological role of the whales in its waters from 1986 to 1989, taking 292 fin whales and 70

---

[34]   International Whaling Commission, *Report of the Scientific Committee*, 48 REP. INT'L WHALING COMMISSION 55, 102 (1998).

[35]   ADOLPHUS S. BLIX, LARS WALLØE & Ø. ULTANG (eds), *Preface*, WHALES SEALS FISH & MAN (1995).

sei whales. The IWC recognized the Icelandic whaling station as an International Whale Research Center.[36]

Since its moratorium decision, the IWC has established detailed guidelines for its Scientific Committee to employ in reviewing and commenting on proposed special permits. These guidelines concern the objectives of the research, and the means through which they relate to the Commission's needs; the methodology and procedures to be instituted, including the whales to be taken, and the feasibility of using non-lethal techniques; the effects of the catch on the stock(s); and the arrangements for participation by scientists from other countries.[37] The IWC may make recommendations to the respective governments, although it has no authority to forbid or rescind the issuance of permits.

In the past 15 years, the IWC has routinely adopted a Resolution inviting Japan to reconsider its research programs, to which the Japanese authorities have declined the offer on an ongoing basis. The IWC has also adopted Resolutions at the annual meetings requesting Norway and Iceland, when it was still a member of the IWC, to reconsider research programs and commercial harvesting operations. Critics of Article VIII continue to suggest that special permit catches are being used as mechanisms to circumvent the prohibition on commercial whaling until the moratorium is lifted. However, the IWC has presented useful information on Japanese operations to its Scientific Committee, although some of Japan's research objectives fall outside IWC research guidelines.[38] The Committee has also held special workshops to review and evaluate the results coming from the Japanese research catches.

10.   WHALE WATCHING

Some proponents of the termination of the harvesting of whales have touted the rapidly growing whale watching industry as fostering a

---

[36]   International Whaling Commission, *Chairman's Report of the Thirty-Second Annual Meeting*, 31 REP. INT'L WHALING COMMISSION 17, 25 (1981).

[37]   International Whaling Commission, *Report of the Scientific Committee*, 39 REP. INT'L WHALING COMMISSION 33, 154 (1989).

[38]   *Infra* note 39, at 95.

benign use of whales that is both economically viable and has enter-
tainment and educational value. In 1993, the IWC invited the Contracting Governments to undertake a preliminary assessment of the
extent of their whale watching activities, as well as an assessment of
their economic and scientific value.[39] A number of NGOs have encouraged this development through sponsorships of workshops and training
courses, including participation by non-IWC member countries.[40]

Whale watching, it is argued, is now more profitable than consumptive use of whales. This new industry is rapidly expanding in many
regions of the world. Some whale watching is conducted from shore,
for example, of gray whales that migrate along the coast of California,
and right whales with their calves in the sheltered bays of South Africa
and Argentina. Additionally, many of these operations use vessels to provide the opportunity to view sperm whales off New Zealand and the
Azores, humpbacks in Alaska, Hawaii and the Caribbean, and minke
whales off Iceland and Norway. The whale watching industry has grown
globally at a remarkable pace in recent years. These activities generate
considerable income for those who conduct the expeditions, as well for
the associated travel and accommodations industries.[41]

The economic benefits of these operations have led the IWC to reaffirm its interest in the subject, while it concurrently encourages scientific work to investigate effects that may arise from these activities. The
Commission has also been working on guidelines to minimize any
potentially harmful impacts that such operations may have on whales.[42]

It may be noted that both whaling and whale watching industries operate in Norway's coastal waters. Thus, the development and sustainability
of one activity does not necessarily preclude the successful conduct of
another. Moreover, the United States has promoted the resumption of

---

[39]   International Whaling Commission, *Chairman's Report of the Forty-Fifth
Annual Meeting*, 44 REP. INT'L WHALING COMMISSION 11, 23 (1994).

[40]   International Whaling Commission, *supra* note 26, at 5; *supra* note 30, at 10.

[41]   International Whaling Commission, *supra* note 3, at 19.

[42]   International Whaling Commission, *Chairman's Report of the Forty-Eighth
Annual Meeting*, 47 REP. INT'L WHALING COMMISSION 17, 21 (1997).

aboriginal whaling by the Makah Indian tribe on gray whales, one of the species often viewed by whale watchers off the North American Pacific coast.

11.   WHICH WAY?

The history of commercial whaling has been marked by the progressive over-exploitation of one species after another. Regulations have not prevented excess harvesting from occurring regardless of the economic ramifications for the industry in the long term. With a pause in commercial whaling now in place, the focus has been on developing procedures to ensure that stocks will not be subject to over-exploitation in the future if the moratorium is lifted, and that there will not be opportunities to conceal illicit activities.

There is no doubt that the removal of an excessive number of whales from the world's oceans has had a profound effect on ocean ecosystems. In the Antarctic, many of the larger animals compete for krill as a source of food. As a consequence, the depletion of the stocks of the great whale species has had observable effects on the growth and reproduction of other species. It also appears that the reproductive rates and numbers of other animal groups such as the penguins, seals and fish may also have changed since the 1930s.[43] What was once a delicate but dynamic equilibrium of competition and coexistence has been profoundly distorted by human intervention. We cannot redress the balance by any direct action now, but only watch and wait to see if the great whales can regain their former position in this ecosystem. The establishment of whale sanctuaries *is* one step designed to provide time and space for that to occur. However, the IWC rejected a proposal put forward by Australia and New Zealand at its Annual Meetings in 2000 and 2001 for a South Pacific Sanctuary.[44]

Right whales, once the primary target of the early global whaling industry, have been severely depleted throughout their ranges. Fortunately, the numbers in the Southern Hemisphere, as seen off Argentina,

---

[43]   Ray Gambell, *Birds and Mammals—Antarctic Whales*, in ANTARCTICA 223, 234 (W.N. Bonner & D.W.H. Walton eds.,1985).

[44]   *Supra* note 27, (in press).

Australia and South Africa, now seem to be increasing. But the stocks in the northern Atlantic Ocean may be on the verge of extinction, unable to recover in the face of their critically small numbers, and further threatened by entanglement (and deaths) in fishing nets, and by ship strikes.[45]

We know that generally the marine environment is becoming ever more hazardous for whales as a consequence of human activities. For example, instances of whale strandings have been attributed to exposure to chemical contaminants.[46] As the IWC and others have noted in recent years, climate change, ozone depletion, underwater noise, and bycatch in fisheries are among the panoply of anthropogenically related threats to whales that further imperil their existence through this century.[47] The role of these animals needs to be understood and appreciated in relation to the other elements in the energy flow of marine ecosystems.[48]

The IWC Scientific Committee has concluded from its testing procedures that the RMP has adequately addressed concerns about the detrimental effects of such environmental changes. However, it went on to state that the species most vulnerable to such threats may be those reduced to levels at which the RMP, even if applied, would result in zero quotas. In addition, the Scientific Committee has held workshops on the effects of chemical pollutants and on the impacts of global warming and

---

[45]   International Whaling Commission, *Report of the Scientific Committee, Anex E. Report of the Sub-Committee on Other Great Whales. Appendix 3. Right Whales —Report of the Cape Town Workshop—Chairman's Summary.* 1 (Suppl.) J. CETACEAN. RES. & MGMT. 139 (1999).

[46]   P.J.H. Reinders *et al* (eds.), *Chemical Pollutants and Cetaceans, Report of the workshop on chemical pollution and cetaceans* (Special Issue 1), J. CETACEAN. RES. & MGMT. 16 (1999).

[47]   50th Meeting of the International Whaling Commission, *Resolution on Environmental Change and Cetaceans*, IWC Resolution 1998-6 (1998). Resolutions from the 50th Meeting are available on the American Society of International Law—Wildlife Interest Group website, at <http://www.interntionalwildlifelaw. org>; *see also* William C.G. Burns, *Climate Change and the International Whaling Commission in the 21st Century*, Ch. 11, *infra*.

[48]   Food and Agriculture Organization, *Mammals in the Seas, Report of the FAO Advisory Committee on Marine Mammal Resources Research Working Party on Marine Mammals*, 5 FAO FISHERIES SERIES 10 (1978).

ozone depletion, and the IWC has reaffirmed its view of the importance of these issues.[49]

Many people, particularly those living in the relative affluence of western societies, believe that whales are unique animals that should not be regarded as resources to be harvested, even as a source of food for human consumption. Documentation of their large brains and gentle and family-oriented behavior is advanced as evidence of intelligence at least equivalent to our own.[50] Emphasis on the aesthetic and sentient values ascribed to whales is the focus of whale watching and educational programs. However, others coming from a background or tradition where hunting is regarded as a normal method of acquiring food find such attitudes difficult to accept. Additionally, in a new century in which there are increasing demands for food to meet the needs of burgeoning human populations, is it justifiable to deliberately deny the use of a particular resource because of one, albeit dominant, cultural approach? It is here that the differing viewpoints, beliefs and traditions of the peoples and nations comprising the IWC come into conflict.

It can be argued, with some justification that a number of whale stocks are abundant, such as minke whales in the Southern Hemisphere, and could sustain carefully regulated and controlled catches at an appropriate level. In addition, certain coastal communities that are prevented from whaling have societal and cultural structures and traditions similar to communities that are allowed to continue hunting whales for subsistence purposes.

It is unlikely that the original drafters, or signatories, of the 1946 International Convention for the Regulation of Whaling either foresaw or intended, through their efforts, to regulate non-consumptive activities such as whale watching, or to monitor or influence environmental factors impacting on whales. However, the United Nations Convention on the Law of the Sea and the 1992 Earth Summit in Rio have enshrined a number of important elements of the ICRW into international wildlife law, particularly in the areas of coastal state sovereignty, sustainable

---

[49]   *Supra* note 14, at 31.

[50]   T. Ogawa, *quoted in* WHALES & WHALING—REPORT OF THE INDEPENDENT INQUIRY CONDUCTED BY THE HON. SIR SIDNEY FROST 202 (1978).

development of natural resources, and the application of the precautionary principle.[51] This development has led to changes in the interpretation and application of the 1946 Convention, and the consequent management policies by which it is implemented.

The original Convention was drafted to control the catching operations of a particular fishery, which throughout its long history had overexploited the resource on which it was dependent. The ICRW is now being interpreted as an instrument to implement a new environmental conservation ethic, but this change of emphasis is not wholly agreed to or accepted by some of the communities most affected by this change in orientation.

This contentious and potentially divisive issue must be resolved soon or the IWC will have forfeited its position as the global authority for the management of whales and whaling. Given the IWC's expertise and its long-term experience in addressing these issues, this would be regrettable. Control could ultimately devolve either to individual states, as is now happening, or perhaps to other international or regional organizations with possibly divergent objectives and expectations.

It is now incumbent upon the IWC to put aside the rancor that has pervaded its deliberations in recent years and find a way to move forward. The IWC must seize control of its affairs within the next few years. Otherwise, there is the danger that its epitaph may ultimately be that it was unable to secure agreement among its members on how to manage the specific resource for which it was established to conserve.

---

[51]   B. Cicin-Sain & R.W. Knecht, *Implications of the Earth Summit for Ocean and Coastal Governance*, 24 OCEAN DEVELOPMENT & INT'L L. 323, 341 (1993).

Chapter 4
Science and Advocacy: A Cautionary Tale from the International Whaling Debate.

William Aron,[1] William T. Burke[2] & Milton M. R. Freeman[3]

1. INTRODUCTION

In 1976, a distinguished scientist suggested use of the term "trans-science" for statements made by scientists that are not supported by established scientific evidence. He wrote:

"[t]he debate on most matters at the intersection of science and society is largely conducted in the public, not the scientific, forum. When scientists express opinions on scientific matters in the public forum they are not subject to the sanctions that regulate opinions expressed in the usual channels of scientific communication. Because these traditional sanctions do not operate, the extra-scientific debate often tends to be irresponsible scientifically: lower standards of proof are demanded in the public than in the professional debate and half-truths are too often perpetrated on the public by scientists."[4]

As we detail below, it appears to be a challenge for some scientists, as specialists in a circumscribed field, to exercise the same skepticism and rigor they employ in their own specialty, when engaging in public

[1]   Director, Alaska Fisheries Science Center, NOAA (retired), Affiliate Professor, School of Marine Affairs, University of Washington, Seattle WA 95108. E-mail: waron@u.washington.edu.

[2]   Professor of Law and Marine Affairs Emeritus, University of Washington, Seattle WA 95108. E-mail: burke@u.washington.edu.

[3]   Senior Research Scholar, Canadian Circumpolar Institute, University of Alberta, Edmonton AB, T6G 0H1 Canada. E-mail: milton.freeman@ualberta.ca.

[4]   Alvin M. Weinberg, *Science in the public forum: keeping it honest.* 191 SCIENCE 431 (1976).

debate involving technical questions falling outside their own areas of competence. As a consequence, errors of judgment arise, leading to dereliction of a scientist's responsibility to avoid misleading the public. A recent commentator has similarly warned scientists of the dangers, to science, of political advocacy, cautioning that "today's scientists need to understand the consequences for science of relying on political advocacy as the primary mechanism of connecting science with policy."[5]

Examples of scientists making errors of judgment, or even committing outright fraud, are not new[6]—but arguably appear to be more common today in those areas of concern considered by the media as especially newsworthy. However, it is irresponsible for respected specialists to allow their passion for some cause to denude their professional responsibility to determine what constitutes the current science on that issue. Such carelessness is particularly prevalent in regard to various contested environmental or resource use issues in which politics, emotion, urban myths, and poor science confound a ready search for answers. Scientists would likely agree that there is no other basis for sound political decisions than the best available scientific information, and most would agree that this is "especially true in the fields of resource management and environmental protection"[7]

Many prominent environmental and animal protection organizations have scientific advisers or may engage scientists—including some from unrelated fields—who, while knowing little of the contemporary science on the specific problems being addressed, nevertheless share public concerns about, e.g., species endangerment, loss of biodiversity, or other environmental issues. While recognizing that scientists are not only scientists and "in donning the white coat at the laboratory door . . . do not step aside from the passions, ambitions, and failings that animate those

---

5   Roger A. Pielke, Jr., *Policy, politics and perspective. The scientific community must distinguish analysis from advocacy.* 416 NATURE 367 (2002).

6   WILLIAM BROAD AND NICHOLAS WADE, BETRAYERS OF THE TRUTH (1982); BROCK K. KILBOURNE AND MARIA T. KILBOURNE (EDS) THE DARK SIDE OF SCIENCE (1983).

7   Gro H. Brundtland, *The scientific underpinnings of policy.* 277 SCIENCE 457 (1997).

in other walks of life,"[8] we are concerned that when scientist-advocates lower their scientific standards and engage in trans-science pronouncements, science itself can be diminished, as can the rights of resource users and competent management of the environment.

Actions by organizations or scientists that increase the level of uncertainty or controversy in environmental or resource management decision-making (*e.g.*, by opposing research on sentimental grounds[9]) have been deemed acts of scientific misconduct.[10] At the present time, when politics and passion intrude into discussion and analysis of many environmental and resource use issues, serious questions are being asked by concerned researchers about the professional responsibility and trustworthiness of those scientists and environmental organizations whose passion or self-interest appears to detract from their responsibility to remain well-informed when operating in a public, as opposed to a scientific, arena.[11]

2.   WHAT CONSTITUTES "CREDIBLE SCIENCE?"

A recent open letter (see Annex) signed by a number of distinguished scientists, that questions the scientific merits of Japan's ongoing whale research program, provides yet another example of scientists' imprudence and irresponsibility in regard to making public pronouncements. The text of this open letter, which was proclaimed to be a "definitive scientific judgment," was published as a full-page advertisement in the

---

8   Broad and Wade, *supra* note 8 at 19.

9   D. S. Butterworth, *Science and sentimentality*. 357 NATURE 532 (1992).

10   NATIONAL ACADEMY OF SCIENCE, RESPONSIBLE SCIENCE: INSURING THE INTEGRITY OF THE RESEARCH PROCESS. VOL. 1 (1992); Tore Schweder, *Distortion of uncertainty in science: antarctic fin whales in the 1950s*. 3 J. INT'L WILDL. L. & POL'Y. 73, 89 (2000).

11   RAYMOND BONNER, AT THE HAND OF MAN: PERIL AND HOPE FOR AFRICA'S WILDLIFE (1993); Nicholas Mrosovsky, *IUCN's credibility critically endangered*. 389 NATURE 436 (1997); see also contributions in, e.g., THE TRUE STATE OF THE PLANET: TEN OF THE WORLD'S PREMIER ENVIRONMENTAL RESEARCHERS IN A MAJOR CHALLENGE TO THE ENVIRONMENTAL MOVEMENT (Ronald Bailey, ed. 1995); TOWARD A SUSTAINABLE WHALING REGIME (Robert L. Friedheim, ed. 2001).

New York Times[12] and can be read on the Internet at <www.wwfus.org>. The letter was signed by 21 scientists, including three Nobel laureates, four members of the U.S. National Academy of Sciences and a former president of the American Association for the Advancement of Science. Sadly, as we indicate below, the letter contains numerous errors of science, fact, and law, and as such reflects poorly on the scientific capability of the instigating environmental organization and the care taken by the scientists overseeing or endorsing the proclaimed "definitive scientific judgment."

The error-filled letter claims that Japan's whale research is bogus because it does not meet "minimum standards for credible science," citing lack of relevancy for management, refusal to release information for independent review, and lack of testable hypotheses. These are serious accusations to level against, *inter alia*, the many leading whale scientists in the IWC Scientific Committee who critically review Japan's research program in order to determine its management relevance and scientific merit. The chair of the IWC Scientific Committee, reporting on the Committee's review of Japan's Antarctic whale research program (JARPA) at the mid-point of this sixteen-year research program, stated "there was general agreement . . . that the data presented on stock structure . . . were important contributions to the objectives of JARPA and stock management"[13] and "JARPA has already made a major contribution to the understanding of certain biological parameters. . . ."[14]

The IWC Scientific Committee's review of Japan's North Pacific minke whale research (JARPN) in 2000 noted that "information obtained . . . had been and will continue to be used . . . and consequently was relevant to their management."[15] With respect to Japanese Antarctic

---

12   World Wildlife Fund, *An Open Letter to the Government of Japan on "Scientific Whaling,"* NEW YORK TIMES (West Coast Edition), May 20, 2002, at A12.

13   1998 REP. INT'L WHALING COMM'N. *Report of the Scientific Committee* 98.

14   *Id.* at 101.

15   *Report of the Workshop to Review the Japanese Whale Research Programme under Special Permit for the North Pacific minke whales (JARPN).* 3 J. CETACEAN RES. MANAGE (Suppl) 389.

research, the IWC Scientific Committee "agreed that [these] studies provided useful information for both the formulation of such hypotheses and for the selection of study areas . . . such studies would be of interest to CCAMLR and Southern Ocean GLOBEC"[16] and at the half-way point of the program "information [from this research] has set the stage for answering many questions about long-term population changes regarding minke whales."[17]

3. PEER REVIEW AND HYPOTHESIS TESTING

The open letter also falsely claims the Japanese "have refused to make information available for independent review."[18] Contrary to these assertions, the research protocols and findings are routinely reviewed by credible international scientists both before and during critical scrutiny by the IWC Scientific Committee. The research data and reports being reviewed, which numbered 133 at the time of the IWC Scientific Committee review of the Antarctic research program,[19] had increased to over 150 research papers and reports one year before the open letter appeared.[20] Information concerning this research is readily available internationally, either on the Internet (e.g. www.icrwhale.org), in some libraries, in peer-reviewed articles, or directly from the laboratories conducting the research. Furthermore, although the scientists' open letter charges "that Japan has refused to make the information it collects available for independent review,"[21] it is normal practice for scientists to withhold preliminary results or unpublished research data prior to their formal presentation at a later time. Nevertheless, this general scientific practice does not prevent full and detailed scrutiny of Japan's research[22]

---

[16]   *Supra* note 13 at 100.

[17]   *Id.* at 101

[18]   see Annex.

[19]   *Supra* note 13, Annex E1.

[20]   1999 REP. INT'L WHALING COMM'N, *Chairman's Report* 28.

[21]   *Supra* note 18.

[22]   *Supra* note 15 at Annex E.

and datasets[23] by members of the IWC Scientific Committee during their annual and intercessional meetings.

The assertion that the Japanese research program lacks a "testable hypothesis" is an indiscriminate broadside without foundation, as may be seen in the authoritative IWC Scientific Committee reviews of the component parts of the overall program. The meeting reviewing the Japanese Antarctic (JARPA) research program noted that the Scientific Committee had commended the program for both the quantity and quality of its scientific work, and called attention to the useful information it provided for the formulation of hypotheses.[24] The IWC JARPA Workshop Report agreed that the data were important contributions to the objectives of JARPA and stock management[25] and specifically that the research "has given valuable information on recruitment, natural mortality, decline in age at sexual maturity and reproductive parameters of minke whales."[26] In regard to the North Pacific minke whale research program (JARPN), as noted above, the Scientific Committee agreed that the information obtained is useful for management and will continue to be used for that purpose.[27]

---

[23]   *Id*. at Annex D.

[24]   *Supra* note 13 at 96, 100. Moreover, we question whether the formulation of hypotheses is significant, or even relevant, in determining the merit of a comprehensive whale research program, one that IWC Scientific Committee members appear to generally approve of, in regard to the program's overall scientific results. Specifically, does it require an explicit hypothesis to justify collecting information on whales' age-specific birth and death rates, or investigate the possible effects of various pollutants on the whales' health, or their quantitative relationships within the North Pacific marine ecosystem?

[25]   *Supra* note 13 at 98.

[26]   *Supra* note 13 at 99.

[27]   *Supra* note 15. For a careful summary of the Scientific Committee's assessment of the overall Japanese research programs, which notes the Commission's misleading characterization (at its 2000 meeting in Adelaide) of the Scientific Committee's discussion of Japan's North Pacific [JARPN II] research program, see Eldon V.C. Greenberg, Paul S. Hoff and Michael I. Goulding, *Japan's whale research program and international law*, 32 CAL. WEST. INT'L L.J. 151, 164–74, 206–07 (2002).

4. COMMERCE, INTERNATIONAL LAW, AND SCIENCE

The open letter is just as erroneous and misleading in its legal statements and interpretations as it is in science. The relevant international law, including treaty law, is both misrepresented and misunderstood. First the relevant treaty, the International Convention for the Regulation of Whaling (ICRW), is simply ignored in the open letter's contention that the research program is a commercial operation because the edible whale products remaining after the scientific samples have been taken are then sold. This non-wasteful disposal of the whale carcass is required by Article VIII (2) of the ICRW, which states "Any whales taken under these special permits shall so far as practicable be processed and the proceeds shall be dealt with in accordance with directions issued by the Government by which the permit was granted."[28] The legally-required disposal by sale of meat taken by scientific permit cannot sensibly be taken to impugn the scientific nature of the undertaking. Certainly the IWC Scientific Committee evaluates the research by the appropriateness of its methodological rigor and its results, and not on the basis of whether the non-sampled tissues are sold, given away, or jettisoned at sea.

Second, the open letter wholly misunderstands general international law concerning whaling. It erroneously charges that Japan claimed an exemption for scientific whaling under international law. However, under general international law, any nation has the right to take whales for food or research purposes and needs no exemption.[29] Further, Japan is a party to the ICRW, Article VIII of which specifically authorizes scientific whaling notwithstanding the ban placed on commercial whaling.[30]

---

[28] International Convention for the Regulation of Whaling with Schedule for Whaling Regulations, Dec. 2 1946, 62 Stat. 1716, T.I.A.S. No. 18\849, 161 U.N.T.S. 361, at Article VIII(2).

[29] Citing both the the 1982 U.N. Convention on the Law of the Sea and the 1958 Convention on the High Seas as reflecting customary international law, Professor Jon L. Jacobson expresses the widely held view that the freedom to hunt whales is protected by such customary law. Jon L. Jacobson, *Whales, the IWC, and the rule of law*, in TOWARD A SUSTAINABLE WHALING REGIME, 84 and n. 18 (Robert L. Friedheim, ed. 2001).

[30] *Supra* note 28.

Under no acceptable circumstances can compliance with a treaty constitute a violation of its provisions.

5.   CONCLUSIONS

Consideration of IWC Scientific Committee members' careful evaluation of Japan's whale research is seemingly discounted or ignored by those signing the open letter. Such behavior draws attention to the ongoing importance of Weinberg's admonition that "the scientist must be beyond reproach in doing his homework thoroughly whenever he makes scientific judgments, and he must delineate as sharply as possible where science ends and . . . trans-science begins."[31]

We are sympathetic to those who for ethical or emotional reasons oppose the killing of whales. However, if scientists' opposition stems from reasons of science, then those scientists should be quite sure of their facts if intending to influence public understanding of an issue that demands scientific input for its resolution. Whaling remains a complex and contentious issue. To understand this complexity, and to form sound judgments about this contested activity, requires careful attention to relevant facts that are readily available from reliable sources. To do otherwise is a dereliction of scientists' professional and ethical responsibility, and is regrettable at a time when public trust in scientists' judgments on a number of environmental and resource use issues is a matter of both national and international importance.

ANNEX

AN OPEN LETTER TO THE GOVERNMENT OF JAPAN ON "SCIENTIFIC WHALING"

Despite its obligation to comply with a global moratorium on commercial whaling, Japan has killed thousands of whales over the past decade, claiming an exemption for "scientific whaling" under international law. We, the undersigned scientists, believe Japan's whale research program fails to meet minimum standards for credible science. In particular:

---

[31]   *Supra* note 4.

We are concerned that Japan's whaling program is not designed to answer scientific questions relevant to the management of whales; that Japan has refused to make the information it collects available for independent review; and that its research program lacks a testable hypothesis or other performance indicators consistent with accepted scientific standards.

Most of the data being gathered by Japan's "scientific whaling" are obtainable by non-lethal means; it is possible, for example, to determine species, gender, population size, migration patterns, stock fidelity, and other key biological information without harming whales. Yet Japan's whale research program kills hundreds of whales each year in the absence of a compelling scientific need.

The commercial nature of Japan's whaling program conflicts with its scientific independence. Japan sells meat from the whales it kills on commercial markets and assigns "scientific whaling" quotas to individual whaling villages. These commercial ties create a profit incentive to kill whales even when no scientific need exists, raising troubling questions about the motives behind Japan's program.

Japan has announced it will soon begin killing sei whales, an internationally listed endangered species, ostensibly to determine the whales' diet. Yet Japan has already analyzed the stomach contents of nearly 20,000 sei whales it killed during the past fifty years. There is no reasonable likelihood that killing additional sei whales now will add to what is already known about their diet.

By continuing to fund and carry out this program, Japan opens itself to serious charges that it is using the pretense of scientific research to evade its commitments to the world community. As scientists, we believe this compromises objective decision-making and undermines public confidence in the role of science to guide policy. Accordingly, we respectfully urge the Japanese government to suspend its "scientific whaling" program.

Frederic Briand, Director General, la Commission Internationale pour l'Exploration Scientifique de la Méditerranée (CIESM), Monaco; Member, UN Experts Group on the Scientific Aspects of Marine Environmental Protection.

Theo Colborn, Senior Scientist, World Wildlife Fund-US; Pew Scholars Award in Environment and Conservation; International Rachel Carson Prize (Norway); Asahi Glass Foundation's International Blue Planet Prize (Japan).

Richard Dawkins, Professor, New College, Oxford University; Royal Society of London Michael Faraday Award; Medal of the Zoological Society of London; Nakayama Prize for Human Science (Japan).

Jared Diamond, Professor of Physiology, University of California, Los Angeles, School of Medicine; Pulitzer Prize; MacArthur Fellow; Britain's Science Book Priz; Fellow, American Ornithologists Union; Coues Award; Burr Award; National Medal of Science (USA); International Cosmos Prize (Japan).

Sylvia Earle, National Geographic Society Explorer-in-Residence; former Chief Scientist, U.S. National Oceanic and Atmospheric Administration; Olguin Marine Environment Award; Knighted by the Netherlands: Order of the Golden Ark.

Edgardo Gomez, Professor, Marine Sciences Institute, University of the Philippines; Pew Fellow in Marine Conservation; Presidential Lingkod Bayan Award, 2000; UNEP Global 500 Roll of Honour.

Roger Guillemin, Distinguished Professor, The Salk Institute; Nobel Prize in Medicine and Physiology;

National Medal of Science (USA); Lasker Foundation Award; Honorary Member, Japan Biochemical Society.

Sir Aaron Klug, Medical Research Council Laboratory of Microbiology, Cambridge University; Nobel Prize in Chemistry; former President, Royal Society (England); Honorary Fellow of Peterhouse and of Trinity College.

Masakazu Konishi, Professor, California Institute of Technology; Member, National Academy of Sciences; Dana Award for Achievement in Health; International Prize for Biology, Japan Society for the Promotion of Science.

Jane Lubchenco, Distinguished Professor of Zoology, Oregon State University; MacArthur Fellow; Pew Scholar in Capital Conservation and the Environment; Heinz Award in the Environment; member, National Academy of Science; former president, Ecological Society of America; former president, American Association for the Advancement of Science

Alan MacDiarmid, Blanchard Professor of Chemistry, University of Pennsylvania; Nobel Prize for Chemistry; Royal Society of Chemistry Centenary Medal and Lectureship (England).

Laurence Mee, Visiting Professor, PlymouthUniversity Environmental Research Center; Pew Fellow; Fellow, Royal Society of Chemistry; Fellow, Georgian Academy of Ecological Sciences.

Elliott Norse, President, Marine Conservation Biology Institute; Pew Fellow; Evergreen Award; Committee on Human Dimensions of Global Change, National Research Council; founding mem-ber, Society for Conservation Biology.

Giuseppe Notarbartolo di Sciara, President, Istituto Centrale per la Ricerca Scientifica e Tecnologica Applicata al Mare; former Pres., European Cetacean Society; Tridente d'Oro Prize.

Gordon Orions, Professor Emeritus of Zoology, University of Washington; Eminent Ecologists Award, Ecological Society of America; Guggenheim Fellow; Chairman, Board of Environmental Studies and Toxicology, National Research Council (USA).

Roger Payne, Founder/President, Ocean Alliance; MacArthur Fellow; Lyndhurst Prize Fellow; UNEP Global 500 Laureate; Knighted by the Netherlands: Order of the Golden Ark.

Carl Safina, Lecturer, Yale University School of Forestry and Environmental Studies; Vice President for Marine Conservation, National Audubon Society; MacArthur Fellow; Board of Governors, Society for Conservation Biology.

David Suzuki, Professor Emeritus, University of British Columbia; Fellow, Royal Society of Canada; Sanford Fleming Medal, Royal Canadian Institute; Medal of Honour, Canadian Medical Association; UNEP Global 500 Roll of Honour; UNESCO Kalinga Prize for Science.

John Terborgh, Director, Duke University Center for Tropical Conservation; MacArthur Fellow; Guggenheim Fellow; Pew Fellow; Member, National Academy of Sciences; Daniel Giraud Elliott Medal, National Academy of Sciences.

Edward O. Wilson, University Research Professor, Harvard University; Pulitzer Prize (twice); National Medal of Science (USA); Tyler Environmental Prize; Prix du Institut de la Vie, Paris; Crafoord Prize, Royal Swedish Academy of Sciences; International Prize for Biology (Japan).

George Woodwell, Director, Woods Hole Research Center; Founder, Ecosystems Center, Marine Biological Laboratory; Heinz Award in the Environment; Member, National Academy of Science; Fellow, American Academy of Arts and Sciences.

Chapter 5
The Framework for Conservation of Whales and other Cetaceans as Components of Marine Biodiversity

Patricia Birnie

1.   INTRODUCTION

Whales and other cetaceans, as is now internationally recognized, are different from fish because of certain special characteristics.[1] These have made them, especially vulnerable to capture and over-exploitation; many species became so grossly over depleted by the early 20th century that some had reached the point of extinction. For the past 55 years commercial whaling has thus been regulated globally by a single global organization, the International Whaling Commission (IWC) established under the 1946 International Convention for Regulation of Whaling[2] (ICRW). The focus of the IWC is almost exclusively on the so-called "great whale" species and it is currently able to only make recommendations to the parties to the ICRW on smaller cetaceans. This chapter re-examines this approach and considers whether it is appropriate in the light of relevant new developments in international law, institutions and treaties following the entry into force of the United Nations Conventions on the Law of the Sea (UNCLOS)[3] and Biological Diversity (CBD)[4] and

---

[1]   *See generally* R. GAMBELL, WHALES 6–45 (1989); C. de Klemm, *Fisheries Conservation and Management and Marine Biodiversity, in* DEVELOPMENTS IN INTERNATIONAL FISHERIES LAW 423–30 (E. Hey ed., 1999); J. Cooke, *Introduction and Overview, in UICN, Dolphins, Porpoises and Whales of the World* 4–18 (1991).

[2]   International Convention for Regulation of Whaling, 1946,190 L.N.T.S. 79; 161 U.N.T.S. 72.

[3]   United Nations Convention on the Law of the Sea 1982. United Nations, New York, 1988; 21 I.L.M. 1245 (1982); Agreement on The Implementation of Part XI thereof 1994, 33 I.L.M. 1309 (1994) (*hereinafter* UNCLOS).

[4]   United Nations Convention on Biological Diversity 1992, 31 I.L.M. 818 (1992).

related agreements, and application of the principles and programs adopted by the United Nations Conference on Environment and Development (UNCED).[5] It examines the problems primarily from the viewpoint of the unique conservatory requirements arising from the CBD, which must now be taken into account both in order to achieve the UNCED goal of sustainable development and to conserve whales and other cetaceans as irreplaceable components of the marine ecosystem. The issues arising from the new conservatory perspectives are considered in five Sections in this chapter. Section 2 addresses the special characteristics of cetaceans relevant to their conservation as components of marine diversity. Section 3 discusses the conservatory requirements of the Biodiversity Convention and their implications concerning whales and other cetaceans. Section 4 details the existing regime for conservation of these species. Section 5 outlines the sectoral nature of the related regimes for conservation of fisheries and of the marine environment. Section 6 current attempts to coordinate and integrate relevant measures taken under various relevant treaties and the adequacy of these actions to achieve the objects of the CBD, before final conclusions are drawn.

2.   SPECIAL CHARACTERISTICS OF WHALES AND OTHER CETACEANS RELEVANT TO THEIR CONSERVATION AS COMPONENTS OF MARINE BIODIVERSITY: THEIR SIGNIFICANCE IN NEW CONSERVATORY PERSPECTIVES

Whales present unique conservatory problems. It is internationally recognized that their differences from fish in many ways and their unique characteristics have historically made them exceptionally vulnerable to capture, while their size and unique properties have made them particularly attractive to hunt for commercial purposes. They are warm-blooded mammals and thus air breathers. The females of the species give birth after up to a year or more of gestation to live young, who are dependent on their mother's milk and require her protection on their first long migration from breeding to feeding grounds. Whales also live in social groupings (pods) and can be identified as individuals. Along with

---

5   United Nations Conference on Environment and Development; Declaration on Environment and Development GA Res. 47/190, (1992), 31 I.L.M. 874 (1992).

the smaller dolphins and porpoises they form the group known as cetaceans. Although modern advances in science, technology and related disciplines have facilitated progress in the acquisition of data and knowledge about them and development of the theoretical bases from which more accurate calculation of stock status and conservatory measures can be based, nonetheless, as an eminent cetologist noted, "much is still mysterious about them"[6] and "in the relationship between whales and humans whales have generally suffered very badly."[7] Thus, substantial scientific research is still being carried out both inside and outside the IWC to learn more about their lives, particularly on identifying individual whales in particular populations.[8] However, the scope of this research is limited by funding constraints, This research takes the form of sightings and observations from coastlines of live whales at sea or those stranded on shore, or from ships, aircraft and satellites,[9] photography, sound recordings, measurement, controversial radio-tagging which can be picked up from shore in some cases, and sampling, though this is also regarded as controversial by those supporting animal rights. Whales can be seen more easily if they swim close inshore as some populations do but details concerning where most whales go are often unknown. Many, especially Blue whales, swim vast distances deep under water, and even if located when they come up to breathe may never be sighted again. Getting close enough to recognize special identifying characteristics is thus very difficult. Mating is rarely observed. It is known that whales create sounds and clicks probably for various echolocation purposes but details of this remain obscure,[10] although advances have been made over time.

Ensuring their continued existence as a major component of the ocean's ecosystems worldwide represents a challenge to humankind not only to provide the data required for effective conservation but also to engender cooperation for their protection and management which

---

[6]   GAMBELL, *supra* note 1, at 6.

[7]   *Id*. at 42.

[8]   *Id*. at 43.

[9]   *Id*. at 44–45.

[10]   *Id*. at 31–33.

necessarily must recognize the different characteristics and vulnerabilities of different species of whales. The largest so-called "great whales," which include the Blue, Humpback and Right whales, are all baleen whales. These species use their fibrous "baleen" plates to filter water from krill and other plankton on which they feed.[11] These species usually travel vast distances between polar feeding grounds and tropical wintering grounds and live in small social groups, though they can form large herds on feeding grounds. The toothed whales, which include the large Sperm whales, the Killer, Beaked, Bottlenose and Beluga whales as well as dolphins and porpoises, vary widely in size and habits. Some travel long distances between feeding and wintering areas, others are largely confined to localized areas. They may live in pods, alone or in large schools moving between groups or staying in their original group, and feed mainly on fish or squid that could also be depleted by over exploitation or habitat damage or loss. Most small cetaceans are currently unregulated by any global institution though some regional commissions have recently been established to conserve them in particular areas and they can also be listed and subjected to special requirements under a few international agreements as indicated in Section 4.

The various populations of whales are distributed all over the globe. Most of the baleen whales are distributed globally, with separate populations found in the Northern and Southern Oceans; whales are present in all the world's oceans, from the edge of the polar ice to the Equator.[12] The distribution and migration of some species varies with the changing seasons as they move to the North and South Polar Seas (whose seasons are opposite) in search of food and return, at different times, to the warmer Equatorial waters in winter to breed and give birth. Those found in shallow coastal waters off the Pacific Coast of North America are the most widely viewed but whale watching is now widespread in countries bordering those areas of the world's seas and oceans where whales regularly appear.

Though the whaling industry, which once posed the greatest threat to the large whales, has greatly declined, stocks are recovering only slowly in most cases, if at all. Moreover, as outlined in Section 4, a number of

---

[11]   *Id.* at 20–23.

[12]   *Id.* at 15–20.

additional threats to their survival have emerged in recent years. Some limited, and steadily increasing, whaling currently takes place, effectively unregulated by the IWC since it is pursued under permitted exceptions to the ICRW, either under nationally issued scientific permits (by Japan) or resort to objection procedures (by Norway). Relevant details of such whaling operations are outlined in Section 4. Under exemptions provided for under the Schedule to the ICRW a few whales are also allowed to be taken from otherwise depleted stocks by indigenous peoples, subject to the condition that the meat and products are used exclusively for local consumption to satisfy aboriginal subsistence needs.[13]

As outlined in Section 4, the IWC has attempted, with limited success, to preserve at least sufficient whales to ensure proper and effective conservation and development of whale stocks albeit in order to make possible the orderly development of the whaling industry,[14] should the commercial moratorium on whaling be lifted in the future. However, the entry into force on December 29, 1993 of the Biodiversity Convention has introduced broader environmental and ecological perspectives into marine species conservation in general at the global level. This necessitates review of state practices both within areas of national jurisdiction and beyond, including the practices of existing, but as yet largely uncoordinated, marine conservatory agreements relevant to the CBD's stated purposes and requirements. The CBD currently has over 180 ratifications, and of the ICRW's 47 parties all but South Korea and the U.S. are also CBD parties. This broader environmental perspective is especially needed in the case of such migratory and vulnerable species such as cetaceans since the seas and oceans that cetaceans inhabit cover 70 percent of the earth's surface and can be said to represent its most extensive but least understood ecosystem.[15] Given their unique characteristics, the conservation of cetaceans thus presents more complex problems

---

[13]   Schedule to International Convention for the Regulation of Whaling; 161 U.N.T.S. 143; amended by Protocol, 2 December 1956, 338 U.N.T.S. 366. For the Schedule, as amended at the 52nd Annual Meeting 2002, *see* Annual Report, IWC (2001).

[14]   Preamble to International Convention for the Regulation of Whaling, *supra* note 2.

[15]   De Klemm, *supra* note 1, at 1. On such ecosystems generally *see* GLOBAL MARINE BIODIVERSITY (E.A. Norse, 1993), *passim*.

of regulation and management than hitherto envisaged. Advances in many modern technologies, accompanied by acceleration in economic development to meet human needs, are accelerating the degradation of many marine ecosystems and the cetacean habitats they sustain.[16] At the same time, other technological advances are enabling hitherto unimaginable and unobtainable exploration of even the deepest ocean habitats and their inhabitants.

3.   THE CONVENTION ON BIOLOGICAL DIVERSITY: ITS APPLICATION TO CONSERVATION OF MARINE BIODIVERSITY[17]

The Convention on Biological Diversity's Preamble expresses, *inter alia*, its parties' recognition of the *intrinsic* value of biological diversity and its components, as well as its instrumental value, including the maintenance of the biosphere's life-sustaining systems. It also acknowledges our serious information deficiencies and thus the urgent need to develop scientific, technical and institutional capacities to facilitate the formulation and implementation of appropriate measures. It notes that the fundamental requirement for conservation of biodiversity is both the *in-situ* conservation of ecosystems and natural habitats and the maintenance and recovery of viable populations of species in their natural surroundings. It thus stresses the importance of and need to promote international, regional and global co-operation among states, inter-governmental

---

[16]   D. Freestone, *The Conservation of Marine Ecosystems Under International Law, in* INTERNATIONAL LAW & THE CONSERVATION OF BIODIVERSITY 91–108 (M. Bowman, C. Redgwell eds., 1996); Norse note 15 *supra* at note 52 *infra*; P. Birnie, *The Conservation and Management of Marine Mammals and Anadromous and Catadromous Species, in* Hey, *supra* note 1, at 357–94.

[17]   On this *see generally*, MARINE BIODIVERSITY, (R.F.G. Ormond, J.D. Gage & M.V. Angel eds., 1997); J. Pullen & L. Warren, *The Biodiversity Convention and its Implementation with Respect to Marine Biodiversity*, 1 J. BIOSCI. & L. 263–69 (1997); C. Joyner, *Biodiversity in the Marine Environment: Resource Implications for the Law of the Sea*, 28 VANDERBILT J. TRANSNAT'L L. 635–687 (1995); De Klemm, *supra* note 1, at 623–99. P. BIRNIE & A. BOYLE, INTERNATIONAL LAW & THE ENVIRONMENT Chs. 11 & 13 & 545–98, 646–98 (2d ed. 2002); C. de Klemm & C. Shine, *Biological Diversity Conservation and the Law*, IUCN, Environmental Policy and Law Paper No. 29 (1993).

organizations (IGOs) the non-governmental sector (NGOs) for the conservation and sustainable use of the components of biodiversity for the benefit of present and future generations. It reinforces this by recognizing that conservation of biodiversity is "a common concern of humankind."

This section, therefore, aims to outline in general terms the CBD's more specific requirements concerning conservation of marine biodiversity and the existing international arrangements for cooperation on the variety of factors involved in conservation of cetaceans, and to consider their appropriateness for achievement of the CBD's purposes in the light of its relevant provisions. These include its definitions of key terms that circumscribe its scope. "Biological diversity," is defined in Article 2 as "the variability among living organisms from all sources including . . . *marine and other aquatic ecosystems and the ecological complexes of which they are part*" (emphasis added). This "includes diversity within species, between species and of ecosystems." "Ecosystem" is defined as "a dynamic complex of plant, animal and microorganism communities and their non-living environment interacting as a functional unit." "Habitat" refers to the place or type of site where an organism or population naturally occurs" and "*in-situ* conservation" means "the conservation of ecosystems and natural habitats and maintenance and recovery of viable populations of species in their natural surroundings." "*In-situ* conditions" are defined as those "where genetic resources exist within ecosystems and natural habitats. . ." Recognising that one of the favored conservation techniques under prevailing treaties and national legislation is establishment of "protected areas" or "reserves," Article 8 of the CBD requires, "as far as possible and as appropriate," the establishment of a *system* of protected areas or areas where special measures need to be taken to conserve biological diversity.[18] These too are defined: "protected area" in the CBD context means a geographically defined area that is designated or regulated or managed

---

[18]   On marine special areas *see* K. Gjerde, *High Seas Marine Issues of Interest to the International Lawyer*, 16(3) INT'L J. MAR. & COMM. LAW, 515–18 (1998); H. Thiel & J. A. Koslow (eds.), *Managing Risks to Biodiversity and the Environment on the High Seas, Including Tools such as Marine Protected Areas—Scientific Requirements and Legal Aspects*, Proceedings of the Expert Workshop held at the International Academy for Nature Conservation, Isle of Vilm, Germany, 2001.

to achieve specific conservation objectives. The alternative provided by the CBD of *ex situ* conservation is defined as conservation of biological diversity components *outside* their natural habitats. This is not discussed further in the present chapter since this is not feasible in the case of the great whales and the life of smaller cetaceans kept in zoos, marine parks and aquaria is known to be shortened since their wild habitats cannot be reproduced. "Sustainable use" is defined as "the use of components of biodiversity in a way and at a rate that does not lead to the long-term decline of biodiversity, thereby maintaining its potential to meet the needs and aspirations of present and future generations," whatever these may prove to be. Sustainable use is thus a permissible option and must be guided in the context of cetaceans by the present generation's perception of the aspirations and needs of future generations vis-à-vis these species.

This analysis concentrates on the conservation objectives laid down in the CBD but it should be noted that in most cases these are coupled with the options not only of sustainable use but of sustainable development and that virtually all obligations are qualified by such amorphous words or phrases as "as appropriate," "as far as possible," "according to particular conditions and capabilities" etc. However, the Convention does not set forth guidelines for reconciling developmental needs with those of conservation to ensure sustainability.

The parties' obligations under Article 4 of the CBD are limited jurisdictionally. The provisions of the Convention apply to each contracting party in the case of "components of biodiversity important for its conservation" (an "indicative list" of which is provided in Annex I to the Convention) in areas within the limits of their national jurisdiction. This list includes "ecosystems and habitats: containing high diversity, large numbers of endemic or threatened species or wilderness; required by migratory species; of sound economic cultural or scientific importance (*inter alia*)." The Convention's provisions also apply in the case of "processes and activities, regardless of where their effects occur," "if carried out under [a Party's] jurisdiction and control" not only within those areas but "beyond those limits," The marine areas potentially within the scope of the CBD thus include, as appropriate, not only those in the first category, *i.e.*, within their internal waters, territorial sea and exclusive economic or fisheries zone (EEZ/EFZ) and archipelagic waters, but also the high seas beyond so far as such processes and activities are concerned. The term "processes and activities" is defined in Article 2 of the Convention, but could potentially apply to vessels and installations registered in states parties or flying their flag, including factory ships and vessels used for waste disposal, to the extent that this is permitted under other relevant conventions. However, apart from the definition provided in Article 2 of a "marine ecosystem," the CBD's only other specific reference to the marine environment is found in Article 22 concerning the relationship of the Convention to other international conventions. This provision saves any effect on the rights and duties of the CBD parties deriving from other international agreements unless their exercise would cause "serious damage or threat to biological diversity." No criteria are provided against which such threats can be measured but they could include harassment or over-exploitation of cetaceans or destruction or degradation of their habitat to the extent that such activities would constitute a "serious damage or threat" to these species under Article 22(1).

The Article adds that parties "must implement the Convention with respect to the marine environment consistently with the rights and obligations of States under the law of the sea." It is noteworthy that the CBD does not specifically refer here to the UN Convention on the Law of the Sea. This is not surprising since, as outlined in Section 4, the Convention does not explicitly require conservation of marine biodiversity and not all states (including some CBD parties) are parties to it. The UNCLOS could, however, be broadly interpreted as mandating attention to biodiversity concerns, as discussed later, in the light of its articles concerning fisheries and marine environment protection, including specific provisions concerning conservation of marine mammals generally and cetaceans in particular. Rather, for reasons of compromise aimed at ensuring consensus, as outlined by Chandler, the CBD relates the issue to the general law of the sea.[19] Since the conclusion of UNCLOS in 1982, this general law has witnessed a rapid growth of relevant declarations, principles and conventions concerning conservation of all forms of fisheries and of the marine environment habitats of all marine species at global, regional and other levels. These, as well as the many pre-existing

---

[19]   M. Chandler, *The Biodiversity Convention: Selected Issues of Interest to the International Lawyer*, 4 COLO. J. INT'L ENVTL. L. & POL'Y 1–175 (1993).

instruments, provide the policy directives and forums through which the CBD parties can cooperate to fulfil their obligations under Article 5 of the CBD which requires them to "cooperate through competent international organisations in respect of areas beyond national jurisdiction and on other matters of mutual interest for the conservation of biological diversity." In the case of such highly migratory transboundary species as whales and many other cetaceans, cooperation is indispensable for the success of conservation programs. An exceptionally wide range of organizations is now potentially involved since Annex I of the CBD indicates that, *inter alia*, for purposes of fulfilling the detailed requirements for identifying and monitoring components of biodiversity under Article 7, attention must be paid, particularly for purposes of *in situ* conservation, to: identifying and monitoring those ecosystems and habitats which contain high diversity, large numbers of endemic or threatened species, or wilderness; or are required by migratory species; are, *inter alia*, of cultural or scientific importance or which are representative, unique or associated with key evolutionary or other biological processes, as well as to species and communities which are threatened or of scientific or cultural importance for research into the conservation and sustainable use of biological diversity. These requirements are clearly applicable to the habitats and ecosystems vital to whales and other cetaceans in general since spatial boundaries cannot be easily or clearly delineated.

The requirements for *in-situ* conservation are outlined in some detail in Article 8 of the Biodiversity Convention. The issue addressed in this chapter is not their sufficiency but the subject matter of their requirements. This includes not only establishing a system of protected areas where special measures need to be taken to conserve biological diversity, but also regulating or managing biological resources important for conservation of biological diversity within and outside such areas. This obligation thus comprehends conservation of all forms of cetacean species; promoting protection of ecosystems, natural habitats and maintenance of viable populations of species in natural surroundings, and environmentally sound development in areas adjacent to such protected areas to further their protection (which would require ensuring protection of cetacean habitats); rehabilitating and restoring degraded ecosystems and promoting recovery of threatened species (which also would apply to many cetacean stocks); controlling the risks associated with use and release, *inter alia* of those alien species that threaten ecosystems, habitats or species. It is not known, so far as the author is aware, whether alien species have yet affected the variability of cetaceans as components of the biodiversity of marine ecosystems but presumably they could do so in certain circumstances if they degrade cetaceans' habitat or food sources or otherwise affect their health. As indicated in Section 6 of this chapter, a Protocol on alien species may well be added to the CBD in due course. Meanwhile the CBD's Conference of the Parties (COP) has established various operational objectives concerning identification of the impacts of such species on ecosystems, habitats and other species.[20]

Cooperation is required under Article 20 in providing financial and other support for all these *in situ* conservation measures. Specifically, Article 17 requires parties to facilitate exchange of public information relevant to conservation of biological diversity, in particular exchanging the results of technical and scientific research, training and surveying programs, specialized knowledge (including that of indigenous peoples) and transfer of relevant technologies. Concerning this last requirement, Article 18 further requires CBD parties to promote this form of cooperation, where necessary, through the appropriate international and national institutions. This must include, *inter alia*, strengthening national capabilities (for which purpose a Clearing House Mechanism (CHM) has now been established[21]); training of personnel and experts and use of technologies, and agreed joint research programs. The CHM has now highlighted the importance of cooperation with other organizations in developing these requirements and instructed the CBD's Executive Secretary to improve synergy regarding information exchange with other biodiversity related conventions.[22] These are identified in general terms in Sections 5 and 6 of this chapter.

---

[20]   Alien species were identified by the CBD's COP as a thematic issue of the CBD's work program in its "Jakarta Mandate on Marine and Coastal Biodiversity"; SECRETARIAT, CONVENTION BIOLOGICAL DIVERSITY, HANDBOOK OF THE CONVENTION ON BIOLOGICAL DIVERSITY 245–48, 336, 472, 541 (2001).

[21]   *Id*. at 173–75.

[22]   *Id*. at 175.

For purposes of facilitating, among other objectives, further cooperation, the Convention established three other institutions, under Articles 23, 24 and 25 respectively: a Conference of the Parties, (the Sixth Meeting of which took place in The Hague from April 7–19, 2002); a Secretariat located in Nairobi (enabling close cooperation with UNEP) charged with the task, among others, of coordination with other concerned international bodies, and a Subsidiary Body on Scientific Technical and Technological Advice (SBSTTA) whose tasks include provision of advice on scientific programs and international cooperation in research and development concerning conservation of biodiversity. The progress to date made by these bodies in the conservation of marine biodiversity is discussed in Section 5.

It remains now to consider what the constituents of the existing system for conservation of cetaceans as vital components of many of the world's marine ecosystems are and how the CBD affects or supplements these. Obviously our first object of study must be the regulation currently provided under the ICRW, which has already been very well documented. As Freestone has pointed out, the precise problems of conservation of marine ecosystems and biodiversity have been largely overlooked by the CBD[23] despite their inclusion within its jurisdictional scope. He observes that it is paradoxical that though their problems make them particularly appropriate for regulation at the international level under a treaty on biological diversity the most important discussions in this context are taking place in other forums, including those relating to land-based sources and straddling fish stocks, or at the regional or sectoral level. Biodiversity issues are not given a high profile at IWC meetings. For example, none of the 22 Annexes to the 2001 Report of its Scientific Committee[24] specifically address this topic, though it is dealt with incidentally under the Annex on Environmental Concerns,[25] which is discussed in Section 5 of this chapter.

---

[23]   *Supra* note 16.

[24]   IWC/53/4.

[25]   *Id.* at Annex J.

4.   THE EXISTING SYSTEM FOR CONSERVATION OF WHALES AND OTHER CETACEANS

4.1   Introduction

Can it be said at present that a global regime (in a broad sense) exists for conservation of whales and other cetaceans, much less for conserving cetaceans as components of biodiversity or even any form of organized system? International attention has focussed to date on the so-called "whaling regime" established by the ICRW and on the manner in which the IWC has exercised its obligation to conserve whales, as set out in the preambular recitals of the ICRW. Writing about the "regime" of the International Whaling Commission (IWC) at the end of the 20th century, Andresen recently exposed, in a different context, the limitations of this system, and concluded that "it would be a tall order for anyone to claim that this regime is characterised by order."[26] He defined "order" in this context of this institution as occurring in two stages; first establishment of "mutually accepted rules and regulations for the management of whales among all the main actors," and secondly the conducting of harvesting, in a "sustainable manner," in line with the best available scientific advice.[27] He went on to discuss the orderliness of management of whales (which he concluded was still lacking), and, more specifically, the contribution of various international regimes to this end. He rightly pointed out that though the IWC is the main regulatory body, a challenge remains "to trace the effects of related international regimes and conferences channelled within or outside the IWC or as the IWC has emerged, as a nested regime."[28] However, he took a narrow view of the scope of the present management regime, including only institutions closely related to whales. Thus the IWC, Andresen suggested, is linked only to a limited number of other bodies and instruments, such as the Delegation of the 1972 UN Conference on the Human

---

[26]   Steiner Andresen, *The International Whaling Regime: Order at the Turn of the Century?, in* ORDER FOR THE OCEANS AT THE TURN OF THE CENTURY 215–28, at 215 (D. Vidas & W. Oestreng eds., 1999). On regime theory and international fisheries management, *see* S. KAYE, INTERNATIONAL FISHERIES MANAGEMENT (2001), Ch. 2, at 15–42.

[27]   *Id.*

[28]   *Id.*

Environment (UNCHE),[29] the 1982 United Nations Convention on the Law of the Sea (UNCLOS),[30] the 1973 Convention on International Trade in Endangered Species (CITES),[31] the 1992 North Atlantic Marine Mammal Commission (NAMMCO)[32] and the various existing international trade regimes. He did not include the 1992 United Nations Conference on Environment and Development (UNCED)[33] and the 1980 Convention for the Conservation of Antarctic Marine Living Resources (CCAMLR)[34] in this list because of, in his view, "their marginal influence on the whaling issue."[35] He also relegated to footnotes several other relevant regimes, including the International Council for the Exploration of the Sea (ICES) and the UN Food and Agriculture Organization (FAO).[36]

This approach begs the question of the purview of the term "whaling regime" and the contours of the "whaling issue." As posited, in the limited context of "management of whaling," the narrowness of this approach is open to debate following the adoption of the UNCED instruments and the CBD. The International Whaling Commission was established in 1946 by the International Convention for the Regulation of Whaling for the specific purpose of internationalizing the regulation of whaling and providing a single global institution through which this activity could be "properly regulated" in what was then the predominant

---

29   UN DOC A/CONF; 48/14/REV 1 (1972); text in P. BIRNIE & A. BOYLE, BASIC DOCUMENTS ON INTERNATIONAL ENVIRONMENTAL LAW 2–8 (1995).

30   *Supra* note 2.

31   12 I.L.M. 1055 (1973).

32   Established by Agreement on Co-operation in Research, Conservation and Management of Marine Mammals in the North Atlantic, Nuuk, 1992, text in 1 THE MARINE MAMMALS COMMISSION COMPENDIUM OF SELECTED TREATIES, INTERNATIONAL AGREEMENTS AND OTHER RELEVANT DOCUMENTS ON MARINE RESOURCES, WILDLIFE, AND ENVIRONMENT (4 vols.) 1618. NAMMCO Annual Reports are issued by NAMMCO, Tromso, Norway.

33   *Supra* note 5

34   19 I.L.M. 841 (1980); BIRNIE & BOYLE, *supra* note 17, at 628–44.

35   Andresen, *supra* note 26, at 216.

36   *Id.* at 216, n.2.

interest of the nations of the world," "the safeguarding for future generations the great natural resources represented by the whale stocks."[37] The original 15 participating governments desired to provide opportunities for stocks then grossly depleted by over-harvesting to recover under "a system ensuring proper and effective conservation."[38] The opportunity provided in the ICRW for the Commission to be brought within the framework of a specialized agency of the then newly established United Nations,[39] was not taken up and has not to this day been further pursued. Although later the UN Food and Agriculture Organization was considered by the IWC as a possible appropriate body for exercise of a supervisory integrating role, no approach, then or subsequently, was ever made to it for these purposes. The U.S. had proposed during the 1945/46 negotiations concerning the establishment of the IWC that it should be incorporated into FAO[40] in order to avoid duplication of functions and to promote conservation. It suggested that it should act as an autonomous body within FAO entering into agreements with it and other relevant international bodies to define its responsibilities, operating methods and inter-relationships.[41] Though, with hindsight, that might, as biodiversity concerns emerged, have today facilitated a broader approach, and closer linkage through FAO to its work on improving coordination and acquisition of data on fisheries bodies, and relevant on-going scientific research, the founding fathers of the IWC rejected the U.S. proposal considering that more than one specialized UN agency would be concerned with conservation and development of "whale fisheries." Thus, they deferred the decision for two years and then decided *not* to bring the IWC under the purview of FAO or any

---

[37]   ICRW, Preamble.

[38]   *Id.*

[39]   Article III (6), ICRW.

[40]   On this *see* P. BIRNIE, INTERNATIONAL REGULATION OF WHALING 182–88 (2 vols) (1983). Article 1 (2) of the FAO Constitution requires it to promote and where appropriate to recommend national and international action with respect to the conservation of natural resources, in which term under Article XVI, fisheries and marine products are included.

[41]   *Id.* at 185. The arguments for and against such incorporation are set out at pp. 186–87 therein, and remain valid.

other organization, though the need for close cooperation with FAO was stressed.[42] The need for close liaison with existing independent scientific research institutions and the need for more scientific research were also recognized by the IWC, as was the need for different regulations in different areas in order to conserve depleted stocks.[43] The importance of preserving marine biodiversity, the subtleties of the role of the whales within the marine ecosystems through which they migrate, and the need to protect these ecosystems in order to sustain viable levels of whale stocks and their food chains was not nearly as fully and widely appreciated as it is today. These broader needs have been more fully comprehended and the Stockholm and Rio Declarations and action programs have called for enhanced international cooperation in the relevant fields at all levels, as have the UNCLOS, the CBD, the UNEP and the FAO. Thus, there is a compelling rationale to move away from the confrontational "regime" or "governance" intensification approaches and to concentrate rather on means of mobilizing a "proper and effective *system*" to ensure conservation of cetaceans. The institutional implications of this new goal requires reconsideration in the context of the much wider range of relevant established marine institutions available in the 21st century. Moreover, we must reassess methods for coordinating the activities of these regimes to engender cooperation and thus effectuate the CBD's objectives. More order is now required in the relevant international organizational system to achieve this purpose, though ultimately the requisite measures will necessarily be taken by the relevant states within areas or vessels subject to their various forms of jurisdiction, in conformity with their various treaty obligations.

4.2   Reconsideration of the Institutional Framework for Conservation of Whales and Other Cetaceans in the Light of the CBD and Other UNCED Instruments

A highly contentious issue is the whether there remains a compelling rationale for continuing to vest exclusive authority for whaling issues,

---

[42]   *Id.* at 213.

[43]   *Id.*

including population assessments and assessment of threats, in a single international organization.[44] Andresen observes that the attention accorded to "protection" in post-1972 agreements is linked to the intense focus placed by the environmental movement on marine mammals as "charismatic megafauna"[45] but that international institutions have now been created to counter this development. He contends that "the key environmental slogan" of . . . "sustainable development" is not easily reconciled with the "slogan of protection."[46] This chapter, however, is not concerned with the different meanings attributed by some commentators to "conservation" and "protection." It focuses on the increasingly urgent questions concerning how to systematize or, at the least, coordinate the now considerable "range" of existing organizations and other institutions relevant to protecting/conserving whales and other cetaceans as components of biodiversity. This issue is salient regardless of whether cetaceans are exploited, throughout their entire migratory range and in their feeding and breeding grounds, and both within areas of national jurisdiction and on the high seas. The IWC's membership remains limited and its research agenda is probably too narrow to encompass all of the critical research necessary to ensure that cetaceans remain an important component of marine ecosystems. Even though the ICRW permits *any* state to become a party, and whales are found in most states' maritime zones, its membership has never risen above the current tally of 47, while the CBD binds over 180 states. Meanwhile, we have become aware that though the oceans contain the world's largest ecosystems much less is known about the 160,000–250,000 species so far identified therein and their interrelationships than about terrestrial species more visible to humans.[47] As a consequence of technology that now affords the world's citizens to view the oceans' depth and food chain relationships, they are becoming increasingly aware of the perils

---

[44] Andresen, *supra* note 26., *passim*, and works cited therein.

[45] He cites in support of this view R. Friedheim, *A Bad Regime Succeeding a Bad Regime, in* Towards a Sustainable Whaling Regime 19 (R. Friedheim ed., 2001).

[46] Andresen, *supra* note 26, at 219.

[47] De Klemm, *supra* note 1, at 424 ff.

facing ocean species and the need for effective measures to protect ecosystems.

The methods of conserving biodiversity have been both enlarged and constrained by the principles laid down in the Declaration adopted by the UNCED in 1992 and the relevant chapters of Agenda 21,[48] (its action program) and the CBD. Identifying the existing institutional framework and its deficiencies when measured against these new and additional international purposes presents a daunting task. Detailed analysis is beyond the scope of this chapter since it requires in-depth scrutiny not only of the role of the IWC, but also the myriad of existing agreements at the international, regional, sub-regional and bilateral level that are germane to management of cetaceans, fisheries and other marine living marine species, as well as habitat protection. The threats posed to cetaceans by environmental change are as important a consideration today as prevention of their over-exploitation was in 1946. No attempt has yet been made, so far as this writer is aware, exhaustively to analyze the relative current *practice* of *all* these organizations, as distinct from their constitutional powers and objectives, in relation to cetaceans as part of the global marine biodiversity conservation framework, though there are frequent references in the literature on this to the relevance of "other bodies." Many of the apposite instruments have now been listed in the numerous marine and environmental treaty compendia that have been recently published. The practice associated with some of these instruments has been examined in the light of the need for better con-servation of fisheries or protection of the marine environment. However, a comprehensive legal analysis of their combined roles and interrela-tionships in conserving cetaceans as unique components of marine bio-diversity has not yet been attempted. There are numerous monographs and ad hoc articles on the practice of particular fisheries or marine envi-ronment protection instruments as well as regional or species specific clusters of these. All that can be done here is to identify the contours of the problem, the gaps in the present system for establishing cooperation and coordination of these agreements, and the nascent international law concerning cooperation and coordination as evidenced by these and by

---

[48]   For the text of the 40 chapters of Agenda 21 *see* S. JOHNSON, THE EARTH SUMMIT 128–205 (1995).

subsequent state and institutional practice.

The question arises whether the present organizational structure is capable of providing sufficient and accurate scientific data to fine tune conservatory ecosystem management of cetaceans in general, given the insufficiencies of data frequently cited by the IWC's Scientific Committee, as well as the scientific bodies for other regimes. Is the IWC, as presently empowered and structured, the body best suited to receive and interpret this data and coordinate the research programs of *all* the concerned bodies? The received wisdom for half a century has been to leave whales, if not all cetaceans to the IWC, though there has been some movement away from this position in the last decade. Before we can answer these questions, we must identify the contours of the existing regime and the principles to which it is now required to give effect.

4.3   Relevant Principles of the Rio Declaration and Agenda 21

Principles germane to the conservation of cetaceans, though not necessarily expressly set forth in the Rio Declaration and not *per se* binding, are now becoming widely accepted in the practice of many states and institutions. This is evidenced in relevant national legislation and new or revised marine-related treaties. This includes principles addressing state responsibility: not to cause damage to the interests of other states or to international areas (para. 2); a duty to cooperate in order to conserve, protect and restore the earth's ecosystem (para.7); a call for strengthening indigenous capacity building by improving scientific understanding through exchanges of scientific and technological knowledge (para. 9); a call for enactment of effective environmental legislation (para. 11); a duty to cooperate to discourage or prevent transfer to other states of activities or substances that cause severe environmental degradation (para.14); adoption of a precautionary approach (para. 15); a call to undertake environmental impact assessments (para. 17); a duty to provide timely notification and information, and to consult with potentially affected states on activities with significant adverse transboundary environmental effects (para. 19). Finally, states are commanded to cooperate in good faith in fulfilling these principles and in "further development of international law in the field of sustainable development" (para. 27). Although these principles are hedged with numerous qualifications respecting states' sovereign rights, as are the CBD's provisions, the intention to enhance environmental protection is clear, and despite the initial lack of a compulsive requirement, they already dominate dialogues concerning the further development of existing commitments of states and the undertaking of new commitments concerning conservation of marine biodiversity. The Rio principles are especially relevant to the obligations arising under the accompanying CBD and to interpretation of relevant UNCLOS articles in its light. Many of these are outlined in more detail in UNCED's Agenda 21, Chapter 17 on "The Oceans and All Kinds of Seas." The IWC has no formal means of learning to what extent these principles are applied in such bodies or enacted in the relevant legislation of member states. It may receive brief reports from these regimes, or from members who have observer status or are otherwise represented at these meetings.

4.4.   Fulfilling the CBD's Requirements for Conservation of Marine Biodiversity: The UNCLOS and Related Conventions

The earth's environment, interdependent ecosystems and living resources, including marine species, were not categorized in the CBD as a "common heritage of mankind," as were the deep seabed's mineral resources in Article 136 of the UNCLOS. However, the CBD did, as noted earlier, declare in its Preamble that "the conservation of biological diversity is a common concern of humankind." Thus, the more than 180 states now party to Convention are accorded the opportunity to make formal representations to those states that fail to observe the obligations they have undertaken.[49]

It must be acknowledged that the CBD is replete with vague terms and amorphous commitments by the parties. This led one critic to describe it as an "amalgam of concepts in unstable and often vaguely

---

[49]   In this it followed the precedent set by the UN in negotiation of the Framework Convention on Climate Change, also adopted at Rio. On the background to the CBD negotiation *see* BIRNIE & BOYLE, *supra* note 17, at Ch. 11, 555–98, Ch. 13, 646–96.

[50]   H. SCHEIBER, *The Biodiversity Convention and Access to Marine Genetic*

defined relationship to one another,"[50] which amounts "to little more that the expression of platitudinous generalities,"[51] depending on the level of party implementation. However, the exceptionally wide participation in the CBD, which the generality of many of its terms has facilitated, is salutary despite its lack of specificity concerning party responsibilities and the ambiguity of many of its provisions. Provisions related to marine conservation need to be elaborated upon and made to work in practice, and there is now no shortage of relevant organizations through which its parties can work to develop both agreed interpretations of relevant provisions and the measures required to execute these provisions. The unresolved question is whether or not they will cooperate sufficiently to ensure better coordination of the growing but diffused institutional structure for conservation of marine biodiversity. As we have seen, this is a particularly serious problem in relation to cetaceans because of the vast range of the migratory routes of the so-called "great whales," such as the Blue, Humpback, Right and beluga whales. These species feed on krill and other plankton, traversing diverse maritime zones of national jurisdiction as well as the high seas and are exposed to the numerous threats outlined earlier in this chapter. The responsibility for confronting these threats lies with a complex array of regional and global institutions.

Traditionally marine conservation has been more concerned with ensuring, on an ad hoc basis, the so-called "rational" or "wise" use of "common property" or "shared" resources such as fish or marine mammals, than with protection of their migratory routes and habitats. Such arguments also focus on preventing over exploitation of specific species of wild fauna and flora through trade and other forms of control. It has only been relatively recently that treaties have been formulated to address the conservation of marine ecosystems, and then only in a few specific areas such as Antarctica, some regions of South-East Asia, the Caribbean and Western Indian Ocean and a few outstanding natural marine sites now listed under UNESCO's World Heritage Convention.[52]

Materials in International Law 1, 187–201, at 193. *Id.*, at 187, *in* ORDER FOR THE OCEANS AT THE TURN OF THE CENTURY 187, 193 (D. Vidas & W. Østreng eds., 1999).

[51]   *Id.* at 187.

[52]   Text in 11 I.L.M. (1972) 1358; for discussion of this and other related conventions *see* PATRICIA BIRNIE & ALAN BOYLE, INTERNATIONAL LAW AND THE ENVI-

All of these treaties can provide vehicles for developing measures to protect various components of marine biodiversity and many do so but in a piecemeal and often limited fashion.[53] Moreover, hitherto with some exceptions, they have seldom considered the effect of their particular activities on, or, (so far as the writer is aware), collected data relevant to the conservation of marine mammals. This is especially so in the case of the numerous fisheries conventions that are mainly, but not exclusively, concerned with regulating the taking of particular species of fish or mixed fishing stocks and are usually focused on high seas fisheries or straddling stocks rather than those found within national jurisdictions.[54]

Unfortunately, as noted earlier, because of the many compromises required to secure consensus on an agreed text, while the Biodiversity Convention does provide a framework within which its parties can take the action it requires for conservation of marine biodiversity, it does not prescribe any explicit measures for doing so. To date, its COP and subsidiary bodies have addressed only limited aspects of these problems in the "Jakarta Mandate" (discussed later). Similarly, as we have seen, the UNCLOS does not explicitly refer to the need for conservation of marine biodiversity although the CBD does require this. Moreover the CBD provides that its marine environment provisions must be implemented "consistently" with the law of the sea. This raises problems of interpretation since the UNCLOS Article 61—on conservation of the

---

RONMENT (2002), at Chs. 7–10 relating to protection of various aspects of the marine and atmospheric environment, and 11–13 on protection of terrestrial and marine biodiversity. On progress under the World Heritage Convention *see*, World Conservation, 3, IUCN BULL. 2001, 5.

53   For critiques of the major environmental conventions relevant to conservation of biodiversity *see* de Klemm, *supra* note 1, at 433–490, P. VAN HEIJNSBERGEN, INTERNATIONAL PROTECTION OF WILD FAUNA AND FLORA, especially Ch. 2 at 9–42, Ch. 6 at 75–140, Ch. 8 at 149–60, Ch. 9 at 161–96; R. Churchill, *The Contribution of Existing Agreements for the Conservation of Terrestrial Species and Habitats to Maintenance of Biodiversity, in* INTERNATIONAL LAW AND CONSERVATION OF BIODIVERSITY 71–89 and works cited in note 17 (M. Bowman & C. Redgwell eds., 1996).

54   On this *see* R. CHURCHILL & A.V. LOWE, THE LAW OF THE SEA 279–327 (3d ed. 1999).

living resources of the Exclusive Economic Zone—requires coastal
States to ensure "though proper conservation and management measures
that their maintenance is not endangered by over exploitation."[55] It
requires that populations of harvested species be maintained at maximum sustainable yield (MSY) levels as qualified, *inter alia*, by relevant
environmental as well as economic factors, interdependence of stocks
and generally recommended international standards, whether subregional, regional or global.[56] It also requires that "available" scientific
information, catch and fishing effort strategies and other data relevant
to the conservation of fish stocks be *contributed and exchanged* on a
regular basis through competent international organizations, whether
sub-regional, regional or global, as appropriate and with participation by
all states concerned. This mandate extends to states whose nationals are
allowed to fish in the exclusive economic zone.[57] Similar requirements
are laid down in Article 119 for conservation of the living resources of
the high seas. However, while Part XIII of UNCLOS recognizes the need
to foster the requisite scientific research to achieve these goals, it imposes
restrictions on freedom of international access for this purpose to waters
under the national jurisdiction of coastal states.[58]

The special problems of conservation presented by fish stocks and
their associated species occurring in the exclusive economic zones
(EEZs) of two or more coastal states or an EEZ and an adjacent area
beyond are addressed in Article 63. In both cases, though in different
terms, the concerned states are obliged "either directly or through appropriate sub-regional or regional organisations to agree on the measures
necessary to co-ordinate and ensure conservation of" these stocks.
Although a number of regional fisheries commissions already exist

---

[55]   UNCLOS, *supra* note 2, at art. 61 (2).

[56]   *Id*. at art. 61 (3).

[57]   *Id*. at art. 61 (5); On marine scientific research and transfer of technology,
see CHURCHILL & LOWE, *supra* note 54, at 400–16, especially 415–16, on cooperation and transfers.

[58]   On this *see* A. Soons, *Marine Scientific Research Provisions in the
Convention on the Law of the Sea: Issues of Interpretation, in* THE UN CONVEN-
TION ON THE LAW OF THE SEA: IMPACT AND IMPLEMENTATION 365–72 (E. D. Brown
& R. R, Churchill eds., 1989).

through which these goals can be achieved, no coherent body of principles and responsibilities for coordinating their activities at the international level existed until the recent entry into force of the 1995 UN Agreement for the Implementation of the Provisions of the 1982 United Nations Convention on the Law of the Sea Relating to the Conservation and Management of Straddling Fish Stocks and Highly Migratory Fish Stocks (hereafter the SFA).[59] The SFA provides a new framework for cooperation, coordination and further development of the bodies relevant for effectuating its objectives. It establishes management principles of general application, providing a blueprint for fisheries conservation and management in general. Additionally, the SFA acknowledges the importance of a cooperative approach to fisheries management regimes and demands compatible conservation and management emphasising the interdependency of stocks. Finally the SFA recognizes that, as holistic management is required, neither coastal nor distant water fishing states can manage stocks in isolation, and explicitly referred to in the ICRW's Preamble. The parallels with the conservation needs of cetaceans are obvious, although the SFA does not specifically apply to these species. There must be a comprehensive global framework within which cooperation in conservation and management (if the latter is required) can occur. Nandan has emphasized that, through the SFA's combination of new mechanisms, clarification of the roles and responsibilities of regional management bodies, and supplementation of existing measures by new ones, a serious gap in effective application of conservation and management measures has now been filled. That the SFA has had an impact even before its entry into force is evidenced, Nandan concludes, by its use as a global standard for review of relevant existing management organizations and establishment of new ones. Many of these agreements reference relevant UNCLOS and SFA provisions in their preambles that require both coastal states and states fishing in the region to cooperate, *inter alia*, to ensure conservation of the relevant stocks.[60]

---

[59]   34 I.L.M. 1542 (1995); in force December 2001; on this *see* M. Hayashi, *The 1995 UN Fish Stocks Agreement and the Law of the Sea, in* Vidas & Ostreng, *supra* note 50, at 37–56; *Id.* in Hey, *supra* note 1 at 55–84, and 577–88.

[60]   Paper given by S. Nandan, *Improvements in Global and Regional, Ocean's Governance,* at the Global Conference on "Oceans and Coasts at Rio," Unesco, Paris, Dec. 3–7, 2001, on file with the author.

Though the SFA does not specifically apply to cetaceans there is no reason why its standards should not be used as a guide to protection of marine biodiversity generally. However, much depends here on subsequent state practice in the IWC and its progress in cooperating to other conventions.

UNCLOS Article 64 similarly requires states exploiting highly migratory species in a region to cooperate to ensure conservation both directly and through appropriate organizations. Although Article 65 of UNCLOS makes specific and separate provision for conservation of marine mammals, this, like Article 61–64 and the SFA, focuses on conservation considerations. It permits more strict limitation or regulation of exploitation of marine mammals, creates a clear obligation on states to cooperate for their conservation and, "in the case of cetaceans in particular to work through the appropriate international organisations for their conservation, management and study."[61] The organizations appropriate to achieve these goals are not identified. Neither the coastal state nor a competent international organization are precluded from prohibiting marine mammal exploitation or regulating it more strictly than is required elsewhere in Part V of the UNCLOS. However, even if they opt not to exploit these species, they are not relieved of the duty to cooperate to conserve and study them, working "in particular through the appropriate organisations." The organizations appropriate for achievement of these goals are not identified in UNCLOS. The choice is left open. Canada, which withdrew from the ICRW at this point in the UNCLOS negotiations, considered that the North West Atlantic Fisheries Organization (NAFO) would best suit its purposes.[62] However, when Iceland, Norway, the Faroes (an autonomous region of Denmark) and Greenland established North Atlantic Marine Mammal Commission (NAMMCO) in 1992,[63] only Iceland withdrew from the ICRW. The IWC remains the central and only global regulatory body for cetaceans. However, the ambiguity of Article 65 leaves open a wide choice of other

---

[61]   On the interpretation of this article and para. 17.62 of Ch. 17 of Agenda 21, *see* P. Birnie, *UNCED and Marine Mammals*, 17 MARINE POL'Y 501–14 (1993); Andresen, *supra* note 24, at 221–22.

[62]   BIRNIE, *supra* note 40.

[63]   *Supra* note 31.

organizations through which states can, and now it is submitted, should, also cooperate to ensure data collection and scientific research vital to effective conservation of all cetaceans under the CBD. For example, Agenda 21 also reflects states' recognition that the IWC is *"responsible"* for conservation and management of "whale stocks" (not, in terms, "cetaceans") under the ICRW, but it also recognizes the *"work* of the IWC's Scientific Committee" in carrying out studies of large whales and other cetaceans, and that of other such existing bodies such, as the Inter-American Tropical Tuna Commission (IATTC) and the Agreement on Small Cetaceans of the Baltic and North Seas (ASCOBANS), in conservation, management and study of cetaceans and other marine mammals.[64] It provides no advice on how to coordinate these studies, or as to which body is responsible for doing so. Nor does it take account of the fact that in the 57 years since conclusion of the ICRW, the 20 since the UNCLOS and the ten since the UNCED and CBD, a large number of fisheries and marine environment conservation conventions have been concluded and that their practices have not only gradually embraced many of the requirements of the UNCED instruments but in doing so, have recognized, by concluding Protocols or Declarations, that fisheries must now acknowledge, and seek to remedy, the adverse effect that some fishing practices (*e.g.*, bottom and pair trawling, discarding bycatch, use of fine filament nets) have not only directly on other species, including cetaceans, but on degradation of the marine environment which provides their habitat. *Vice versa*, several marine environment protection conventions have recognized the adverse effects that pollution of the marine environment from point and multiple sources of pollution under their control have, on marine habitats and the species the seas' food chains sustain. Among examples too numerous to cite here, the post UNCED revision in 1992 of the Paris Convention for the Protection of the Marine Environment of the N.E. Atlantic to establish the OSPAR Commission by combining the 1974 Paris Convention on Land-Based Pollution and the 1972 Oslo Convention on Prevention of Marine Pollution by dumping from Ships, as well as the revision in the same year of the Helsinki Convention on Protection of the Marine Environment of the Baltic Sea Area; both recognized the significance

---

[64]   Agenda 21, Ch. 17, para. 17–62; NAMMCO not having been concluded at that date.

of these activities in improving protection of fisheries habitat and now aim to work closely with the relevant fisheries conventions, especially in seeking scientific advice, collecting relevant data and adopting UNCED's precautionary approach.[65]

Since the conclusion of the ICRW, the UNCLOS and since the CBD and other UNCED instruments, an increasingly large number of conventions or conservation of fisheries and marine environment have been concluded dealing with the multiplicity of related issues. Practice under these conventions has embraced many of the Rio Declaration principles including the need to restore the earth's ecosystems, improve scientific knowledge, apply a precautionary approach, undertake environmental impact assessment etc.

The time is thus ripe for a review of state practice in relation to UNCED's injunction in Agenda 21's para. 17.63 that *all* states (at the least) *should* cooperate for the conservation, management and study of cetaceans. There is, a need to identify the organisations which have played or could play a leading role in these fields, as well as the extent to which technological and other advances in all forms of marine scientific research and data collection made since conclusion of the ICRW are being used by the IWC and its member states.

5. THE SECTORAL NATURE OF THE EXISTING REGIME FOR CONSERVING FISHERIES AND THE MARINE ENVIRONMENT

Potentially, since cetaceans' survival depends not only on protection from over-exploitation and other threats but also preservation of their habitats, all fisheries,[66] marine environment

---

[65]   On these *see* for case studies of North and Baltic Sea marine environment protection regimes see 13 I.J.M.L.C. 299–472 (1998). *See also* L. de La Fayette, *The London Convention 1972: Preparing for the Future*, 13 I.J.M.L.C. 515–36 (1998); L. De La Fállete, *The OSPAR Convention Comes into Force, Continuity and Progress*, 14 I.J.M.L.C. 247–98 (1999).

[66]   For details of these, *see* S.H. Marashi, *The Role of FAO Regional Fishery Bodies in the Light of New Developments in World Fisheries*, FAO (1994); *Summary Information on the Role of International Fishery and Other Bodies with regard to*

protection[67] and scientific research bodies[68] are relevant to their conservation since some cetaceans are likely to migrate through maritime areas within their substantive or jurisdictional scope. A recent UNEP study concluded that there are at least 502 international treaties related to the environment (all establishing institutions of various kinds), 60 percent of which date from 1972. The largest ad hoc cluster before that date consisted of biodiversity or species-related agreements, including the ICRW; another large early cluster addressed marine environmental issues, particularly the IMO Conventions on marine pollution; a third cluster related to nuclear energy, nuclear weapons testing and radiation. Of the 300 conventions concluded after 1972, 70 percent were regional. The greatest impact derives from the 17 multi-sectoral regional seas conventions and actions plans, encompassing 416 conventions, protocols and related agreements; the largest of these—40 percent of the total—concern the marine environment. They include UNCLOS, new IMO Conventions and Protocols, UNEP's 1995 Global

---

the Conservation and Management of Living Resources of the High Seas, FAO Fisheries Circular No. 908, FIPL/C908; Ecosystem-Based Management of Fisheries: Opportunities and Challenges for Co-ordination between Marine Regional Fishery Bodies and Regional Seas Conventions, report on Second Meeting of FAO and Non-FAO Regional Fishery Bodies or Arrangements, RFB/II/2001/7 (hereafter FN/FRFB report); J. Swan & B.P. Sadia, Contribution of the Committee on Fisheries to Global Fisheries Governance 1977–1997, FAO, Rome, 1999, (hereafter Contribution of COFI Report).

67   For example, the International Council for the Exploration of the Sea (ICES) and the Pacific International Council for the Exploration of the Sea (PICES) and the International Oceanographic Commission (IOC), see P. Ehlers, The Intergovernmental Oceanographic Commission: An International Organisation for the Promotion of Marine Research, 15 I.J.M.C.L. 2000, 533–54, which critiques IOC's global role in the context of its new statutes and relation to UNESCO.the Para. 17,62; NAMMCO had not then been concluded.

68   See Multilateral Environmental Agreements: A Summary, Background Paper prepared by UNEP for First Meeting of Open Ended Intergovernmental Group of Ministers or Their Representatives on International Environmental Governance, New York, 2001, (I.L.M. 3); UNEP/I.L.M./1/INF, Mar. 30, 2001, based on information supplied by 20 Secretariats of Multilateral Environmental Agreements (MEAs). For analysis of these and many other related conventions, see BIRNIE & BOYLE, supra note 17, at Chs. 6–13.

Plan of Action for the Protection of the Marine Environment from Land-Based Activities, the UNEP and other regional seas conventions and action plans, and the numerous regional fisheries conventions and protocols. The specifically biodiversity related conventions referred to earlier in this chapter (the WHC; CITES, CMS and CBD) comprise the second largest cluster; nuclear related agreements are also important, with nine new global conventions and protocols and several regional agreements. New chemicals-related conventions have now emerged including the 2001 Stockholm Convention on Persistent Organic Pollutants (POPS).[69] Several atmospheric/energy related conventions are also potentially relevant to marine environmental habitats status: these include the 1985 Vienna Convention for the Protection of the Ozone Layer and its Montreal Protocol[70] and the 1992 Climate Change Convention (FCCC).[71] Yet, as others have pointed out, despite these external developments the ICRW itself has not really been "greened" over these years and other wildlife conservation oriented treaties are providing more effective tools for this purpose.[72] The ICRW does not prohibit its Commission, Secretariat and Scientific Committee from establishing wider cooperation and collaboration with other institutions that would better meet the requirement to conserve whales and other cetaceans as components of the biodiversity of the marine areas through which the migrate. As illustrated in the section that follows, these bodies have not yet, however, made full use of the resources of these organizations, though several cooperative agreements have been concluded.

As Schram and Tahindro have pointed out,[73] the UNCLOS conservation and management principle, as set out in Article 119 (1)(a), requiring

---

[69]   40 I.L.M. 532 (2001).

[70]   26 I.L.M. 1529 (1987).

[71]   3 I.L.M. 851 (1992).

[72]   G. Rose & S. Crane, *The Evolution of International Whaling Law, in* THE GREENING OF INTERNATIONAL LAW 180 (P. Sands, ed. 1993).

[73]   G.G. Schram & A. Tahindro, *Developments in Principles for the Adoption of Fisheries Measures, in* Hey, *supra* note 1, at 255–56.

that measures be based on the "best scientific evidence available" to the states concerned, has recently been incorporated into Article 5 (b) of the SFA. The section stipulates that "the requirement of the best scientific evidence available shall guide states in adopting measures for these stocks," and signals that the need for precaution does not exempt the states and management authorities concerned from their responsibilities to acquire the requisite scientific information to facilitate effect management. As the 1989 UNGA Resolution 44/225 on large scale pelagic driftnet fishing indicated, conservation and management measures concerning marine living resources should take account of the best available scientific "data and analysis," apparently an attempt to clarify the "best evidence" concept by equating it with "statistically sound evidence."[74] Article 5 (j) and (k) of the SFA also require that coastal states as well as high seas fishing states collect and share timely, complete, accurate detailed data on their fishing activities, promote and conduct scientific research and develop appropriate technologies to support conservation and management measures.[75] The SFA's Annex I, an integral part of the convention, lists standard requirements to effectuate these objectives and outlines the information required for conservation as well as identifying the principles for data collection and compilation, reporting, data verification and exchange, all of which is vital for conservation measures. Fisheries experts now believe that new areas of biological and ecological research should be developed, enabling better appreciation of the respective roles of environmental variability and spawning stock size, thus enhancing the quality of scientific information for conservation and management.[76] While there are some major differences between theories and methods for conservation of fisheries and of cetaceans, management regimes in both sectors require the possible data. Given the limited membership of the IWC and the extensive global and regional migrations of cetaceans, especially whales, the problem of

---

[74]   *Id.*, citing The Precautionary Approach to Fisheries with Reference to Straddling Fish Stocks and Highly Migratory Fish Stocks, UN DOC. A/CONF. 164/INF/8, 1994, par. 5, *reproduced in* UNITED NATIONS CONFERENCE ON STRADDLING AND HIGHLY MIGRATORY FISH STOCKS: SELECTED DOCUMENTS 555 (J-P Levy & G. Schram eds., 1996).

[75]   Schram & Tahindro, *supra* note 73, at 261.

[76]   *Id.* at 262.

obtaining the best and most comprehensive data on all aspects of cetacean conservation would be best served by casting the data net beyond this limited membership. Ideally, it would be salutary to obtain data from all CBD parties and other relevant fisheries and marine environment protection agreements in a more ecosystem related approach. Progress in meeting this goal is described below.

6. CURRENT PRACTICE OF THE IWC IN ADDRESSING ENVIRONMENTAL ECOLOGICAL CONCERNS AND COOPERATING WITH INSTITUTIONS AND TREATY REGIMES RELEVANT TO CONSERVATION AND STUDY OF CETACEANS

6.1 The IWC and Environmental/Ecological Concerns

6.1.1 Role of the IWC and Its Scientific Committee

The IWC can, of course, decide for itself with which relevant organizations it wishes to cooperate on environmental matters. In so doing it can seek the advice and recommendations of its Scientific Committee (SC) but does not necessarily have to. In case of disagreement the Commission can resolve the issue by resort to voting (adoption of resolutions require only a simple majority of members present and voting)[77] but it prefers to act by consensus, though this is often elusive given the present political impasse. However, it has not always proved possible in this field, even on scientific issues, given the possible implications of these decisions for the resumption of commercial whaling.

The rules governing membership of the SC are important since its advice is crucial to the functioning of the Commission, its decisions being required by Article V (2) (b) of the ICRW to be based on "scientific findings." The SC consists of scientists nominated by each contracting government that wishes to be represented on it.[78] Representatives of intergovernmental organizations with particular relevance to the SC's work can participate as non-voting members, if they are acceptable to the parties, as can the WCU (IUCN). Non-member governments can be observers

---

[77] For its Rules of Procedure, *see* IWC Annual Rep. 2000, 106–109, at E 99.

[78] For details *see*, Rules of Procedure of the Scientific Committee. IWC Ann.Rep.2000, 106–109 at M (Committees).

and any other international organisations that send observers to the IWC can also nominate a scientifically qualified observer. Finally the Chairman can invite other (non-voting) qualified scientists ("invited participants") that might contribute to the Commission's work and they can present papers.[79] The SC thus, as presently composed, can be said to include almost all the world's leading cetologists. It can, *inter alia*, request special reports on relevant matters as necessary. There is thus considerable scope for introducing discussion in this forum of all the environmental/ecological issues raised in this chapter. The number of invited experts in the relevant fields has grown correspondingly. At the 54th Meeting of the IWC, held in Japan in May 2002, 47 parties sent delegations, the highest number in the history of IWC meetings. Iceland attended under the rubric of a new category crafted by the parties, a "Government invited to assist as an Observer," and there were five non-member government observers[80] but only eight inter-governmental observers[81] and 100 NGOs.[82] The size of the Scientific Committee and its range of expertise has grown correspondingly. The participation of an Icelandic scientist was noted specifically as "being without prejudice to the positions of individual members of the Commission on Iceland's attempted reservation on the commercial whaling moratorium."

The Scientific Committee's mandate under its Rules of Procedure (ROP) includes review of: scientific and statistical information regarding whales and whaling; current research programs of governments, other international or private organizations; and the scientific permits and programs planned by Contracting Governments. It can also consider other matters referred to it by the IWC or its Chairman. The Committee meets before the IWC's Annual Meeting and reports on these, make recommendations. Commissioners *should* also, under the SC's ROP, transmit to the IWC any reports on whaling published in their own countries. It works through standing sub-committees and working groups related to areas and species, including on the identification, status and trends

---

[79]   *Id.*, at 106, 107, at A.

[80]   IWC/54/3 (Draft 2), at 1–3.

[81]   Canada, Cape Verde, Ivory Coast, Nicaragua, Suriname, *Id.* 4.

[82]   *Id.* at 4–7.

of stocks, biological parameters, and related matters. It concentrates on large cetaceans, particularly those exploited or considered for exploitation, but a committee on small cetaceans also exists.[83] It can also convene workshops and has done so on such topics as climate change and the status of small cetaceans around the world. Decisions at its meetings are made by a simple majority.

Specific topics of current concern include comprehensive assessment of whale stocks, implementations of the Revised Management Procedure (RMP) and the effects of environmental change on cetaceans.

6.2.   Environmental and Ecological Issues of the IWC's Agenda

The scope of the IWC's and the SC's agenda has grown considerably over the years, reflecting not only the issues on which its members hold conflicting views but also rising international concern about the environmental issues addressed in this chapter. Thus, items on its agenda include not only such hardy annuals as aboriginal subsistence whaling, comprehensive assessment of whale stocks, the RMP and related International Observer Scheme issue, but also whale watching (including its environmental impact), proposals for the creation of sanctuaries covering vast areas of the oceans, a range of environmental concerns and appropriate forms of cooperation with at least a few specific related organisations. We shall concentrate in this section on the environmental and health issues and cooperation with other organizations.

6.2.1.   Scientific Permits and Environmental and So-Called "Health" Issues

The SC regularly reviews various relevant issues. According to its report to the 53rd Annual Meeting of the IWC, covering the year 2000–2001,[84] it reviewed, *inter alia*, intersessional work carried out to improve its abil-

---

[83]   IWC Rules of Procedure, Annual Rep. 2000, at M. 101 and ROP of SC at C, 107.

[84]   IWC Chair's Report to the 53rd Annual Meeting, July 23–27, 2001, at 40–45.

ity to assess the impact of scientific permit catches on stocks. In particular, it reviewed the documentation concerning permits to kill certain
whales for purposes of scientific research, issued by Japan under its so-
called JARPA and JARPN II research programmes.[85] As to the former,
the SC took a cautious stance, concluding that further modelling
approaches needed to be examined and that stock abundance estimates
had to be agreed to before it could give advice on the effects of JARPA
on Antarctic minke whale stocks. Japan's JARPA II program is aimed
at obtaining information to aid conservation and sustainable use of
marine living resources in the North Pacific, including studies based on
ecosystem modelling, the feeding ecology and ecosystems of minke,
Bryde's and sperm whales, their stock structure and environmental
effects, especially of pollution, on cetaceans and the marine ecosystem. There was considerable disagreement in the SC concerning most
of the aspects of this program, including on its objectives, methodology, and the quality of data derived using non-lethal research techniques. As a result, the SC established a steering group to establish
approaches that might facilitate assessment of the scientific benefit of
the research program. The IWC merely noted the SC's report and
accepted its recommendation.

The SC itself has also developed two major research proposals of its
own, on environmental and health issues. The first is a multi-national,
multi-disciplinary research proposal—POLLUTION 2000+ with the
twin aims of establishing, *inter alia*, whether certain relationships exist
between biomarkers (of exposure of whales to and/or the effects thereon
of PCBs) and establishing PCB levels in certain tissues. This proposal
was strongly endorsed by ASCOBANS and the ICES Working Group
on Marine Mammal Habitats. However, the IWC only allocated £57,000
for the program in 1999/2000, half of the sum required to carry it out.
A revised budget was approved for 2000/01, scaling back the program
to focus only on pollution impacts on bottlenose dolphins and on harbor porpoises.

Though progress was made on these sub-projects, further progress
on pollutant issues at the 53rd IWC Meeting was limited because of the
shortfall in funds. Japan meanwhile contended that chemical pollution

---

[85]   *Id.* at 11 Scientific Permits.

issues were of secondary importance, especially as the program related only to small cetaceans, and it argued that the funding should thus come from the IWC's limited Small Cetaceans Fund. However, the IWC accepted the SC's recommendations.

The SC also reported on its other multidisciplinary program— SOWER 2000—which examined the influence of temporal and spatial variability in the physical and biological Antarctic environment on the distribution, abundance and migration of whales, and also on progress on the SOWER 2000 program in general, particularly concerning the Southern Ocean GLOBEC and preliminary results of earlier collaboration with CCAMLR. It sought validation of the data collected on the IWC/CCAMLR cruises, to enable collaborative analysis. Noteworthy is the fact that the inter-disciplinary approach of these cooperative studies had benefited both organizations and it was regarded as important to maintain this cooperation to ensure timely results on analysis and verification of the now very large amount of whale data that has been collected. However, while Japan commended the accumulation of knowledge on Antarctic ecosystems and geology and the conduct of surveys and research, it doubted the value of according priority to the CCAMLR/ SO-GLOBEC programs and environmental work related to the management of small cetaceans, which it thought had expanded inappropriately. It also argued that the results may be negatively biased by the research methods used and urged caution in, *inter alia*, the use of data in ecosystem modelling. The IWC, however, noted the report and accepted its recommendations.

The question of competition between cetaceans and fisheries was also on the agenda. The SC agreed on the importance of using models to answer questions concerning whether removing marine mammals from an ecosystem increases fish yields and, conversely, whether reducing fish yields would accelerate the recovery rate of depleted cetaceans stocks. It proposed a brief workshop to consider the questions involved. Although the U.S. thought the view that whales precipitated fish stock declines was oversimplified and biologically unsound and was concerned that the issue was put before a body not recognized as having competence in management of whale stocks, it accepted that these were appropriate issues for consideration by the SC. It thus approved the workshop proposal and also proposed, jointly with Japan, a Resolution on Interactions Between Whales and Fish Stocks. Norway noted that the issue was already being addressed in a workshop convened by NAMMCO with Canadian participation. New Zealand proposed review also of the impact of fisheries on cetaceans through by-catch and prey depletion. The SC Chairman stressed that in answering these broad questions, the Workshop would take a full ecosystem approach, looking at the full range of interactions between fisheries and cetaceans. It would also interpret "fishing" in a broad sense, to include study of other marine resources such as krill, and would consider whatever data were appropriate to examining the ecosystem models. The U.K. noted that any multispecies ecosystem approach should include consideration of environmental threats and concerns. The unusual joint U.S./Japan resolution addressing the vexed questions of interactions between whales and fish stocks gave notice (with current FAO initiatives in mind) that the IWC was according priority to this issue and that any studies on ecosystem based fisheries management undertaken by FAO must be holistic and balanced in approach. The resolution endorsed the proposed SC Workshop and asked the IWC Secretary to seek cooperation with FAO.[86]

Other environmental health issues raised in this IWC Meeting resulted in adoption of a Resolution proposed by New Zealand and others, suggesting that the IWC Secretary send the resolution on pollution to the Secretariat of the Stockholm Convention on Persistent Organic Pollutants (POPs) that had just come into force. This idea was approved by a majority. Another Resolution, brought by New Zealand and other concerned governments, noted the importance of habitat protection and integrated coastal zone management, FAO having reported that up to 34 percent of the world's coastal zones were at high risk and 17 percent at moderate risk possibly from climate change effects, ubiquitous POPs and polluted river outflows exacerbated by lack of control of land-based pollution, poor planning an emissions of untreated sewage. While accepting that the IWC could not itself resolve all these threats, the co-sponsors of this Resolution noted the responsibilities imposed on states by the UNCLOS concerning protection and preservation of the marine

---

[86]   *Id.* at 12.32, 43.

environment,[87] Norway supported placing conservation and management of whale stocks in a broader context and the Resolution (as revised) attracted consensus. Though Japan opined that the issues were beyond the IWC's competence it did not block the consensus.[88]

The complexity of the issues had given rise to a number of problems. Thus, discussion also centred on the State of Cetacean Environment Report (SOCER) produced by a SC Working Group. Though the Working Group's report was approved by the SC, it expressed concern that it might be misinterpreted as expressing the SC's view. Therefore, the report was not appended to the SC report, but it was agreed that it could be made available to the IWC under its editors' names and that the process of developing and refining such a report should continue. A Workshop on Habitat Degradation and its assessment was approved in order to develop, *inter alia*, a methodology for quantifying the relationship between environmental variables and the health of a given population. The proposed SOCER was not without its critics in the IWC, however, it was regarded by some as being still at the prototype stage needing refinement.

6.3   The IWC's Cooperation with Other Organizations[89]

At the 53rd Meeting, the IWC's SC also received reports on its cooperation with 12 organisations: the CMS; ASCOBANS; ACCOBAMS; ICES; CCAMLR; GLOBE; NAMMCO; FAO (COFI); PICES; CITES; IUCN and ECCO (East Caribbean Cetacean Organization). It noted the report and accepted its recommendations without further discussion. No details concerning the discussion on or the content of the reports are mentioned in the IWC Chair's Report (details of the SC's discussion of such items are now relegated to a new publication of the IWC, the *Journal of Cetacean Research and Management*,[90] which replaces the

---

[87]   UNCLOS Part XII.

[88]   Chair's Report of 53rd Meeting, Item 12.4 at 44.

[89]   Id/.13, Cooperation with Other Organizations, at 45–46.

[90]   *Id*. at 13.2., 46, *see also* Annual Report of the IWC, 2000, 17.5 Action Arising, 49.

former more detailed scientific sections of the IWC Reports). It appears
from the brief statements presented at the IWC Meetings by its observers
at the meetings of the few related bodies that they have attended that
"cooperation" is somewhat formalistic. It consists primarily of non-com-
mittal reportage of events occurring at the respective Annual Meetings
of contracting parties of these other regimes, rather than reflecting
exchanges of opinion, data, analysis or inter-sessional visits to the secretariats of the other conventions. The time may be ripe for review of the
methodology of "observer status." IWC "observers" are already delegates at the meetings concerned representing their states. That said, it
clearly remains important that regular links exist through which vital relevant information can immediately be passed on without bureaucratic
delay (such as on CITES listings). Furthermore, consistent with the provisions of the CBD, the IWC should seek to foster co-operate and coordination with fisheries management organisations and conventions
concerning protection of the marine environment. It is disappointing that
a recent attempt to extend the scope of the IWC's cooperation with the
IMO was frustrated. The Secretary of the IWC requested that the IMO
Secretary-General (S-G) distribute the IWC's Resolution 2000-8 on protection of the critically endangered Western North Atlantic Right Whale
from ship strikes. Since registered vessels of the U.S., Canada and others use shipping lanes that transverse the habitats of Western North
Atlantic Right Whales, the IWC regarded cooperation with the IMO as
necessary to protect them. However, the IMO rejected the request
because the IWC has no formal cooperative arrangement with the IMO
such as the Memorandum of Understanding (MOU) that it has with the
CMS.[91] The IMO Secretary General suggested that the Resolution be
submitted by a country having membership of both organizations; this
Sweden did subsequently. Following further enquiries by the IWC
Secretary on the benefits and procedures relating to a formal cooperative
arrangement, the IMO Secretary General, while expressing IMO's support for the IWC's whale conservation objectives, concluded that this
issue was not central to IMO activities. This seems a surprising decision
by IMO as it is responsible for among other issues, ship safety, and prevention of oil release and other forms of pollution of the marine envi-

---

[91]   IWC, Chairman's Report of the 53rd Annual Meeting, 3–6 July 2000, at
16.2, 49.

ronment from vessels, and has standing committees on both issues. Furthermore, it has already issued a Notice of Avoidance of certain whale species. New Zealand and Mexico suggested that concerned IWC members that are members of both organizations could pursue the matter further and the IWC concurred.[92]

7.   CONCLUSIONS

In establishing the ICRW in 1946, its parties recognized the "interests of the world in safeguarding for the future generations the great natural resources represented by the whale stocks." The UNCLOS set forth some of the conservatory responsibilities of states fishing the world's marine living resources and established their responsibility for protecting the marine environment. It also provided that neither states nor competent international organizations were precluded from prohibiting or regulating marine mammals more strictly than otherwise provided and that they *must cooperate* in conserving them through appropriate international organizations. The CBD, recognizing the intrinsic value of biodiversity and, *inter alia*, the ecological value of its components, affirmed that its conservation is a "common concern of humankind." Not surprisingly the CBD parties have accorded priority to conservation of marine and coastal biodiversity. Cetaceans, however, legally remain "common property resources" which cannot in practice be exploited without effective institutional arrangements to ensure cooperation. Questions of sufficiency of participation of states in the relevant bodies and regulation and the adequacy of the scientific advice provided by such bodies arise. As Nandan has pointed out in relation to straddling stocks and highly migratory fish stocks, governments must not insist on the inappropriate "old rules of the game;"[93] more ecological approaches are necessary to conserve biodiversity. In the absence of any overriding World Fisheries Authority,[94] in the case of cetaceans globally as components much now depends on whether states cooperate effectively, not

---

[92]   *Id.*

[93]   S. Nanden, 24 E.P.L. 144 (1994).

[94]   On a draft for such a body, *see* A. KOERS, THE INTERNATIONAL REGULATION OF MARINE FISHERIES 307–24, 331–39 (1973).

only through the ICRW but in relationship with the many relevant international and regional conventions referred to in this chapter as well as other relevant bodies and coordinate the activities of these bodies and the implementation nationally of the measures required by them. The SFA addresses this problem; states parties must affect their duty to cooperate by becoming members of the appropriate bodies, or lose their right to fish on the high seas. It has been suggested that once the SFA is widely in force a new conception of common property will emerge.[95] An instant problem arising in the ICRW/CBD relationship is the lack of specificity concerning this in either of these texts and the number of CBD parties that are not parties to the ICRW and related conventions. In the interim, the best use has to be made of the possibility of exchange of well-qualified observers among these conventions, joint studies and workshops, etc., and close involvement of their Secretariats in exchange of information and organization of such workshops. The solution adopted for some of these conventions of co-locating their Secretariats is not likely to be feasible for the ICRW. Meanwhile it should be remarked that often not all conventions and international bodies with observer status at the IWC are in practice represented at its meetings,[96] and that the IWC's observers at meetings of other concerned bodies are frequently already members of these bodies coming from states party to them.[97] The IWC was represented by a member of the Secretariat only at the first meeting of the parties to ACCOBAMS meeting. The report of the IWC observers is informative but brief and hardly represents or results in the required coordination. Meanwhile the IWC still has no regular item addressing marine biodiversity on its agenda. For this establishment of more formal Memoranda of Understanding is required, setting out the required cooperative and coordinate activities. In the context of the CBD goals; recognizing the intrinsic value of biodiversity and the common interest in its conservation, the IWC should work in closer concert not only with the bodies involved directly or indirectly in

---

[95]   P. BIRNIE & A. BOYLE, INTERNATIONAL LAW & THE ENVIRONMENT 684 (2d ed. 2002).

[96]   *See e.g.*, IWC/54.

[97]   *See e.g.*, IWC/54/10, *Cooperation with Other Organizations, Reports of Observers to the IATTC, ICES, CCAMLR, NAMMCO, ICCAT, ACCOBAMS, COFI.*

takes of all kinds of cetaceans, but, also with the growing number of fisheries organizations and bodies concerned with the protection of the marine habitat from all sources of degradation. Since scientific bodies, such as ICES, PICES and the IOC, continually regret the lack of relevant data to ensure the soundness of their advice perhaps efforts could now be made to utilize the opportunities provided by the growth in IFOs and RFOs to involve the observers now required by the SFA to be on board them in a new form of "whale watching" by training them to gather data on the numbers, locations and species of cetaceans observed. This might benefit ICES studies on trophic interactions.

PART II

The North Atlantic Marine Mammal Commission and the World Council of Whalers

Chapter 6
NAMMCO—Regional Cooperation, Sustainable Use, Sustainable Communities

Grete Hovelsrud-Broda[1]

1. INTRODUCTION

The North Atlantic Marine Mammal Commission (NAMMCO) is an international intergovernmental body established to foster, strengthen, and further develop cooperation on conservation, management and study of marine mammals in the North Atlantic. The four parties to NAMMCO (Faroe Islands, Greenland, Iceland and Norway) formulated the NAMMCO Agreement to foster marine ecosystem research and to better understand the role of marine mammals in this system. Such measures are based on the best available scientific evidence, taking into account both the complexity and vulnerability of the marine ecosystem, and the rights and the needs of residents of coastal communities to make a sustainable living from what the sea can provide. NAMMCO provides a mechanism for the conservation and management of all species of cetaceans (whales, dolphins, and porpoises) and pinnipeds (seals and walruses) in the region, many of which were not previously included in international agreements. This chapter will focus on NAMMCO's efforts in the context of cetaceans.

Section 2 of the chapter outlines the negotiating history for the establishment of NAMMCO. Section 3 describes the organization's structure and scope. Section 4 addresses the issues of sustainable use of marine mammals and efforts to foster sustainable communities, and the role of NAMMCO in furthering these objectives. The final section reviews the work to date of the NAMMCO Scientific Committee and Management Committee in the context of cetaceans.

[1]   Secretary to NAMMCO, Polar Environmental Centre 9296 Tromso, Norway, gretehb@nammco.no.

2.   THE ESTABLISHMENT OF NAMMCO

2.1.   Background

The process of establishing NAMMCO began in 1988 with a conference that addressed common concerns about rational utilization of marine mammals in the North Atlantic. Five additional annual international conferences were convened, culminating in the establishment of NAMMCO. The nations attending these conferences were interested in fostering multi-species and ecosystem management approaches in the North Atlantic. This emphasis on conservation was partly in response to, and in pointed contrast with, the increased support for a preservationist approach[2] by many parties to the International Whaling Commission (IWC). In 1990, the four current parties to NAMMCO signed a Memorandum of Understanding (MoU) establishing the North Atlantic Committee for Coordination of Marine Mammals Research (NAC). The signatories to NAC agreed that the MoU could serve not only as an alternative to other organizations such as IWC, but that it would also fill a gap with respect to regional intergovernmental cooperation on the study and management of pinnipeds. In addition, the signatories agreed that a focus on ecosystems and multi-species interaction was lacking in other marine mammal management organizations and that NAC would adopt such an approach. The NAC signatories also considered it important to focus on the interaction with local communities that utilized the marine mammals. The NAC was the first step towards establishment of a regional international organization that would address issues of conservation and management of marine mammals. In 1992, NAMMCO was established as an Agreement between the four signatories to the NAC.

The increased interest in the 1980s in regional cooperation for management of living marine resources was integrally related to the establishment of the 200-mile exclusive economic zones (EEZs) beginning in the late 1970s. The establishment of EEZs and the challenges asso-

---

2   In wildlife management there is a marked distinction between "preservation" and "conservation" approaches. In this context, the preservationist perspective holds that whales should not be managed for utilisation, but rather should be preserved in their present or pre-harvesting states. The conservationist perspective advocates conservation of the resource by rational and sustainable management, which may include commercial and/or subsistence use of the resource.

ciated with management of species that migrate between national and international waters created a strong exigency for regional cooperation to manage these resources.[3] All of the countries involved in establishing NAMMCO utilized marine resources in the North Atlantic and understood the importance of cooperating to manage these resources.

It is often claimed that NAMMCO was established as an alternative to the IWC. However, an examination of the context in which NAMMCO was created evinces a more complex situation. NAMMCO is concerned not only with large and small cetaceans, but also pinnipeds and marine ecosystem and multi-species interaction issues, thereby expanding its management purview well beyond the jurisdiction of the IWC. It is therefore not appropriate to consider NAMMCO solely in the context of, and as an alternative to, the IWC. There were six substantive factors that led to the formation of NAMMCO.[4] First, as mentioned above, the establishment of EEZs led to an increased awareness of the importance of regional cooperation in managing shared stocks and species. While the IWC's imposition of a moratorium on commercial whaling in the 1980s was problematic for Norway and Iceland, this was not the only concern. Questions also arose as to the rights of coastal states to utilize their resources in their exclusive economic zones.[5] In this context, the increased sway of the preservationist outlook in the IWC compelled whaling nations, especially Norway and Iceland, to explore alternative ways to manage the harvesting of whales. Secondly, the parties to NAMMCO needed an international forum for cooperation in the management and conservation of small cetaceans. The IWC was not viewed as a useful or competent mechanism for this purpose. Moreover, there was growing concern that the IWC appeared to be seeking to "expand its sphere of influence" by exerting "creeping jurisdiction" over small cetaceans.[6] Third, the parties to NAMMCO also needed an international organization that could address issues associated with marine mammals

---

[3]   Alf Håkon Hoel, *Regionalization of International Whale Management: The Case of North Atlantic Marine Mammals Commission*, 46 Arctic 116, 117 (1993).

[4]   Guðmundur Eiriksson, *The North Atlantic Marine Mammal Commission* (NAMMCO) unpublished 1, 1 (1992), and Hoel, *supra* note 3 at 118.

[5]   Hoel, *supra* note 3 at 118.

[6]   *Id.* at 121.

other than whales, such as seals and walruses. Fourth, given the heavy dependence of all NAMMCO members on fishing, they recognized the need for a multi-species ecosystem approach to the management of the interaction of whales, seals and fish stocks.[7] Fifth, the parties to NAMMCO saw a need for a forum that could disseminate objective information about marine mammals.[8] Finally, the parties to NAMMCO recognized the importance of establishing an investigative and management body that would address the needs of coastal communities and indigenous people.

2.2.   Overview of NAMMCO

NAMMCO was established by the Agreement on Cooperation in Research, Conservation and Management of Marine Mammals in the North Atlantic (referred to here as the NAMMCO Agreement), signed in Nuuk, Greenland on April 9, 1992, by the Faroe Islands, Greenland, Iceland and Norway.[9] The Agreement is entered into by the ministries of fisheries in each country rather than between the respective governments, ensuring greater autonomy for the Faroe Islands and Greenland.[10]

The North Atlantic Marine Mammal Commission is comprised of four major elements:

- *The Council* is the highest authority of the Commission. Member countries meet annually to exchange information, discuss matters of mutual interest and make decisions related to the aims of the organization. Observers from the governments of Canada, Denmark, Japan, Russia and St. Lucia also regularly attend council meetings;

---

[7]   Eiriksson, *supra* note 4, at 2; Hoel, *supra* note 3, at 121.

[8]   Hoel, *supra* note 3, at 121.

[9]   Agreement on Cooperation in Research, Conservation and Management of Marine Mammals in the North Atlantic. Apr. 9, 1992, Nuuk Greenland. Faroe Islands—Greenland—Iceland—Norway (hereinafter NAMMCO Agreement), available at <www.nammco.no>.

[10]   The Faroe Islands and Greenland have Home Rule governments and have full local autonomy, while foreign relations remain under the control of Denmark.

- *Management Committees*, species-specific or general, propose measures for conservation and management and make recommendations to the Council concerning scientific research;
- *The Scientific Committee* consisting of three scientists from each member country is responsible for providing scientific advice in response to requests from the NAMMCO Council. This forms the basis for the advice on conservation and management decisions forwarded to the member countries. The Scientific Committee provides advice to the Council based on the best available scientific findings, and also draws upon analysis by external experts;
- *The Secretariat*, which serves the Council and its Committees in their general work and meetings, is hosted by Norway at the University of Tromsø. The Secretariat also compiles data on species relevant to the specific conservation and management interests of the organization, and provides information for the general public on the work of NAMMCO. The Secretary to the Commission represents NAMMCO in international meetings and fora.

In addition to these fundamental elements, ad hoc working groups may also be convened by the Council to address specific areas of inquiry, such as the exchange of technical advice on hunting methods and the development of a reciprocal observer scheme for coastal whaling and sealing.[11] The Joint Control Scheme for the Hunting of Marine Mammals, implemented in 1998, developed out of one of these working groups and provides for an exchange of international observers appointed by NAMMCO.

Where appropriate, the Scientific Committee establishes its own specialist working groups to address requests for advice from the NAMMCO Council. Such working groups may include experts from both NAMMCO member and non-member countries.[12]

The NAMMCO Fund provides financial support for projects that contribute to the knowledge and understanding of marine mammal conservation and sustainable use of marine mammal resources. The Fund is

---

[11]   NAMMCO Agreement, *supra* note 9, at arts. 4, 2a.

[12]   *Id.* at arts. 6, 2.

operated by a Board that consists of a representative from each NAMMCO member country, and is administered by the Secretariat.[13] The Fund has provided funding for projects including the production of films, books, posters, brochures and web sites, and the holding of symposia and meetings.

The NAMMCO Agreement is registered in accordance with Article 102 of the UN Charter and principles of international law as reflected in the 1982 UN Convention on the Law of the Sea (UNCLOS).[14] In the case of cetaceans, UNCLOS requires cooperation through the appropriate international organizations for the conservation, management and study of cetaceans.[15] This is reiterated in Agenda 21, adopted by the UN Conference on Environment and Development in 1992, which, in addition to the International Whaling Commission, also recognizes the work of other international organizations in the conservation, management and study of cetaceans and other marine mammals.[16]

The NAMMCO Agreement applies to all marine mammals in the North Atlantic, without prejudice to obligations of the parties under other international agreements. This includes smaller species of whales, as well as seals and walruses. This is highly significant because the Agreement represents the first international mechanism for cooperation on conservation and management of these species in this region. Through NAMMCO, member countries also aim to enhance research on the role and relationship of marine mammals in the ecosystem, as well as on the effects of marine pollution and other human activities on these marine mammals. In accordance with UNCLOS, NAMMCO

---

[13]   North Atlantic Marine Mammal Commission 1993, *Report of the Third Meeting of the Council 1993* at 7. North Atlantic Marine Mammal Commission, Tromsø, Norway (1993).

[14]   United Nations Convention on the Law of the Sea, Dec. 10, 1982, U.N. Doc. A/CONF.62/122), 21 I.L.M. 1261 (entered into force on Nov. 16, 1994) (hereinafter UNCLOS).

[15]   *Id.* at art. 65.

[16]   Agenda 21, Chapter 17, Sec. 17.61(c), <http://www.un.org/esa/sustdev/agenda21chapter17.htm>.

ensures regional cooperation in an area where ecosystems and resources cross boundaries of several states and the high seas.[17]

3.   SUSTAINABLE USE—SUSTAINABLE COMMUNITIES

In the NAMMCO Agreement Preamble the parties refer to "the general principles of conservation and sustainable use of natural resources as reflected in the report of the World Commission on Environment and Development."[18] The Preamble further states that the parties are "Convinced that regional bodies in the North Atlantic can ensure effective conservation, sustainable marine resource utilisation and development with due regard to the needs of coastal communities and indigenous people."[19]

NAMMCO considers both the role of marine mammals in the ecosystem and the sustainable use of these resources. These conservation efforts are closely linked to efforts to ensure the (long-term) viability of coastal communities. The sustainability of one species must be considered in conjunction with the ecosystem in which it lives. This must include human communities, in particular where humans actively utilize and are dependent upon renewable resources for their livelihood. This consideration applies principally to the coastal and indigenous communities of the NAMMCO countries that often harvest more than one species of marine mammal. It is the position of NAMMCO member countries that it is appropriate to assess the interaction of species in an ecosystem.[20] Application of a multi-species approach poses significant challenges, the difficulties of which will be further discussed below.

Sustainable use of renewable resources and the sustainability of small coastal communities are closely linked in the North Atlantic region.

---

[17]   NAMMCO Agreement, *supra* note 9, at Preamble.

[18]   *Id.*

[19]   *Id.*

[20]   *Id*; North Atlantic Marine Mammal Commission 1999, *NAMMCO Annual Report 1998*, 13 North Atlantic Marine Mammal Commission, Tromsø, Norway (1999); North Atlantic Marine Mammal Commission 2000, *NAMMCO Annual Report 1999*, 13 North Atlantic Marine Mammal Commission, Tromsø, Norway (2000).

There is a strong interdependence between the human enclaves and the adjacent marine environment that comprise the local ecosystem. Countries or communities that are dependent upon natural renewable resources wish to ensure their sustainable utilization. This objective is successfully reflected in the various management schemes that have been formulated by different countries and cultural groups in the region. Such management schemes have been based on western scientific knowledge, on customary practices with deep prehistoric roots, or a combination of the two. National governments establish catch quotas and seasonal regulations on hunting to ensure sustainable levels of harvesting. When adhered to, these provisions or sanctions greatly contribute to the conservation of species and ecosystems, protecting the interests of the communities dependent upon the resources.

3.1.   Resource Use in NAMMCO Member Countries

Greenland can be generally characterized as more economically, socially and culturally dependent upon marine mammal harvesting than the three other members. In the Faroe Islands, the primary targeted marine mammal is the pilot whale, while Iceland focuses on the harvesting of larger whales. In Norway, many small coastal communities are engaged in both sealing and whaling. However, regardless of the different forms of resource use and extent of dependency, it is necessary to determine the level at which a species or stock can be sustainably harvested. This question can only be answered through accurate estimates of marine mammal abundance and knowledge of migratory patterns. Since NAMMCO's creation, the Scientific Committee has been engaged in developing abundance estimates and calculating sustainable levels of harvest for several marine mammal stocks and species. The latter issue is particularly complex. There is not a clear-cut relationship between the socio-economic status of a community, the utilization of marine mammals and the sustainability of these resources. Most communities have multiple sources of income and utilize more than one species as food sources. Nevertheless, it is important that they can assess the level at which a resource *can* be used sustainably and factor that information into analysis of the socio-economy of the community in question.

Within the framework of sustainable use developed by the Brundtland Commission, it is assumed that natural resources are used to further community development, but only at levels that do not compromise the ability of future generations to meet their own needs. Questions related to the sustainability of renewable resource use and community development lead to questions regarding markets, cash and income. All contemporary societies are dependent upon a steady cash flow, and in, for example, many small communities in Greenland, marine mammal products (mainly from seals) remain an important source of cash. On the other hand, the pilot whale hunt drives in the Faroe Islands do not generate cash directly, but are an important source of meat for households. In this sense, the monetary replacement value is probably high. The income generated by harvesting marine mammals in Norway is also integral to the economies of the small whaling and sealing communities. If Iceland resumes whaling operations, income and the nutritional value of the harvest will be its most important aspects. Also, in all these nations the social and cultural values of marine mammal hunting, and the communal aspects associated with consumption and distribution of the foodstuffs, must not be overlooked.

3.2. Marine Mammal Use in the NAMMCO Member Countries

The four current parties to NAMMCO have a long history of, and are currently utilizing (or have until recently utilized), marine mammals within national coastal waters as economic, nutritional and community resources. NAMMCO represents an important mechanism to assist communities in ensuring proper regional cooperation and management of shared stocks and species, and those that are of particular interest to each member country.

The Vikings probably introduced pilot whale hunting in the Faroe Islands from the Norwegian West Coast. As early as 1298, a legal document outlined rights to both driven and stranded whales. Pilot whale hunting in the Faroe Islands is opportunistic because the animals are driven towards and killed on the beach only when observed along the coast from boats or from land. Each pilot whale drive hunt must be authorized by government officials. The pilot whale drive hunt is not a commercial activity, and the edible meat and the blubber are distributed to the hunters and their families, and to local inhabitants. The annual average harvest of pilot whales in the Faroe Islands between 1990 and 1996 was 956 animals. The pilot whale is an important source of food, constituting about one-third of the meat supply in the Faroes.[21] Meat from domestic livestock is quite expensive in the Faroe Islands and therefore the monetary replacement value is high for pilot whale meat. Pilot whale hunt drives are also instrumental in maintaining a way of life and important cultural values in the Faroe Islands.

Whale species, including fin, minke, pilot and narwhal, together with various species of seals, remain important components of the diet of inhabitants of Greenland, and also retain cultural and economic importance. Marine mammal hunting is steeped in tradition, extending back to when the first Inuit came to Greenland thousands of years ago. Greenlandic hunters use both traditional hunting methods, such as qayaqs equipped with traditional harpoons, and imported technologies, such as motor powered vessels and harpoon cannons. Greenlanders draw on both traditional and contemporary methods of harvesting and processing, and operate in a mixed cash/subsistence economy.[22]

Iceland ceased whaling in 1986 as a result of the IWC's imposition of a moratorium on commercial whaling, and continued with scientific whaling until 1989. Prior to the moratorium, Iceland primarily hunted fin whales, but also sei, sperm, and minke whales from vessels equipped with 90-mm and 50-mm harpoon cannons. In March 1999, the Icelandic *Althingi* (Parliament) voted to resume whaling operations. In June 2001, Iceland sought to rejoin the IWC,[23] signaling that whaling could com-

---

21   Anonymous, WHALES AND WHALING IN THE FAROE ISLANDS 1, 3 Tórshavn, Faroe Islands: Department of Fisheries (1999).

22   Grete K. Hovelsrud-Broda, *The Integrative Role of Seals in an East Greenlandic Hunting Village*, 36(1–2) ARCTIC ANTHROPOLOGY 37–50 (1999).

23   Iceland's application to rejoin the IWC with a reservation to the moratorium on commercial whaling was rejected at the 53rd and 54th Meetings of the Parties. IWC, *Final Press Release*, 54th Meeting of the International Whaling Commission, <http://www.iwcoffice.org/2002PressRelease.htm> (2002); IWC, *Final Press Release*, 53rd Meeting of the International Whaling Commission, <http://www.iwcoffice.org/pressrelease2001.htm> (2001).

mence at any time. It is anticipated that minke and fin whales will be targeted should operations resume.

Currently only minke whales are hunted by Norwegians, from small fishing boats (between 15 and 40 meters long). The Norwegian government established a quota of 549 minke whales for 2001. Whaling takes place annually between May and July; Norwegian whalers usually engage in fishing during other times of the year. Many of the coastal communities in Norway, as in Greenland, are dependent upon whaling for jobs and income, both from hunting and employment at the processing plants, and whales play an important role in the social and cultural activities of these communities.

4.   SCIENTIFIC AND MANAGEMENT ISSUES THAT HAVE BEEN CONSIDERED BY THE NAMMCO SCIENTIFIC COMMITTEE AND MANAGEMENT COMMITTEE

The NAMMCO member countries have divided the North Atlantic into stock areas for various marine mammal species. Management recommendations of the Commission pertain to the stocks that are of interest to one or more member countries at any given time. The NAMMCO Scientific Committee will have assessed all available scientific findings and presented its analysis and results to the Council before the Management Committee prepares advice for the Council and member countries to institute. In the following section, I will briefly outline some of the questions posed to the Scientific Committee and the subsequent management advice provided by the Management Committee.

4.1. Long-Finned Pilot Whales

In 1993, the Council requested the NAMMCO Scientific Committee to analyze the effects of the pilot whale (drive) hunt in the Faroe Islands on North Atlantic long-finned pilot whales. In particular, the Council asked the Scientific Committee to determine whether the number of whales harvested were consistent with sustainable utilization. The Scientific Committee reviewed the ICES Study Group Report on Long-Finned Pilot Whales, which had used information sampled from the Faroese drive fishery, analyzed distributional, genetic and morphometric evidence and consulted studies on social structure and behavioral factors. In addition, the Scientific Committee relied on three separate abundance estimates, historical catches, population dynamics and population models from the NASS-Surveys from 1987, 1989 and 1995. Based on these sets of data the Scientific Committee concluded that:

> the effects of historic and present catches in the Faroe Islands have had a negligible effect on the long-term trends in the pilot whale stock. The Scientific Committee also noted that the annual catch of 2,000 individuals in the eastern Atlantic corresponds to an exploitation rate of 0.26 per cent of the present best estimate of the abundance of pilot whales in the Northeast Atlantic (778,000 pilot whales from NASS-89).[24]

At its Seventh Meeting in the Faroe Islands, the NAMMCO Council noted the Management Committee's conclusion, based on the Scientific Committee report, that the drive hunt of pilot whales in the Faroe Islands, as it has been practised for centuries, is sustainable.[25]

4.2.   Beluga and Narwhal

In 1999, the Scientific Committee, at the request of the NAMMCO Council, conducted a broad assessment of the status of all known stocks of beluga and narwhal in the North Atlantic and adjacent seas. In addition to members of the Committee, invited experts from Canada and the Russian Federation participated in the assessment. Based on recent evidence from genetics, and contaminants and tracking studies, it can be concluded that both narwhal and beluga are strongly philopatric (*i.e.*, returning to the same areas year after year). Management of these species must therefore be based on aggregations that are seasonally but regularly present at specific fjords, coastlines, promontories or estuaries. While many aggregations of narwhal and beluga are either not harvested, or harvested at or below sustainable levels, some, such as the

---

24   North Atlantic Marine Mammal Commission 1998, *NAMMCO Annual Report 1997*, 99 North Atlantic Marine Mammal Commission, Tromsø, Norway (1998).

25   North Atlantic Marine Mammal Commission 1998, *NAMMCO Annual Report 1997*, 26 North Atlantic Marine Mammal Commission, Tromsø, Norway (1998).

West Greenland beluga, appeared to be declining as a consequence of excessive harvesting. The Scientific Committee noted that index surveys conducted in the West Greenland beluga wintering area since 1982 indicated a decline of more than 60 percent in abundance, and that the aggregation was likely declining due to over-exploitation.

In response, the Council asked the Scientific Committee to provide advice on the level of sustainable utilization of West Greenland beluga in different areas and under different management objectives. The Scientific Committee concluded that the stock is substantially depleted and that present harvests are several times the sustainable yield. The Committee projected that extinction of local stock would likely occur within 20 years if current levels of harvesting continue.[26] The Committee determined that harvesting of the stock must be reduced to approximately 100 animals per year to achieve a sustainable level of the population over the next decade. Greenland is presently considering this advice and will likely take action in the near future to bring its beluga harvest within sustainable levels.

4.3. Fin Whales

In 1999, the Council asked the Scientific Committee to examine the stock structure of fin whales throughout the North Atlantic, and to advise them on sustainable harvest levels of the East Greenland and Iceland stock. The Scientific Committee concluded that there is genetic evidence of large-scale stock differentiation among North Atlantic fin whales, but that much more information is needed before stock boundaries can be delineated with any certainty. For the East Greenland and Iceland stock area, the Committee concluded that a catch of up to 200 fin whales per year would be sustainable under very conservative assumptions. However, the Committee concluded that such a catch should be spread throughout the area, roughly in proportion to observed density. This advice was accepted by the Management Committee of NAMMCO and conveyed to the relevant member countries. To date, no fin whales have been harvested in this area since Iceland ceased whaling in 1989.

---

[26]   While a change in beluga distribution might also explain the observed reduction in abundance, there was no evidence to support this hypothesis.

In 2000, the Scientific Committee conducted an assessment of fin whales in Faroese waters. The Committee found that there was insufficient information about the stock relationships of fin whales in the area, and therefore was unable to complete the assessment. The Committee's analysis of historical catch records and recent abundance estimates indicate that fin whales in the area have likely been substantially depleted by past harvests, although these results may be unreliable. As a result, the Committee recommended to the Faroese that research be conducted to resolve the stock structure of fin whales in the area. In 2000, the Committee initiated a fin whale biopsy sampling and satellite telemetry program.

4.4.   Marine Mammal Fisheries Interaction

The NAMMCO Agreement Preamble expresses the desire of member states "to enhance their co-operation in research on marine mammals and their role in the ecosystem, including, where appropriate, multi-species approaches. . . ."[27] Understanding the role of marine mammals in the marine ecosystem, and the extent to which they can be affected by fisheries, or have an impact on commercially important fish stocks, is of special interest to those North Atlantic nations with heavy economic dependence on fisheries. Therefore, the Council has requested the Scientific Committee to periodically review and update available information in this field. The Scientific Committee's work in this context has focussed on harp seals, hooded seals and minke whales, major predators in the North Atlantic ecosystem. In 1998, the Scientific Committee concluded that these animals might have substantial direct and/or indirect effects on commercial fish stocks.[28] It is clear that significant uncertainties remain in the calculation of consumption of fish stocks by marine mammals, and this is the most important factor hindering the development of multi-species fishery models. The Scientific Committee concluded that more data is needed in a number of areas, including abundance estimates and seasonal distribution for all relevant

---

[27]   NAMMCO Agreement, *supra* note 9, at Preamble.

[28]   North Atlantic Marine Mammal Commission 1999, *NAMMCO Annual Report 1998*, 91 North Atlantic Marine Mammal Commission, Tromsø, Norway (1999).

species, information about the diet composition of the predator species, the link between diet composition and prey abundance, and the seasonal variation of energy consumption.

In multi-species fishery models, it is necessary to explicitly describe the uncertainty inherent in consumption estimates, but this is not possible with the data currently available. The quality of the data necessary to estimate consumption is generally highest for minke whales and harp seals in the Barents and Norwegian Seas, for pilot whales around the Faroes, and harp, hooded and grey seals off southeastern Canada. Folkow *et al.* recently conducted research on the food consumption of minke whales in the region and estimated that the "stock of 85,000 minke whales appeared to have consumed more than 1.8 million tonnes [herring, krill, cod, capelin, haddock and other fish species] in the coastal waters off northern Norway, in the Barents Sea and around Spitsbergen during a six-month period.[29] Schweder *et al.* constructed a food web model in which minke whales prey on cod herring and capelin; cod on herring, capelin, and young cod; while herring prey on young capelin. They suggested that in general, "catches of cod, herring and capelin are reduced with an increased abundance of minke whales."[30]

Consumption by marine mammals of fish stocks approaches or exceeds fish harvesting in some areas.[31] In the coming years, NAMMCO will continue to facilitate the development of multi-species models to assess the impact of marine mammal consumption on fisheries.

---

[29] L.P. Folkow *et al.*, *Estimated Food Consumption of Minke Whales Balaenoptera acutorostrata in Northeast Atlantic Waters in 1992–1995, in* 2 MINKE WHALES, HARP & HOODED SEALS: MAJOR PREDATORS IN THE NORTH ATLANTIC ECOSYSTEM, North Atlantic Marine Mammal Commission 65, 70–72 (Gísli A. Vikingsson & Finn O. Kapel eds., 2000).

[30] Tore Schweder, Gro S. Hagen & Einar Hatlebakk, *Direct and indirect effects of minke whale abundance on cod and herring fisheries: A scenario experiment for the Greater Barents Sea, in* Vikingsson & Kapel, *supra* note 29, at 120, 129.

[31] North Atlantic Marine Mammal Commission 2001, *NAMMCO Annual Report 2000*, 126 North Atlantic Marine Mammal Commission, Tromsø, Norway (2001).

5.   CONCLUSION

The North Atlantic Marine Mammal Commission represents an important forum for its member countries to engender cooperation in the study and management of marine mammals in the North Atlantic region. The member countries have interests in shared stocks and species, and those that are found only within national waters. Regardless of the geographical range (within the North Atlantic) of the species, concerns regarding management can be raised in NAMMCO. Each member country has the opportunity to consult a larger and more diverse scientific group than those within their national borders. In a regional and intergovernmental organization such as NAMMCO, sustainable use must be considered within the framework of the common interests in the region and in its resources, combined with a need to address the socio-economic and cultural differences among the member countries with respect to marine mammal harvesting. Sustainable use programs must foster the objectives of local communities, while also being solidly grounded in objective scientific principles, embrace an ecosystem approach and allow for integration into more far-reaching management systems. This is a delicate balancing act that NAMMCO has been able to maintain to this point. NAMMCO continues to strive for refinement and increased understanding of the cultural, socio-economic and scientific basis for sustainable use of marine mammals.

Chapter 7
The Competence of Pro-Consumptive International Organizations to Regulate Cetacean Resources

Howard S. Schiffman[1]

1.   INTRODUCTION

In the last ten years, a number of international marine management organizations have been created. While some of these organizations focus on the conservation of marine species, others seek to foster consumptive use of these species, including cetaceans. The latter organizations include among their members whaling states, some of which retain membership in the International Whaling Commission (IWC), organizations representing communities with traditional whaling histories and other international organizations committed to a resurgence of whaling. These organizations claim to further the goal of marine mammal conservation, but there can be little doubt that they also seek to undermine the current conservation scheme of the IWC. The effective regulation of whaling practices by these pro-whaling organizations is currently limited. The potential for these organizations, the North Atlantic Marine Mammal Commission (NAMMCO), in particular, to undertake such responsibilities in the future, however, is worthy of consideration. Potential conflicts with the IWC also raise important questions about the role and status of international organizations in international environmental law.

The issues addressed in this chapter reflect both legal and policy considerations and are informed by the instruments and scholarly writings of both perspectives. The dearth of primary and secondary sources directly germane to these matters, however, is a limitation that should be noted at the outset. The questions presented concern possible future

[1]   Howard S. Schiffman, J.D., LL.M., Adjunct Assistant Professor, International Programs, New York University School of Continuing and Professional Studies.

actions by international organizations; as such, they are necessarily speculative. Much of the analysis and some of the conclusions contained herein are also necessarily derivative and in some cases analogical. At the same time, the relative novelty of these questions in the field of marine mammal conservation underscores the importance of the analysis.

2.   DESCRIPTIONS OF SOME NEW "CONSUMPTIVE-FRIENDLY" INTERNATIONAL ACTORS IN THE REALM OF MARINE MAMMAL MANAGEMENT

This section will provide a brief overview of the structure and function of key international organizations that have arisen in recent years to foster whaling. An understanding of the structure and function that is provided for by the constitutive instruments of an international organization is a necessary first step to understand the legality of its actions in international law in general, and in relation to other international organizations, in particular.

2.1.   The North Atlantic Marine Mammal Commission (NAMMCO)

NAMMCO came into being in 1992 after the signing of the Agreement on Cooperation in Research, Conservation and Management of Marine Mammals in the North Atlantic (NAMMCO Agreement) between Norway, Iceland, Greenland and the Faroe Islands.[2] The stated objective

---

2   Agreement on Cooperation in Research, Conservation and Management of Marine Mammals in the North Atlantic. 1945 U.N.T.S. (Treaty No. 33321) [hereafter NAMMCO Agreement]. The original signatories of the NAMMCO Agreement are the states of Norway and Iceland, and two self-governing territories of Denmark: Greenland and the Faroe Islands. Other parties are free to join NAMMCO with the consent of the original members. *See id.* at art. 10. Thus far, no other party has done so although the governments of Canada and Russia have been invited to join by the NAMMCO Council. Japan has also participated in NAMMCO conferences. One observer traces the political process leading to the development of NAMMCO to 1988, ". . . when the first of a series of annual international conferences on the rational utilization of marine mammals was hosted by the Icelandic minister of fisheries at the time. . . ." Kate Sanderson, *The North Atlantic Marine Mammal Commission—in principle and practice, in* WHALING IN

of NAMMCO, as set forth in Article 2 of the Agreement, is "to contribute through regional consultation, rational management and study of marine mammals in the North Atlantic."[3] NAMMCO is comprised of a Council,[4] Management Committees,[5] a Scientific Committee[6] and a Secretariat.[7]

Each NAMMCO member is represented on the Council.[8] The Council provides a forum for the study, analysis and exchange of information among the parties on matters concerning marine mammals in the North Atlantic;[9] establishes appropriate Management Committees and coordinates their activities;[10] establishes guidelines and objectives for the work of the Management Committees;[11] establishes working arrangements with the International Council for the Exploration of the Sea and other appropriate organizations;[12] coordinates requests for scientific advice[13] and establishes cooperation with states not parties to NAMMCO in order to further the objective set out in Article 2.[14]

THE NORTH ATLANTIC: ECONOMIC AND POLICY PERSPECTIVES 68 (Gudrun Petursdottir ed., 1997). (This book reports the proceedings of a conference held in Reykjavik on Mar. 1, 1997.)

[3]   NAMMCO Agreement, *supra* note 2, at art. 2.

[4]   *Id.* at art. 3(a).

[5]   *Id.* at art. 3(b).

[6]   *Id.* at art. 3(c).

[7]   *Id.* at art. 3(d).

[8]   *Id.* at art. 4.

[9]   *Id.* at art. 4(a).

[10]   *Id.* at art. 4(b).

[11]   *Id.* at art. 4(c).

[12]   *Id.* at art. 4(d). The International Council for the Exploration of the Sea is an intergovernmental organization concerned with marine and fisheries science. Founded in 1902, it is the oldest such organization in the world. *See* Website of the International Council for the Exploration of the Sea (visited Nov. 10, 2002 <http://www.ices.dk/hl/About_ICES.htm>.

[13]   NAMMCO Agreement, *supra* note 2, at art. 4(e).

[14]   *Id.* at art. 4(f). For a description of Article 2, *see supra* text accompanying note 3.

The Management Committees of NAMMCO are tasked with proposing measures for the conservation and management of marine mammal stocks[15] and making recommendations to the Council concerning scientific research.[16] The Scientific Committee of NAMMCO consists of experts appointed by the parties to the Agreement.[17] The Scientific Committee may invite other experts to participate in the conduct of its work subject to the approval of the Council.[18] The Scientific Committee is also mandated to provide scientific advice in response to requests from the Council, utilizing available scientific information.[19] Finally, the Secretariat performs functions set forth by the Council under the NAMMCO Agreement.[20]

NAMMCO publishes papers and sponsors meetings and conferences on matters of interest to marine mammal management.[21] These conferences focus on scientific and technical matters such as population status, fisheries interactions, and hunting methods.[22] The NAMMCO Scientific Committee is particularly active in sponsoring events.[23]

2.2.   The World Council of Whalers (WCW)

Unlike NAMMCO, which is an intergovernmental organization, the World Council of Whalers (WCW) is a non-governmental organization. The overarching objective of WCW is:

---

[15]   *See* NAMMCO Agreement, *supra* note 2, at art. 5(a).

[16]   *Id*. at art. 5(b).

[17]   *Id*. at art. 6.

[18]   *Id*.

[19]   *Id*.

[20]   *Id*. at art. 8.

[21]   *NAMMCO Meetings and Conferences*, NAMMCO Website (visited Nov. 10, 2002) <http://www.nammco.no>.

[22]   *Id*.

[23]   *Id*. In 2000, examples of Scientific Committee events included meetings of the working groups on Economic Aspects of Marine Mammal-Fisheries Interactions; North Atlantic Fin Whales and the Population Status of Narwhal and Beluga in the North Atlantic. *Id*.

. . . to provide a forum for whaling peoples around the world, both aboriginal and non-aboriginal. Its mission is to promote their continued sustainable use of marine living resources, to protect their cultural, social, economic and dietary rights, and to address their concerns.[24]

Its more specific goals are:

- To support communities engaged in sustainable whaling by providing a cooperative forum for whalers, governments, researchers and managers to discuss and reach informed decisions regarding the conservation of whales and whaling societies;
- To provide a collective informed voice for whaling peoples around the world;
- To support and encourage the customary relationships between whaling people, their communities and the whale;
- To encourage respect for cultural, social, and economic needs and concerns of whaling communities;
- To promote sustainable and equitable resource use by incorporating the needs, knowledge and teachings of whaling peoples, and including them in the decision-making process;
- To act as an information gathering and dissemination centre to assist whalers and their communities to remain aware of the issues that may affect them.[25]

The WCW was developed after ad hoc meetings in Glasgow, Scotland, in 1992, Kyoto, Japan, in 1993 and Berkeley, California, in 1996.[26] The WCW was formally founded in 1997 and acknowledges participation by whalers and sympathetic observers from a variety of geographic regions and over a dozen countries.[27] The WCW claims no

---

[24]   *History of the World Council of Whalers*, World Council of Whalers Website (visited Nov. 10, 2002) <http://www.worldcouncilofwhalers.com/profileframe.htm>.

[25]   *Objectives of the World Council of Whalers*, World Council of Whalers Website (visited Nov. 10, 2002) <http://www.worldcouncilofwhalers.com/profileframe.htm>.

[26]   *Id.*

[27]   *Id.* At the first annual WCW General Assembly held in Victoria, British Columbia, from March 2–6, 1998, there were over 100 delegates representing 19

direct role in the actual management of whale species but functions as a forum to support and encourage the practice of whaling between and among aboriginal and non-aboriginal peoples.[28] WCW publishes papers and newsletters and maintains a website to help accomplish its mission.[29]

2.3. Rules of International Organizations Applied to NAMMCO and WCW

Although the United Nations Convention on the Law of the Sea (UNC-LOS) does not define the term "competence," we can look to general rules of international law that are typically applied to international organizations to determine their capacity to promulgate a regulatory scheme and even their relationship to other international organizations. International organizations have a long history in international relations and have proliferated since the end of World War II.[30] They serve a variety of functions and represent a multitude of interests in international affairs. In the 20th century, international organizations progressively acquired legal personality in international law.[31] Where the organizations are established by international agreement between governments,

---

countries: Antigua and Barbuda, Australia, Canada, Dominica, the Faroe Islands, Greenland, Grenada, Iceland, Indonesia, Japan, New Zealand, Norway, the Philippines, Russia, St. Kitts and Nevis, St. Lucia, St. Vincent and the Grenadines, Tonga and the United States.

[28]   *Id.*

[29]   *Id.*

[30]   Participation in international organizations at global, regional, and subregional levels, as well as for functional, issue-related purposes, has been a major manifestation of international cooperation since the early 19th century. The 1815 Congress of Vienna and others that followed it served as precursors to the UN system of today where international organizations play a prominent role. *See* BIRNIE & BOYLE, *infra* note 71, at 32.

[31]   PETER MALANCZUK, AKEHURST'S MODERN INTRODUCTION TO INTERNATIONAL LAW, at 91 (7th rev. ed. 1997). Evidence of the growing legal personality of international organizations can be found in their ability to enter into treaties. *See* Vienna Convention on the Law of Treaties Between States and International Organizations or Between International Organizations, *opened for signature*, Mar. 21, 1986, A/CONF. 129/15, 25 I.L.M. 543 (not yet in force). The right to conclude treaties was previously reserved for sovereigns alone.

and the members are primarily sovereign states, they are termed "inter-governmental organizations" (IGOs). The right of states to form and join intergovernmental organizations for any purpose consistent with international law must be presumed to be an element of sovereignty.

Where the organization is not established by international agreement between two or more sovereign states but rather is comprised of other organizations, individuals or other interested parties, they are called "non-governmental organizations" (NGOs). NGOs play an increasingly important role in the world of international affairs; this is certainly so in the area of environmental conservation. By concentrating resources and raising public awareness about particular issues they are often well situated to influence policy and law-making at various levels. IGOs, on the other hand, are directly responsible for law making and the promulgation of management frameworks. By definition, NGOs lack the legal capacity to bind sovereign states. IGOs, however, may do so depending upon the powers its member states have vested in it.

As an IGO, NAMMCO's organizing document is a treaty and two of its founding members are sovereign states.[32] As an NGO, the WCW was not established by an international agreement and has no sovereign states as members.[33] Claiming no direct authority to regulate cetacean resources, the "forum" provided by the WCW[34] can be said to exist more for informational and advocacy purposes on behalf of whaling peoples. If, in the future, the WCW chooses to disseminate recommendations to whaling peoples on the technical aspects of cetacean management from a consumptive standpoint or otherwise, it would not be binding. States, however, would be free to voluntarily implement such recommendations provided they were consistent with any and all obligations in law.

Thus, despite the fact that the WCW plays a supportive role for individuals, communities and states that either conduct whaling operations or hope to see a resumption of whaling on a larger scale in the future, the organization's ability to influence the practice of governments or

---

[32]   *See supra* text Section 2.1.

[33]   *See supra* text Section 2.2.

[34]   *See id.*

other international organizations is limited. In light of this fact, NAMMCO, and not the WCW, is the primary focus of this work.

3.   THE IWC'S COMPETENCE TO REGULATE CETACEAN CONSERVATION AND MANAGEMENT

Any discussion of the powers of new international organizations to participate in the management and regulation of cetacean resources must include an examination of the IWC. The IWC is by far the oldest and most visible intergovernmental organization in the world engaged in marine mammal conservation and management. The IWC was created under the International Convention for the Regulation of Whaling (ICRW) in 1946.[35] The ICRW entered into force in 1948 and was the first significant attempt at an international level to confront the crisis of declining whale stocks precipitated by over-exploitation.

The IWC is empowered—either in collaboration with or through independent agencies of the contracting governments or other public or private agencies, establishments, or organizations, or independently— to encourage, recommend, or if necessary, organize studies and investigations relating to whales and whaling;[36] collect and analyze statistical information concerning the current condition and trend of the whale stocks and the effects of whaling activities thereon;[37] study, appraise and disseminate information concerning methods of maintaining and increasing the populations of whale stocks.[38]

---

[35]   International Convention for the Regulation of Whaling, Dec. 2, 1946, 62 Stat. 1716, 161 U.N.T.S. 72 (entered into force Nov. 10, 1948) [hereafter ICRW]. The original signatories of the ICRW were the United States, Argentina, Australia, Brazil, Canada, Chile, Denmark, France, the Netherlands, New Zealand, Norway, Peru, the Soviet Union, the United Kingdom and South Africa. Each member of the ICRW is represented on the IWC. *Id*. at art. III(1).

[36]   *Id*. at art. IV(a).

[37]   *Id*. at art. IV(b).

[38]   *Id*. at art. IV(c). A Scientific Committee that studies technical and scientific issues pertinent to cetacean populations assists the IWC in its work.

The IWC is also permitted to amend the provisions of its Schedule by adopting regulations with respect to the conservation and utilization of whale resources, fixing protected and unprotected species;[39] open and closed seasons;[40] open and closed waters, including the designation of sanctuary areas;[41] size limits for each species;[42] time, methods, and intensity of whaling (including the maximum catch of whales to be taken in any one season);[43] types and specifications of gear and apparatus and appliances which may be used;[44] methods of measurement;[45] and catch returns and other statistical and biological records.[46] The ICRW provides that amendments to the Schedule "shall be such as are necessary to provide for the conservation, development, and optimum utilization of the whale resources[.]"[47]

Part of the reason we are now seeing the development of international organizations that may play a meaningful role in future cetacean management is the belief among whalers that the IWC has failed to achieve its mandate to provide for the long-term health of the whaling industry.[48] In fact, the history of the IWC clearly reflects concern for the industry as it declined concurrently with the whale stocks. The preamble of the ICRW is quite revealing in this regard. The preamble states:

[r]ecognizing that the whale stocks are susceptible of natural increases if whaling is properly regulated, and that increases in

---

[39]   *Id.* at art. V(1)(a). The ICRW Schedule contains the quota for given whale species as well as aboriginal quotas. At the present time, a moratorium or zero catch limit is in effect for commercial whaling, although several aboriginal quotas are listed.

[40]   *Id.* at art. V(1)(b).

[41]   *Id.* at art. V(1)(c).

[42]   *Id.* at art. V(1)(d).

[43]   *Id.* at art. V(1)(e).

[44]   *Id.* at art. V(1)(f).

[45]   *Id.* at art. V(1)(g).

[46]   *Id.* at art. V(1)(h).

[47]   *Id.* at art. V(2)(a).

[48]   *See* William Burke, *Whaling and International Law, in* Petursdottir *supra* note 2, at 114–17.

the size of whale stocks will permit increases in the number of whales which may be captured without endangering these natural resources.[49]

The initial parties to the ICRW were all whaling states. The ICRW embodies their efforts to save the whaling industry after generations of over-exploitation of the species that it targeted.[50] At the time of its adoption, the ICRW had 15 members; today it has 49. Before the moratorium was adopted in 1982 the IWC experienced a growth phase. Most of the nations that joined at that time focused more on protecting whales than the whaling industry. This may have been a reflection of the growing global concern for degradation of the environment in the past three decades, including states' responses to pressure from increasingly influential environmental groups.[51] More recently, however, some newer members have been more open to renewed prospects for utilization. The approach of the IWC since the early 1980s has emphasized conservation at the expense of utilization of cetacean resources. Therefore, if the few remaining pro-whaling states and other like-minded organizations and individuals are unsuccessful in reversing the agenda of the IWC, it would not be surprising that they seek to undermine the IWC in favor of newer international organizations specifically organized to support whaling.

4.   LEGAL FACTORS INFLUENCING THE CAPACITY OF INTERNATIONAL ORGANIZATIONS TO REGULATE MARINE MAMMALS: UNCLOS AND THE MODERN LAW OF THE SEA

An analysis of the capacity of international organizations to engage in regulatory activity requires a review of several legal factors. First, what

---

[49]   ICRW, *supra* note 35, Preamble at para. 4.

[50]   One need only look at another key assertion in the Preamble. Considering that the history of whaling has seen over-fishing of one area after another and of one species of whale after another to such degree that it is essential to protect all species of whales from further over-fishing[.]

ICRW, *supra* note 35, Preamble at para. 2.

[51]   *See* R.R. CHURCHILL & A.V. LOWE, THE LAW OF THE SEA 317 (3d ed. 1999).

is the legal framework that governs in the area that the organization seeks to regulate? Second, what is the authority of an international organization to manage a marine resource that falls within a state's jurisdiction? Third, what defines the scope of an organization's authority and may it under any circumstances bind non-member states? Finally, how should we evaluate real and potential conflicts between the management schemes of competing international organizations?

4.1. National Jurisdiction and Sovereignty Over Resources

A logical point of departure for a discussion about legal competence to regulate any marine resource is national jurisdiction, even for a discussion of the competence of international organizations. We may start with the proposition, embodied in Principle 21 of the Stockholm Declaration that states are free and indeed obligated to manage the resources that fall within their jurisdiction.[52] In the case of marine resources, the concept of national sovereignty is recognized in the 1982 United Nations Convention on the Law of the Sea (UNCLOS).[53] An innovation of

---

[52]   There is strong support for the concept of national sovereignty over national resources. Perhaps the best example of this is found in Principle 21 of the Stockholm Declaration produced by the 1972 United Nations Conference on the Human Environment. Declaration on the Human Environment, June 16, 1972, U.N. Doc. A/Conf. 48/14/ Rev. (1972), 11 I.L.M. 1416, at Principle 21. The Stockholm Principle 21 provides:

> States have, in accordance with the Charter of the United Nations and the principles of international law, the sovereign right to exploit their own resources pursuant to their own environmental policies, and the responsibility to ensure that activities within their jurisdiction or control do not cause damage to the environment of other States or of areas beyond the limits of national jurisdiction.

Id. The Stockholm Conference is widely regarded as a starting point for international environmental law in general. *See* LAKSHMAN GURUSWAMY & BRENT HENDRICKS, INTERNATIONAL ENVIRONMENTAL LAW IN A NUTSHELL 3 (1997). "The 1972 Stockholm Conference on the Human Environment, . . . may well have been the chrysalis from which international environmental law emerged as a legal subject in its own right, . . ." *Id.*

[53]   United Nations Convention on the Law of the Sea, Dec. 10, 1982, 21 I.L.M. 1261 (entered into force Nov. 16, 1994) [hereinafter UNCLOS], at arts. 61–62.

UNCLOS is its establishment of the Exclusive Economic Zone (EEZ).[54] The importance of the EEZ to conservation and management schemes is enormous. The regime of the EEZ subjects large segments of the ocean to national jurisdiction where prescriptive and enforcement jurisdiction is the strongest. For example, Article 61 of UNCLOS provides that the coastal state shall determine the allowable catch of the living resources in its EEZ.[55] This provision requires action based upon scientific evidence and clearly contemplates the participation of "competent international organizations" in the pursuit of this objective.[56]

Article 62 declares that "the coastal State shall promote the objective of optimum utilization of living resources in the [EEZ] without prejudice to Article 61."[57] This provision, interestingly enough, mentions neither scientific evidence nor participation of international organizations to promote utilization. Despite this, the applicability of appropriate sci-

---

Articles 61 and 62 of UNCLOS are entitled, "Conservation of living resources" and "Utilization of living resources" respectively. For scholarly commentary on Articles 61 and 62 see UNITED NATIONS CONVENTION ON THE LAW OF THE SEA: A COMMENTARY 594–638 (SATYA N. NANDAN & SHABTAI ROSENNE eds., 2d vol., 1995).

[54]   UNCLOS, *supra* note 53, at Part V.
The exclusive economic zone (EEZ) is an area beyond and adjacent to the territorial sea, subject to the specific legal regime established in this Part, under which the rights and jurisdiction of the coastal State and the rights and freedoms of other States are governed by the relevant provisions of this Convention.

Id. at art. 55. The EEZ may extend up to 200 nautical miles from the baselines of the coastal state. *Id.* at art. 57. The area seaward of the EEZ is the "High Seas," or international waters. *See id.* at art. 86; *see also infra* note 62.

[55]   UNCLOS, *supra* note 53, at art. 61.

[56]   *Id.*
The coastal State, taking into account the best scientific evidence available to it, shall ensure through proper conservation and management measures that the maintenance of the living resources in the [EEZ] is not endangered by over-exploitation. As appropriate, the coastal State and competent international organizations, whether sub-regional, regional or global, shall co-operate to this end.

Id. at art. 61(2).

[57]   *Id.* at art. 62(1).

entific information and input of relevant, if not legally empowered international organizations can probably be inferred. Ultimately, the legality of states' actions with regard to the exploitation of their own marine resources may be presumed absent a treaty limitation to the contrary or clear evidence that such exploitation is interfering with the rights of other states. Whether or not a particular case of utilization contravenes conservation objectives is another question.

Under most circumstances, UNCLOS appears to attempt to effectuate a balance between the objectives of conservation and utilization in EEZs. For example, in the case of highly migratory species (including some cetaceans), UNCLOS refers to both objectives.[58] In addition, UNCLOS refers to the role of "appropriate international organizations" to "ensur[e] conservation and promot[e] the objective of optimum utilization."[59] In the particular case of cetaceans, however, the balance of UNCLOS' objectives may tip in favor of conservation. Articles 65 of UNCLOS, entitled "Marine mammals" and applicable to states' conduct in their EEZ's, supports this position, providing:[60]

---

[58]   *Id.* at art. 64. Article 64 is entitled, "Highly migratory species." Article 64 provides:

>   1. The coastal State and other States whose nationals fish in the region for the highly migratory species listed in Annex I shall co-operate directly or *through the appropriate international organizations with a view to ensuring conservation and promoting the objective of optimum utilization of such species throughout the region, both within and beyond the [EEZ]. In regions for which no appropriate international organization exists, the coastal State and other States whose nationals harvest these species in the region shall co-operate to establish such an organization and participate in its work.*

Id. at art. 64(1) (emphasis added). Annex I of UNCLOS lists 17 specific highly migratory covered by Article 64. These include seven Families of cetaceans. They are: Family *Physeteridae*; Family *Balaenopteridae*; Family *Balaenidae*; Family *Eschrichtiidae*; Family *Monodontidae*; Family *Ziphiidae*; Family *Delphinidae*. *Id.* at Annex I.

[59]   *Id.* at art. 64(1). For the entire text of Article 64(1) *see supra* note 58.

[60]   UNCLOS, *supra* note 53, at art. 65. For scholarly commentary on Article 65 see NANDAN & ROSENNE, *supra* note 53, at 659–664.

[n]othing in this Part restricts the right of the coastal State or the competence of an international organization, as appropriate, to prohibit, limit or regulate the exploitation of marine mammals *more strictly* than provided for in this Part. States shall co-operate with a view to the conservation of marine mammals and in the case of cetaceans shall in particular work through the appropriate international organizations for their conservation, management and study.[61]

The plain language of Article 65 suggests that conservation concerns are paramount in the context of cetaceans, permitting stricter management standards than for other marine species. A precautionary approach is clearly justified given the historic record of over-exploitation of many cetacean species. Article 120 reinforces this point, where the drafters saw fit to extend the protections of Article 65 to cetaceans found on the High Seas.[62] The fact that Articles 65 and 120 single out cetaceans for greater conservation is problematic for states and organizations that argue cetaceans are a resource that may be utilized and consumed as any other marine resource. To argue otherwise would ignore the legal maxim and general principle of law—*lex specialis derogat legi generali*—a specific law prevails over a general law.

4.2.   The High Seas

Beyond national jurisdiction, equally important is a discussion of the High Seas, or that segment of the ocean where no state may exercise jurisdiction at the expense of another state.[63] Since no state may exercise its sovereignty in this area, the principles of UNCLOS, applicable to every member, become all the more important in the management context. Each state that utilizes the High Seas must exercise its jurisdiction over the ships flying its flag.[64] As discussed above, Article 120

---

[61]   UNCLOS, *supra* note 53,. at art. 65 (emphasis added).

[62]   *Id.* at art. 120. Article 120 provides: "Article 65 applies to the conservation and management of marine mammals in the high seas." *Id.* The High Seas is that segment of the ocean over which no state may exercise sovereignty. *See id.* at Part VII.

[63]   *Id.* at Part VII.

[64]   *Id.* at art. 94(1).

extends the expansive cetacean conservation provisions of Article 65 to the High Seas. Similar to the standards established for management of cetaceans within EEZs, Article 120 contemplates the possibility of stricter protective measures for cetaceans than for other marine species found in the High Seas.

While UNCLOS emphasizes "freedom of fishing" on the High Seas, this provision does not negate Article 120's provisions.[65] UNCLOS provides that High Seas freedoms must be exercised with "due regard" for the interests of other states on the High Seas.[66] In addition, this freedom must be exercised consistently with "their treaty obligations."[67] Furthermore, freedom of fishing is specifically made subject to the requirements of a section entitled "[c]onservation and management of the living resources of the high seas,"[68] which includes reference to Article 120. Interestingly, Article 120 follows immediately after the provision recognizing the concept of "maximum sustainable yield" in the conservation scheme of living resources found in the High Seas.[69] From an interpretive standpoint, Articles 65 and 120 seem to modify similar provisions with regard to the conservation of resources found in the EEZ and the High Seas, respectively. Therefore, the freedom to harvest cetaceans, if any, must be exercised with the interests of other states in mind and subject to possibly stricter conservation mandates laid down by competent international organizations.

4.3.   The Duty to Cooperate

Under UNCLOS, one factor resonates loudly in the management of living resources: the duty to cooperate with other states directly or through international organizations.[70] It is widely recognized that international

---

[65]   *Id.* at art. 87(1)(e).

[66]   *Id.* at art. 87(2).

[67]   *Id.* at art. 116(a).

[68]   *Id.* at Part VII, Section 2.

[69]   *Id.* at art. 19(1)(a).

[70]   Most significantly, the "duty to cooperate" is discussed in UNCLOS Articles 61, 63, 64, 65, 66, 117, 118 and 119 to name just a few. UNCLOS, *supra* note 53.

organizations provide an essential forum for international cooperation in relation to a spectrum of environmental issues.[71] Accordingly, international organizations can provide such a forum for cooperation in the realm of marine mammal management. Through these organizations states can discuss conservation and utilization measures. Under optimal circumstances, the organizations can initiate scientific research and issue recommendations or binding regulations that may be implemented by members undertaking such conservation and utilization. In short, such organizations should have a prominent place in an overall management scheme. On the other hand, in the case of cetaceans, the relevant articles of UNCLOS suggest that such collective action should occur with conservation as the more prominent objective.

UNCLOS certainly contemplates a role for regional organizations in the conservation of living resources in the EEZ and High Seas. Sound resource management principles suggest that some issues are best addressed at the regional level. Regional organizations can address a management problem in a more focused way and require a balancing of fewer interests than at the global level. In addition, the states of a region often share objectives and incentives for managing common resources. Finally, they are typically engaged in other areas of cooperation, such as trade and tourism, which are conducive to common resource management. The Agreement on the Conservation of Small Cetaceans of the Baltic and North Seas (ASCOBANS),[72] and the Agreement on the

---

In addition to the provisions of UNCLOS, Agenda 21, a product of the 1992 United Nations Conference on Environment and Development, calls for cooperation on matters of conservation of marine living resources. *See* Report of the United Nations Conference on Environment and Development, Rio de Janeiro (June 3–14, 1992), U.N. Doc. A/CONF.151/26, Agenda 21, Aug. 12, 1992; Rio Declaration on Environment and Development, *reprinted in* 31 I.L.M 874 (1992), at Chapter 17.

[71]   PATRICIA W. BIRNIE & ALAN E. BOYLE, INTERNATIONAL LAW AND THE ENVIRONMENT 32 (1992).

[72]   Agreement on the Conservation of Small Cetaceans of the Baltic and North Seas, *reprinted in* II THE MARINE MAMMAL COMMISSION COMPENDIUM OF SELECTED TREATIES, INTERNATIONAL AGREEMENTS, AND OTHER RELEVANT DOCUMENTS ON MARINE RESOURCES, WILDLIFE AND THE ENVIRONMENT 1612 (1994) (entered into force Mar. 29, 1994) [hereinafter ASCOBANS]. ASCOBANS currently has eight parties: Belgium, Denmark, Finland, Germany, the Netherlands, Poland, Sweden and the United Kingdom. Although a full discussion of the

Conservation of Cetaceans of the Black Sea, Mediterranean Sea and the Contiguous Atlantic Area (ACCOBAMS)[73] are examples of cetacean conservation agreements that have been heralded for their regional focus.[74] NAMMCO as a regional organization may also benefit from regional synergies. Advocates of consumptive uses of cetaceans may contend that the establishment of additional regional organizations will help to fully and effectively implement UNCLOS marine species management provisions. However, unlike ASCOBANS and ACCOBAMS, which ban the consumptive use of cetaceans within their respective regions, the pro-consumptive IGOs raise issues about the management decisions that might be made by these organizations.

Whether or not NAMMCO, the World Council of Whalers, or any other pro-consumptive organization in the marine mammal arena will satisfy the role that UNCLOS contemplates for international or regional organizations is an open question at this point. Thus far, no organization

---

ASCOBANS treaty regime is beyond the scope of this work, it should be noted that the language of the ASCOBANS treaty emphasizes the conservation, not consumption, of small cetaceans in the Baltic and North Seas. It is noteworthy that ASCOBANS and NAMMCO have developed something of a cooperative relationship. For example, they have cooperated on the monitoring of harbor porpoises in the Baltic Sea. *See* Robin Churchill, *Sustaining Small Cetaceans: A Preliminary Evaluation of the ASCOBANS and ACCOBAMS Agreements, in* INTERNATIONAL LAW AND SUSTAINABLE DEVELOPMENT: PAST ACHIEVEMENTS AND FUTURE CHALLENGES 240 (Alan Boyle & David Freestone eds., 1999).

[73]   Agreement on the Conservation of Cetaceans of the Black Sea, Mediterranean Sea and Contiguous Atlantic Area, [hereinafter ACCOBAMS] *reprinted in* 36 I.L.M. 777 (1997) (entered into force June 1, 2001). Currently, 16 countries have signed ACCOBAMS. They are: Albania, Croatia, Cyprus, France, Georgia, Greece, Italy, Monaco, Portugal, Spain, Tunisia, Morocco, Romania, Libya, Malta and Bulgaria. Thus far, Monaco, Spain, Morocco, Bulgaria, Croatia, Albania, Georgia, Malta and Romania have ratified.

[74]   For an overview of these two agreements established under the Convention on the Conservation of Migratory Species of Wild Animals, 19 I.L.M. 15 (1980), *see* William C.G. Burns, *The Agreement on the Conservation of Cetaceans of the Black Sea, Mediterranean Sea and Contiguous Atlantic Area (ACCOBAMS): A Regional Response to the Threats Facing Cetaceans*, 1(1) J. INT'L WILDLIFE L. & POL'Y 113–33 (1998); Hugo Nukamp & Andre Nollkaemper, *The Protection of Small Cetaceans in the Face of Uncertainty: An Analysis of the ASCOBANS Agreement*, 9 GEO. INT'L ENVTL. L. REV. 281, 302 (1997).

other than the IWC has put forth regulations for the taking of cetaceans.[75] NAMMCO has only established a "scientific, knowledge-building, and procedural" framework, and there has been very little discussion of possible management schemes.[76] So long as NAMMCO activities remain so restricted, they seem wholly consistent with UNCLOS. If, on the other hand, NAMMCO moves toward regulation of consumptive utilization in the future, a crucial issue will be whether its scheme comports with all of the requirements of UNCLOS, as well as other key obligations of international law. In addition, it will be an issue whether NAMMCO's action complements that of the IWC, contradicts it or is wholly separate and distinct. Another critical question will be whether NAMMCO seeks to bind member states to a regulatory framework or merely offer recommendations for a consumptive management scheme. At that moment, the critical issue of competence will be forced to the surface.

4.4. Regulations by a Pro-Consumptive Organization

If, in the future, NAMMCO chooses to put forth a regulatory scheme seeking to bind member states, three critical questions must be answered. First, does the NAMMCO Agreement confer the authority on NAMMCO to promulgate such a scheme? Second, would the proffered regulatory scheme be consistent with key obligations of the law of the sea? Third, would the proffered regulatory scheme be consistent with IWC obligations incumbent upon NAMMCO member states that are also members of the IWC? These three questions would require affirmative answers before any regulatory scheme promulgated by NAMMCO could withstand scrutiny.

4.4.1. The NAMMCO Agreement: What Powers Does it Confer?

As discussed in Section 2.1, the NAMMCO Agreement establishes four

---

[75]   Steinar Andresen, *NAMMCO, IWC and the Nordic Countries, in* Petursdottir *supra* note 2, at 80.

[76]   *Id.*

components within the regime, including the Management Committees;[77] yet, there is no indication that the NAMMCO Agreement intends to create a framework where binding management obligations are established for member states. Under the Agreement, Management Committees are to "propose to their members measures for conservation or management[,]"[78] and decisions of Management Committees are taken by a unanimous vote of present members.[79] The key word "propose" suggests that the Management Committee's authority is limited to making recommendations that the parties have the discretion to accept or reject.

By contrast, under Article 5 of the ICRW, amendments by the IWC to the ICRW's Schedule with respect to "conservation and utilization of whale resources" are binding on the parties unless they expressly object within 90 days.[80] The provisions of the NAMMCO agreement evince the intent of the parties not to confer commensurate authority on its Management Committees.

4.4.2. Duty to Conserve vs. the Right to Consume

The UNCLOS articles discussed in Section 4 set forth the balance between conservation and utilization that UNCLOS seeks to achieve with regard to living marine resources. As a matter of treaty construction, all relevant provisions must be read together. Significantly, in the case of cetaceans, conservation, not utilization, would seem to be the *lex specialis* in light of Articles 65 and 120. Even if the purpose of Articles 65 and 120 were not to shift the balance in favor of conservation so that cetaceans should be treated as any other marine resource, one must ask whether a consumptive management scheme would satisfy other important requirements under UNCLOS and international environmental law in general.

---

[77]   *See supra* text accompanying notes 4 to 7.

[78]   *See supra* text accompanying note 15. "Management Committees shall with respect to stocks of marine mammals within their respective mandates: (a) propose to their members measures for conservation and management[.]" NAMMCO Agreement, *supra* note 2, at art. 5(a).

[79]   NAMMCO Agreement, *supra* note 2, at art. 5.

[80]   ICRW, *supra* note 35, at art. 5.

First, the "duty to cooperate" in the conservation of marine resources (examined in Section 4.3) might disfavor a consumptive management scheme by a small number of states where many more states have rejected such a scheme for those same species. In other words, UNCLOS encourages wide participation and collective action as an effective management plan, but competing management plans would arguably undercut the spirit of cooperation. To the extent states disagree about management goals and the technical aspects of their implementation, they should be resolved in a cooperative not competitive context.[81] In fact, an attempt to utilize protected species in a consumptive manner in the face of much more stringent conservation measures might be interpreted as a *per se* violation of the duty to cooperate.

Second, the application of the "precautionary approach" would set the bar rather high for any state that wants to consume a resource that other states have subjected to a strict conservation scheme. The precautionary approach has emerged as a key doctrine of international environmental law[82] and is clearly contemplated by UNCLOS.[83] The fact that a large number of individual states and other international organi-

---

[81]   The famous dispute, *Fisheries Jurisdiction Case (United Kingdom v. Iceland)* predated UNCLOS, but the holding of the International Court of Justice clearly endorsed the concept that states with competing fishery interests and different interpretations of the law of the sea should work together to negotiate an equitable solution that would reconcile those competing interests. *See* Fisheries Jurisdiction Case (U.K. v. Ice.) 1974 I.C.J. 3 (July 25).

[82]   For a discussion of the role of the precautionary approach in marine fisheries management *see* Grant J. Hewison, *The Precautionary Approach to Fisheries Management: An Environmental Perspective*, 11 INT'L J. MARINE & COASTAL L. 301 (1996); John M. MacDonald, *Appreciating the Precautionary Principle and Ethical Evolution in Ocean Management*, 26 OCEAN DEV'T & INT'L L. 255 (1995). In substance, the precautionary approach as applied in this context requires states to be more cautious in their management of resources when faced with scientific uncertainty. Numerous multilateral environmental agreements in force today adopt the precautionary approach.

[83]   *See* Separate opinion of Judge Laing, *Southern Bluefin Tuna Cases, Requests for Provisional Measures* (visited Dec. 30, 2000), <http://www.un.org/Depts/los/ITLOS/3Laing.htm>. Judge Laing traces the history of the precautionary approach and states that it cannot be denied that UNCLOS adopts a precautionary approach. *Id.* at para. 17.

zations have found it appropriate to proceed with greater caution in their management of identical species should be given great weight. Practically speaking, divergent management schemes are evidence of scientific disagreement that the precautionary approach was meant to address.

4.4.3. NAMMCO and the IWC

From the perspective of its relationship to other treaty regimes, one of the most important provisions of the NAMMCO Agreement is Article 9: "[t]his Agreement is without prejudice to obligations of the [p]arties under other international agreements."[84] NAMMCO and the IWC differ in some important details. First, NAMMCO is a regional organization dedicated to marine mammal conservation, management and optimum utilization in the North Atlantic. The IWC's mandate, on the other hand, is global. Second, NAMMCO claims within its purview all marine mammals, not just cetaceans. While NAMMCO has so far given advice to members on matters of pinnipeds and small cetaceans it has not done so for species included in the IWC quotas. There is some contact and exchange between NAMMCO and the IWC. Both the IWC and NAMMCO exchange observers to their annual meetings and there is contact between their Scientific Committees.[85]

One of the most critical issues that will need to be addressed in the future, should NAMMCO take substantive management actions, will be

---

[84]   NAMMCO Agreement, *supra* note 2, at art. 9. This provision alone may prove potent and decisive in any direct conflict between NAMMCO action and the IWC. Article 30 of the Vienna Convention on the Law of Treaties Between States and International Organizations or Between International Organizations contemplates such provisions as conclusive in favor of the earlier treaty. *See* Vienna Convention on the Law of Treaties Between States and International Organizations or Between International Organizations, *supra* note 31, at art. 30(2).

> When a treaty specifies that it is subject to, or that it is not to be considered as incompatible with, an earlier or later treaty, *the provisions of that other treaty shall prevail.*

Id. (emphasis added). Although this Vienna Convention is not yet in force its provisions are persuasive authority.

[85]   NAMMCO 1998 Annual Report, § 9.1 (Cooperation with other international organizations/IWC). *See also* Churchill, *supra* note 72.

the extent to which such actions conflict with the conservation scheme of the IWC and the legal implications of such conflict. As a threshold consideration, are there any legal principles that would favor one regime over the other? After all, Article 65 of UNCLOS does not specify which international organizations are competent to manage cetaceans.[86] There is, however, some collateral evidence of the primacy of the IWC in cetacean management. Agenda 21's Oceans Chapter recognizes the IWC in this role, although it does not preclude other organizations.[87] In addition, the Convention on International Trade in Endangered Species of Wild Fauna and Flora (CITES), with 160 state parties, also recognizes the primacy of the IWC on matters of cetacean conservation.[88] Realistically, the existence of an established, ongoing conservation regime with a greater than 50-year history, established by treaty and sustained by a membership of almost 50 states, presents formidable obstacles for any other organization seeking to regulate the same resource. The fact that two NAMMCO members, Norway and Iceland, are also members of the IWC presents additional problems.

Examining a hypothetical scenario, let us assume that NAMMCO, in objection to the failure of the IWC to repeal its moratorium on commercial whaling, purports to authorize a limited commercial hunt by its own members. It would attempt to justify this action with the assertion

---

[86] UNCLOS, *supra* note 53, at art. 65. For the full text of Article 65 *see* text accompanying note 61. It has even been suggested by one observer that Article 65 is unclear as to whether it contemplates more than one international organization for cetacean conservation and management. *See* Ray Gambell, *The International Whaling Commission Today, in* Petursdottir, *supra* note 2. The second sentence of Article 65, however, uses the plural "organizations." UNCLOS, *supra* note 53, at art. 65.

[87] Agenda 21, *supra* note 70, at para. 17.90. Gambell highlights this point. *See* Gambell *supra* note 86, at 58.

[88] Convention on International Trade in Endangered Species of Wild Fauna and Flora, Mar. 13, 1973, 27 U.S.T. 1087, 993 U.N.T.S. 243 [hereinafter CITES]. Various resolutions passed by the CITES Conference of the Parties (COP) established the practice for the CITES COP to defer to the IWC on matters of cetacean conservation. For the complete texts of all CITES resolutions *see* CITES Website (visited Nov. 11, 2002) <http://www.cites.org/eng/resols/index.shtml>. For a discussion of the key CITES resolutions that invoke the IWC *see* David S. Favre, INTERNATIONAL TRADE IN ENDANGERED SPECIES: A GUIDE TO CITES 91–93 (1989).

that the IWC has failed in its stewardship of the whaling industry.[89] In addition, it would argue that a limited hunt would be scientifically supportable as a sustainable utilization of cetacean resources consistent with all UNCLOS obligations.

Should such a scenario arise, proponents of the NAMMCO scheme will need to contend with several factors. First, as discussed in Section 4.4.1, it is questionable whether NAMMCO has the power to establish an obligatory management scheme in the first instance. The IWC does not suffer from this limitation. Therefore, there may not be a legal conflict between a binding obligation and a non-binding regulatory proposal. Second, even if NAMMCO had the power to establish binding regulations, it is important to remember that international organizations can only bind member states. NAMMCO cannot be seen to compete with the IWC in any real sense unless it was to share a substantial number of members in common. This is not so at the present time. Currently, only Norway and Iceland are members of both organizations. While Norway has consistently been a member of both the IWC and NAMMCO, Iceland withdrew from the IWC in 1992 only to rejoin in late 2002.[90] Although Greenland and the Faroe Islands lend legitimacy to NAMMCO, it does not change the basic fact that NAMMCO only has two state parties.[91] Therefore, unless this fact changes, NAMMCO probably cannot be considered a viable management alternative.[92]

---

[89]   See Gambell, *supra* note 86. Iceland and Norway have stated in several meetings of the IWC that since that body seems reluctant to allow the resumption of commercial whaling, it might be necessary for them to turn to alternative forums where the issues of catch limits can be discussed. *Id.* at 58.

[90]   See Gambell, *supra* note 86, at 58. The matter of Iceland's re-entry into the IWC in 2002 was quite controversial because Iceland successfully sought to rejoin with a reservation to the moratorium. *See Iceland and her re-adherence to the Convention after leaving in 1992*, IWC Website (visited Nov. 11, 2002) <http://www.iwcoffice.org/Iceland.htm>. *See also* Walter Gibbs, *Iceland Joins Whale Panel, Giving Whalers Stronger Say*, N.Y. TIMES, Oct. 21, 2002, at A2.

[91]   See Andresen, *supra* note 75, at 80–81.

[92]   See id. Andresen avoids a discussion of the legal intricacies of the relationship between UNCLOS, the IWC and NAMMCO by proclaiming he is not a lawyer. *Id.* at 80. His point, however, that NAMMCO's limited membership undercuts the possibility for direct competition between the IWC and NAMMCO is well

Even if NAMMCO were able to coax a number of other like-minded IWC members to join its ranks,[93] those states would be limited by their obligations under the ICRW. As with any treaty, members are required to perform their obligations in good faith.[94] The NAMMCO Agreement also specifies: "[t]his Agreement is without prejudice to obligations of the Parties under other international agreements."[95] Thus, the NAMMCO Agreement itself recognizes the difficulties that would be posed by conflicting regulatory schemes. The two state members of NAMMCO, Norway and Iceland, demonstrate in a microcosm the potential difficulties for states that are members in both organizations. Norway immediately chose to straddle the fence by participating in both organizations. Iceland, by contrast, initially decided to withdraw from the IWC and put all of its eggs in the NAMMCO basket.[96] Iceland's decision to rejoin the IWC after 10 years of absence and Norway's long-standing membership in both the IWC and NAMMCO may ultimately require those countries to engage in a delicate balancing act should the two organizations set forth conflicting regulatory requirements.

Legally speaking, if NAMMCO wants to implement regulatory measures in the future it would cause fewer problems if it recruited mem-

---

taken from both a legal and policy standpoint. One could fancy an argument that proponents of a NAMMCO regulatory scheme could rely upon the general legal principle providing that later obligations prevail over earlier ones, but this would only apply, at best, to the extent of the shared membership. In addition, the IWC moratorium is reviewed at annual meetings so it would be difficult to characterize as a "prior obligation."

93   As discussed in note 2, Russia and Canada have been mentioned as possible new members. *See* note 2, *supra*. *See also* Andresen, *supra* note 75, at 81–82. Although Russia is a member of the IWC, Canada is not. *Id*. Canada does, however, retain strong interests in sealing and small-scale aboriginal whaling. *Id*.

94   *See* Vienna Convention on the Law of Treaties, May 23, 1969, 1155 U.N.T.S. 330, at art. 26. This section of the Vienna Convention recognizes one of the most important doctrines of international law—*pacta sunt servanda*—that a treaty in force is binding upon the parties to it and must be performed by them in good faith. *Id*. Significantly, the Vienna Convention specifies that it "applies to any treaty which is the constituent instrument of an international organization." *Id*. at art. 5.

95   NAMMCO Agreement, *supra* note 2, at art. 9.

96   *See* Andresen, *supra* note 75, at 83.

bers that are either not members of the IWC or are states willing to surrender their IWC membership. Additionally, if NAMMCO wants to test the bounds of its legal and political capacity to regulate cetaceans it will probably first attempt to do so for small cetaceans in the North Atlantic, because the IWC has not asserted jurisdiction over these species[97] (although even in this scenario, NAMMCO would need to concern itself with any potential conflict with actions taken by ASCOBANS). Finally, any NAMMCO member would be on the firmest ground, so to speak, if it implemented a NAMMCO scheme within its own EEZ (to the extent it coincides with the NAMMCO regulatory area) where its sovereign rights to exploit its resources are the strongest.

As noted at the beginning of this section, it would be in NAMMCO's interest to avoid a direct regulatory conflict with the IWC, but that does not mean NAMMCO cannot play an active role to further its objectives. The more likely scenario is one where Norway, Iceland or any other state that might hold membership in both organizations, uses its NAMMCO membership as political leverage to undercut the IWC's more preservationist agenda.[98] Importantly, however, states that choose to walk that fine line may find it difficult to act in good faith on all fronts. Where this occurs against the backdrop of the UNCLOS "duty to cooperate," the scenario becomes confusing indeed. In any event, Iceland's re-entry into the IWC, while most likely based on several factors, might well signify that NAMMCO will not be offering a competing model of cetacean regulation anytime soon.

If NAMMCO continues on its present course, focusing mainly on scientific study, and does not venture directly into the realm of consumptive

---

[97]   *See* William C. Burns, *The International Whaling Commission and the Regulation of the Consumptive and Non-Consumptive Uses of Small Cetaceans: The Critical Agenda for the 1990s*, 13 WIS. INT'L L.J. 105 (1994).

[98]   *See* Andresen, *supra* note 75, at 83–85. The relationship between IGOs is an intriguing, but as yet underdeveloped subject of international law. Some scholars have considered the possibility that one state would use its membership in one IGO to influence the activities and policies of another IGO. *See* WERNER J. FELD, ROBERT S. JORDAN WITH LEON HURWITZ, INTERNATIONAL ORGANIZATIONS: A COMPARATIVE APPROACH 188–89 (1983); *see also* FELICE MORGENSTERN, LEGAL PROBLEMS OF INTERNATIONAL ORGANIZATIONS 19–29 (1986); IAN BROWNLIE, PRINCIPLES OF PUBLIC INTERNATIONAL LAW 692–94 (1979).

cetacean management, cooperation between the organizations may be engendered on the scientific, if not policy, level. On the other hand, if it chooses to offer an alternative management scheme for cetacean resources it would be advisable to attempt to harmonize and coordinate its activities with the IWC to the greatest extent possible.[99] Ultimately, the decision to work with the IWC, or against it, may prove decisive for NAMMCO's future. From a political perspective, should NAMMCO choose to offer an alternative management strategy and position itself in clear opposition to the IWC, it would place its legitimacy in jeopardy. Many prominent IWC members would not regard a conflicting regulatory framework with kindness.[100] On the other hand, the recent accession to membership in the IWC by states sympathetic to the cause of whaling, along with Iceland's re-entry into the organization, may indicate that the IWC, and not any other intergovernmental organization, will be the venue for future battles between pro and anti-whaling advocates.[101]

5.   CONCLUSIONS

The pro-consumptive international organizations that have developed since the early 1990s have the potential to change the landscape of cetacean conservation. At present, NAMMCO is the only intergovernmental organization with a credible possibility of offering a consumptive model of cetacean management. Should it attempt to do so, however, it will need to navigate a number of important legal obstacles set forth by the law of the sea and international environmental law. In particular, the provisions of UNCLOS addressing the protection of living resources emphasize the conservation, not consumption, of cetaceans. Similarly, UNCLOS also calls for cooperation in the discharge of conservation and management obligations. An organization with few members offering an alternative management strategy will have difficulty satisfying that obligation.

---

[99]   *See* BROWNLIE, *supra* note 98, at 692–93 (discussing the coordination of the activities of international organizations).

[100]   *See* Andresen, *supra* note 75, at 84.

[101]   *See* Gibbs, *supra* note 90. Recent IWC members Benin, Gabon, Mongolia, Palau and Iceland all appear to stand solidly with the whalers. *Id.*

The IWC remains the most prominent and capable international organization to discharge these responsibilities and provide a forum for cooperation. Its large membership guarantees participation by states with differing perspectives on cetacean management. In contrast to the IWC, an analysis of the NAMMCO Agreement indicates that it may not have the power to establish a binding management scheme even under the best of circumstances. At the same time, NAMMCO could promulgate a regulatory scheme that its members might voluntarily implement, particularly in their own EEZs. Of course, such actions would need to satisfy all obligations of UNCLOS and other relevant treaties.

Even if the NAMMCO Agreement established the power to institute a binding management scheme upon its members, a conflict with the IWC sets up potential problems for parties that share membership in both organizations. As mentioned, presently, only Norway and Iceland are members of both the IWC and NAMMCO. Should NAMMCO offer an alternative management scheme, Norway, Iceland and any other state that may hold membership in both organizations in the future, will need to walk a fine line to reconcile their obligations. The UNCLOS obligations of conservation and cooperation must be respected as well.

There is nothing to prevent NAMMCO from fulfilling its mandate as an independent forum and organization devoted to the scientific study of marine mammals. To the extent both organizations are so inclined, it may do so in full cooperation with the IWC. NAMMCO is a regional organization, like ASCOBANS and ACCOBAMS, and as such can play a role in future management schemes. Regional organizations are contemplated and even favored by UNCLOS. The role NAMMCO, or any other pro-consumptive IGO chooses to play in the future, will need to be concordant with the modern framework of international law. Their real challenge is that in the case of cetaceans, the legal and political context, exemplified by UNCLOS and the IWC, has evolved along the lines of conservation and not consumption. Ultimately, the actions of pro-consumptive organizations will be judged from that context.

Part III

The Threat to Small Cetaceans and Institutional Responses

Chapter 8
Small Cetaceans: Status, Threats, and Management

Kieran Mulvaney[1] & Bruce McKay[2]

1. INTRODUCTION

Approximately 70 species of cetaceans are often referred to collectively as "small cetaceans." Since these 70 species constitute around 85 percent of all known cetacean species, it is not surprising that they vary considerably in size. The smallest, the vaquita or Gulf of California porpoise (*Phocoena sinus*), has a reported size range of 90–143.5 centimeters. The largest, the Baird's beaked whale (*Berardius bairdii*), may attain a length as great as 13 meters. In fact, membership in the club of small cetaceans is, to some extent, an arbitrary matter. Size is not, in and of itself, either a prerequisite or a disqualifying feature. For example, the Baird's beaked whale, widely classified as a small cetacean, is actually larger, on average, than the minke whale (*Balaenoptera acutorostrata*), which is not classified as such.[3]

The reason for such confusion is that 'small cetacean' is not a strictly biological term but rather a political construct. Its genesis lies in a seemingly innocuous list of species appended to the 1946 International Convention for the Regulation of Whaling (ICRW).[4] This "Annex of

---

[1] Editor, *Ocean Update*, 1219 W. 6th Ave., Anchorage AK 99501, USA, Kieran@alaska.net.

[2] Senior Researcher, SeaWeb, 1731 Connecticut Ave., NW, Washington, DC, USA, bmckay@seaweb.org.

[3] Although some reviewers do indeed designate it as such. *See, for example*, ED MITCHELL, PORPOISE, DOLPHIN AND SMALL WHALE FISHERIES OF THE WORLD: STATUS AND PROBLEMS (1975).

[4] International Convention for the Regulation of Whaling with Schedule of Whaling Regulations, Dec. 2, 1946, 62 Stat. 1716, T.I.A.S No. 19\849, 161 U.N.T.S.

Nomenclature" was drawn up at the Conference at which the ICRW was signed and from which the International Whaling Commission (IWC) was spawned.

The Annex simply listed the species that were most likely to be targeted by the whaling industry of that time. Not surprisingly, this included most of the larger whales but almost none of the smaller ones. As a result, the minke whale is listed, as is the northern bottlenose whale (*Hyperoodon ampullatus*), which was being hunted in the North Atlantic at the time the ICRW was signed. However, the Baird's beaked whale, the larger cousin of the bottlenose whale, is excluded simply because it was not then being targeted by the whaling industry. So the northern bottlenose whale is not a "small cetacean," but the larger Baird's beaked whale is.

2.   THE SMALL CETACEANS

Given that we must draw a line somewhere, we seek to classify small cetaceans in as consistent and logical a manner as possible. There are two suborders of cetaceans: the *mysticeti*, or baleen whales, and the *odontoceti*, or toothed whales. All the species covered in this chapter are odontocetes; indeed, all odontocete species except one—the sperm whale (*Physeter macrocephalus*)—are included here. That means this chapter will cover 70 species of toothed whales in nine families.[5]

Four of the species are the so-called river dolphins: the Indian river dolphin (*Platanista gangetica*);[6] the baiji, or Chinese river dolphin

---

361. Available online at: <http://ourworld.compuserve.com/homepages/iwcoffice/Convention.htm>.

   5   Classifications follow those in DALE W. RICE, MARINE MAMMALS OF THE WORLD: SYSTEMATICS AND DISTRIBUTION (1998).

   6   In keeping with the authoritative classification put forward by Rice, we have here designated just one species of Indian river dolphin; Rice proposes two subspecies, the Indus and Ganges river dolphins, respectively *Platanista gangetica minor* and *P. g. gangetica*. Many authorities, however, argue for two separate species—*P. minor* and *P. gangetica. See, for example*, Hugo P. Castello, *An Introduction to the Whales and*

(*Lipotes vexillifer*); the boto, or Amazon river dolphin (*Inia geoffrensis*); and the franciscana, or La Plata river dolphin (*Pontoporia blainvillei*). Each of these species is the sole member of its family. The first three are the only truly freshwater cetaceans; the franciscana, although exclusively marine, is considered an "honorary" river dolphin because it inhabits the saltwater estuary of the Rio La Plata in Argentina, as well as coastal waters along the mid-eastern shores of South America.

The fifth family, the Ziphiidae, contains the approximately 20 species of beaked and bottlenose whales. Little is known about most of these species and they are rarely seen. Some are known only from isolated sightings, stranded animals, and skulls. Not all species have even been fully described. They range in length from approximately 4.5–6.5 meters for the members of the *Mesoplodon* genus to 13 meters for the Baird's and Arnoux's beaked whales.

The Kogiidae and the Monodontidae families are comprised of only two species each: the pygmy (*Kogia breviceps*) and dwarf (*Kogia sima*) sperm whales in the Kogiidae family; and the narwhal (*Monodon monoceros*) and beluga (*Delphinapterus leucas*), both confined to the waters of the Arctic and sub-Arctic, in the Monodontidae.

By contrast, the Delphinidae contains the most species of all marine mammal families, including these familiar ones:

(1)   The bottlenose dolphin (*Tursiops truncatus*), made famous by the television show "Flipper" and various live dolphin shows;
(2)   The common dolphin (*Delphinus delphis*);
(3)   The long- and short-finned pilot whales (*Globicephala melas* and *G. macrorhynchus*); and
(4)   The orca or killer whale (*Orcinus orca*).

The final small cetacean family is the Phocoenidae, composed of the six species of "true" porpoises, all easily distinguishable from other cetaceans by their rounded bodies, snub noses, and small size.

Dolphins, in THE CONSERVATION OF WHALES AND DOLPHINS: SCIENCE AND PRACTICE 1, 2 (Mark P. Simmonds & Judith D. Hutchinson eds., 1996).

3. THREATS TO SMALL CETACEANS[7]

3.1 Directed Hunts[8]

3.1.1. Commercial Hunts

The small cetacean hunt that most closely resembles commercial whaling operations is in Japan, where the Baird's beaked whale is targeted by small vessels equipped with harpoon guns armed with harpoons tipped with explosive grenades—the same type of vessel and equipment used in coastal minke whaling operations.[9] However, the Baird's beaked whale hunt is not presently regulated by the IWC, and Japan has consistently resisted any such regulation. Japan presently assigns itself an annual quota of 54 whales. A similar Japanese operation conducts hunts for short-finned pilot whales and Risso's dolphins (*Grampus griseus*), with quotas of 50 and 30, respectively.[10]

Other commercial hunts of small cetaceans in Japanese waters generally use either drive fisheries or hand harpoons. The majority of species are killed in so-called drive fisheries, in which a number of small boats drive a herd of dolphins toward shore and into a bay or inlet, where they are later killed. Eleven species are taken this way. The most frequently caught species are striped dolphins (*Stenella coeruleoalba*), short-finned pilot whales, bottlenose dolphins, false killer whales (*Pseudorca crassidens*), Risso's dolphins, and spotted dolphins (*Stenella attenuata*). By far the highest catches have been of striped dolphins, although catches have declined in recent decades. Prior to 1963, annual catches were as high as 10,000–20,000, although records are incom-

---

7   In addition to the specific sources cited in this chapter, readers may also want to see our own overview of small cetacean issues, an expanded version of this chapter: Kieran Mulvaney & Bruce McKay, *Small Cetaceans: Small Whales, Dolphins and Porpoises, in* SEAS AT THE MILLENNIUM: AN ENVIRONMENTAL ASSESSMENT, 89–103 (Charles Sheppard ed., 2000).

8   *See also*, Kieran Mulvaney, *Directed Kills of Small Cetaceans Worldwide, in* Simmonds & Hutchinson, *supra* note 6, at 89–108.

9   K. Balcomb & C.A. Goebel, *Some Information on a Berardius bairdii fishery in Japan*, 27 REP. INT'L WHALING COMMISSION 485, 486 (1977).

10   *Report of the Sub-Committee on Small Cetaceans*, 43 REP. INT'L WHALING COMMISSION 130, 135 (1993).

plete; after 1963, the catch dropped to a mean average of 7,350, until 1980 when, following a peak year of 16,237, the mean over the following decade dropped to 2,390.[11] The average has fallen further since then; 596 were reported killed in 1999.[12]

Two species, the Dall's porpoise (*Phocoenoides dalli*) and Pacific white-sided dolphin (*Lagenorhynchus obliquidens*), are killed with hand harpoons, with concern especially acute in recent years over the size of the catch of the Dall's porpoise. Prior to 1988, the annual take was approximately 10,000; however, in 1988, the recorded take soared to more than 40,000. The catch immediately declined again, to around 29,000 in 1989 and 22,000 in 1990, but was still substantially higher than historical levels. Following intense international criticism, the government of Japan established an annual quota of 17,700; the reported takes in 1998 and 1999 were 11,385[13] and 14,807,[14] respectively.

In Peru, small cetaceans are harpooned and also caught in fishing nets, either directly or as bycatch. They are sold primarily for human consumption. Principal species include dusky dolphins (*Lagenorhynchus obscurus*), long-beaked common dolphins (*Delphinus capensis*), Burmeister's porpoises (*Phocoena spinipinnis*), and bottlenose dolphins. The most recent estimates place the number of small cetaceans killed at around 17,600 annually, even though the Peruvian government passed a law prohibiting cetacean hunting a decade ago.[15] In the Magellan region of Chile, small cetaceans have been taken for use as fish bait since the mid-1970s. This catch now seems to be declining considerably,

---

[11]   *Id.* at 136

[12]   IWC/52/4/Annex K, *Report of the Scientific Committee, Report of the Standing Sub-Committee on Small Cetaceans*, IWC/52/4, 52nd Meeting of the International Whaling Commission, Adelaide, Australia, May 2000, Appendix 4, at 32.

[13]   IWC 51/4/Annex 1, *Report of the Scientific Committee, Report of the sub-committee on Small Cetaceans*, 51st Meeting of the International Whaling Commission, Grenada, May 1999, Appendix 2, at 5.

[14]   IWC/52/4/Annex K, *supra* note 12.

[15]   Report of the Sub-Committee on Small Cetaceans. 45 REP. INT. WHAL. COMMISSION. 165, 170 (1995).

although as many as 600 dolphins may still be killed per year.[16]

In Sri Lanka, large numbers of small cetaceans are caught in gill nets, deliberately as well as incidentally. At least 17 species are involved, including a few large cetaceans. Some estimates suggest that the total number of all cetaceans killed each year may be as high as 40,000.[17]

3.1.2.   Traditional and Subsistence Hunts

There are two principal so-called "traditional" hunts of small cetaceans: for belugas and narwhals in the Arctic and sub-Arctic; and for long-finned pilot whales in the Faroe Islands. Historically, the hunters have used the catch for subsistence rather than selling it.

Belugas and narwhals have long been hunted by native peoples in the Arctic and sub-Arctic. These hunters have used the whales' blubber oil for lighting and cooking, and they eat the *muktuk* (the whales' skin and adhering blubber) and meat. In addition, because of the value of the spiraled tusk of mature males, a commercial element has been introduced to the narwhal hunt. Accurate overall catch statistics are difficult to gather, although landings of narwhals in Canada and Greenland are estimated to be around 1,000, and for belugas in Canada, Greenland, and Alaska to be more than that.

Pilot whales have been the target of a drive fishery for hundreds of years in the Faroe Islands, a Danish protectorate in the North Atlantic. In a process known as the *grindadrap*, pods of pilot whales are herded toward shore by small boats and then killed in the shallows or on the shore with spears and knives. The meat is distributed free to local inhabitants. The hunt has become the target of animal protection groups, who criticize the inhumane killing method, question the impact on pilot whale numbers, and challenge the need for the hunt in a society which is now relatively wealthy and westernized.

Hunting statistics date back to 1584, and largely unbroken records exist from 1709 to the present. Since then, the total recorded number of

---

[16]   *Id.* at 169.

[17]   *Report of the Workshop on Mortality of Cetaceans in Passive Fishing Nets and Traps*, 15 REP. INT'L WHALING COMMISSION. (Special Issue), 15 (1994).

whales killed is somewhat more than 240,000, with annual kills in 1998, 1999, and 2000 each less than 1,000.[18]

3.2. Conflicts with Fisheries

Almost without exception, conflicts between fisheries and marine mammals of all kinds take place whenever and wherever the two co-exist.[19] Individual animals may be shot or otherwise killed by fishermen who perceive them to be a threat to their livelihood. On occasion, attempts may even be made to eradicate or at least reduce local cetacean populations for the same reason. For instance, around 4,750 small cetaceans were killed at Iki Island in Japan from 1976 to 1982 by fishermen who believed the dolphins were responsible for declines in their catches of yellowtail.[20]

Around the world, cetaceans and other marine mammals become entangled in fishing nets and gear.[21] Sometimes, particularly in less-developed countries where food and money are at a premium, these incidentally caught cetaceans may be consumed or sold, which in turn may lead to a directed commercial hunt of cetaceans. For example, this was the impetus for the development of the commercial small cetacean hunts in Sri Lanka and Peru. Because a number of such interactions take place in parts of the world where there is little attempt to record such figures, the total number of cetaceans killed is unknown. However, it is safe to

---

[18]   Faroes Islands authorities no longer submit annual catch data to the International Whaling Commission. A full annual listing of all cetaceans killed in the Faroes since 1584 is available at <http://www.highnorth.no/statistik/faroe-whale.htm>.

[19]   For an overview *see, for example*, Simon P. Northridge & Robert J. Hofman, *Marine Mammal Interactions with Fisheries, in* CONSERVATION & MANAGEMENT OF MARINE MAMMALS 99–119 (John R. Twiss, Jr. & Randall R. Reeves eds., 1999).

[20]   For an overview, *see* Toshio Kasuya, *Fishery-dolphin conflict in the Iki Island area of Japan, in* MARINE MAMMALS & FISHERIES 253–72 (J.R. Beddington, R.J.H. Beverton & D.M. Lavigne eds., 1985).

[21]   For an overview, *see* the *Report of the Workshop on Mortality of Cetaceans in Passive Fishing Nets and Traps, supra* note 17, at 1–72.

say that, wherever coastal and, for some species, riverine gill nets are utilized, there are likely to be incidental catches of small cetaceans.

Among the documented instances of incidental catches are river dolphins in the Indus, Ganges, and Yangtze rivers; harbor porpoises and other species in gill-net fisheries off the coast of California; bottlenose dolphins and other species in shark nets off South Africa and Australia; Hector's dolphins (*Cephalorhynchus hectori*) off New Zealand; franciscana in Uruguay and Brazil; vaquita in the Gulf of California; and around 10,000 harbor porpoises (*Phocoena phocoena*) each year in the North Atlantic.

During the 1980s, there was considerable concern about the extremely high levels of mortality of a number of species in large-scale surface gill nets, widely known as drift nets. These nets often extended more than 50 kilometers in length. Japanese, Taiwanese, and Korean squid fisheries in the North Pacific were among the fisheries using these drift nets. Smaller drift nets were also used by, for example, French fishermen to catch albacore tuna in the North Atlantic and Italian fleets to catch swordfish in the Mediterranean. A 1991 review by the Food and Agriculture Organization (FAO) of the United Nations[22] catalogued the literally tens of thousands of small cetaceans—and millions of seabirds, non-target fish, and other marine wildlife—that were killed in these nets each year. The following year, the UN General Assembly passed a resolution[23] calling for a moratorium on the use of these nets. That moratorium has largely held, although smaller-scale drift nets continue to be widely used in coastal waters and there remains some concern over possible illegal use on the high seas.

But it is the purse-seine fishery for yellowfin tuna in the Eastern Tropical Pacific (ETP) that has been responsible for more cetacean deaths than any other human activity.[24] In this region, herds of certain dolphin species swim in association with schools of large yellowfin tuna.

---

[22]   Simon P. Northridge, Driftnet Fisheries and Their Impacts on Non-Target Species (1991).

[23]   U.N. Gen. Ass. Res. 46/215.

[24]   For a recent overview, *see* Michael L. Gosliner, *The Tuna-Dolphin Controversy, in* Twiss & Reeves, *supra* note 19, at 120–55.

(The reasons for this are not completely understood.) Beginning in the 1950s, tuna seiners used this association to their advantage by deliberately setting nets around dolphin herds and thus trapping both dolphins and tuna. Many dolphins were drowned or crushed in the nets themselves or in the winch that was used to haul the nets aboard. Close to seven million dolphins are estimated to have died in this way since 1959, with over a half million killed in both 1960 and 1961. Catches began declining substantially in 1972, after the U.S. Marine Mammal Protection Act took effect. Catches took a further downward turn following an agreement negotiated in 1992 by the Inter-American Tropical Tuna Commission (IATTC).[25] Today, approximately two to three thousand dolphins are killed annually. Many environmental campaigners acknowledge this substantial improvement over past years but still insist they will not rest until annual mortality is zero. Others, however, argue that dolphin mortality in the fishery is "no longer . . . biologically significant." They suggest that "we are finally approaching the point at which further reductions in dolphin mortality using traditional fishing methods are unlikely."[26]

3.3.   Pollution[27]

While there is still relatively little information on how much pollutants affect cetaceans, there is general agreement that various health-related impacts are likely occurring in some populations.[28] The majority of contaminant studies on marine mammals have focused on pinnipeds, not cetaceans; still, the highest levels of persistent organic pollutants (POPs), such as PCBs, in mammals are found in odontocetes, notably

---

[25]   *Id.* at 124, 151.

[26]   *Id.* at 151.

[27]   For an overview *see, for example*, Thomas J. O'Shea, *Environmental Contaminants and Marine Mammals, in* BIOLOGY OF MARINE MAMMALS 296–310 (John E. Reynolds & Sentiel A. Rommel eds., 1999).

[28]   *See generally*, William C. Burns, *From the Lance to the Laboratory: The Impact of Anthropogenic Environmental Degradation on Cetacean Species*, 10(2) SEA WIND 2–7 (1997).

St. Lawrence River beluga whales,[29] U.S. east coast bottlenose dolphins,[30] and orcas from the Puget Sound region in the Pacific northwest.[31]

Correlations between environmental contaminants and impacts on health-related components in small cetaceans have been noted or proposed for a number of species and populations, including bottlenose dolphins in the Gulf of Mexico and along the U.S. east coast,[32] St. Lawrence River beluga whales,[33] North Sea and Baltic Sea harbor porpoise,[34] North Pacific Dall's porpoise,[35] Mediterranean Sea striped dolphins,[36] and har-

---

[29]   Daniel Martineau *et al.*, *Levels of Organochlorine Chemicals in Tissues of Beluga Whales (Delphinapterus leucas) from the St. Lawrence Estuary, Quebec, Canada*, 16 ARCHIVE ENVTL. CONTAMINATION CANADA 137 (1987).

[30]   Joseph R. Geraci, *Clinical Investigation of the 1987–88 Mass Mortality of Bottlenose Dolphins along the U.S. Central and South Atlantic Coast*. Final Report to National Marine Fisheries Service, U.S. Navy (Office of Naval Research) and Marine Mammal Commission, at 36, 37 (1989).

[31]   Peter S. Ross *et al.*, *High PCB Concentrations in Free-Ranging Pacific Killer Whales, Orcinus orca: Effects of Age, Sex and Dietary Preference*, 40 MAR. POLLUTION BULL. 504 (2000).

[32]   K. Kannan *et al.*, *Elevated Accumulation of Tributyltin and its Breakdown Products in Bottlenose Dolphins (Tursiops truncatus) Found Stranded along the US Atlantic and Gulf Coasts*, 31 ENVTL. SCI. TOXICOLOGY 296 (1997).

[33]   Sylvain De Guise *et al.*, *Effects of In Vitro Exposure of Beluga Whale Leukocytes to Selected Organochlorines*, 55 J. TOXICOLOGY ENVTL. HEALTH 479 (1998); Sylvain De Guise, Andre Lagace & Pierre Beland, *Tumors in St. Lawrence Beluga Whales* (Delphinapterus leucas), 31 VETERINARY PATHOLOGY 444 (1994).

[34]   P.D. Jepson *et al.*, *Investigating Potential Associations between Chronic Exposure to Polychlorinated Biphenyls and Infectious Disease Mortality in Harbour Porpoises from England and Wales*, 244 SCI. TOTAL ENV'T. 339 (1999); U. Siebert *et al.*, *Potential Relation between Mercury Concentrations and Necropsy Findings in Cetaceans from German Waters of the North and Baltic Seas*, 38 MAR. POLLUTION BULL. 285 (1999).

[35]   A.N. Subramaniam *et al.*, *Reductions in the Testosterone Levels by PCBs and DDE in Dall's Porpoises of Northwestern North Pacific*, 18 MAR. POLLUTION BULL. 643 (1987).

[36]   A. Aguilar & A. Borrell, *Abnormally High Polychlorinated Biphenyl Levels in Striped Dolphins (Stenella coeruleoalba) Affected by the 1990–1992 Mediterranean Epizootic*, 154 SCI. TOTAL ENV'T 237 (1994).

bor porpoise in the Black Sea,[37] among many others. Definitive evidence for any cause-and-effect relationship between contaminants and the fitness of an odontocete population is, however, unlikely to be demonstrated, at least not in the near term. This is due largely to ethical and logistical considerations in marine mammal research and also to the substantial difficulties in understanding the dynamics of naturally chaotic populations and in disentangling the impacts of other stresses such as reduced forage, pathogens, climatic fluxes, algal toxins, and human disturbance. In addition, numerous uncertainties exist with regard to the interactive effects among chemicals, species-specific susceptibilities, and the potential for endocrine system disruption and other pathologies. Indeed, little is known about the actual toxicity of the vast majority of the chemicals released into the environment.

The proven, suspected, and potential impacts of chemical contaminants, along with the many uncertainties associated with them, have led many in the environmental protection and public health communities to propose a radically different approach for their use and regulation. This "precautionary approach" advocates, among other things, a focus on clean production, action even before scientific proof of deleterious effects, and a shift in the burden-of-proof (for example, of a chemical's safety) from the public to the proponent. Efforts by a variety of public interest groups and scientists have recently provided the impetus for an international agreement to ban or severely restrict the use of 12 globally pervasive and persistent organic pollutants,[38] all of which have been found in marine mammals. However, the persistence of the many highly toxic and bioaccumulating chemicals already in the environment, not to mention the still-ongoing pollution, suggests that high—and potentially dangerous—levels of environmental contaminants will continue to circulate in a host of marine mammal populations for the foreseeable future.

---

[37]   Shinsuke Tanabe *et al., Isomer-Specific Analysis of Polychlorinated Biphenyls in Harbour Porpoise (Phocoena phocoena) from the Black Sea*, 34 MAR POLLUTION BULL. 712 (1997).

[38]   Press release from the United Nations Environment Program (UNEP) available at <http://irptc.unep.ch/pops/princ5.htm>.

3.4. Environmental Change

Cetaceans potentially face an additional range of consequences from broader environmental change, ranging from habitat alteration and destruction to atmospheric impacts caused by ozone depletion and global climate change.

Because their ranges are limited and directly overlap areas of human habitation, river dolphins are considered most likely to be affected by habitat alteration. There are, for example, concerns that construction of the Three Gorges Dam on the Yangtze River will cause fundamental changes to the ecosystem and thus seriously impact the already-threatened baiji.[39] Dam construction and habitat alteration has already fragmented and diminished populations of Indian river dolphins.[40]

Marine species and populations with restricted coastal distribution may also be at particular risk from habitat change. The Chinese white dolphin (*Sousa chinensis*) is one notable example. Parts of its range have apparently been adversely affected by habitat loss caused by construction and increased shipping traffic.[41]

A wide variety of long-term changes to the marine environment may also be affecting cetaceans in ways that we are only now beginning to understand. Of increasing importance are the disturbances, particularly noise, associated with activities such as shipping, seismic exploration, geophysical surveys, dredging, mining, oil and gas drilling, military activities, and even climate monitoring programs (such as the Acoustic Thermometry of Ocean Climate, or ATOC). While it is known that undersea noise is escalating significantly and while the sources are relatively well understood, there is little information on rates of increase or the effects of this increase on marine wildlife.

---

[39]   A.R. Topping, *Ecological Roulette: Damming the Yangtze*, 74 FOREIGN AFF. 132 (1995).

[40]   Alison Smith, *The River Dolphins: The Road to Extinction, in* Simmonds & Hutchinson *supra* note 6, at 355, 376.

[41]   Nicola Kemp, *Habitat Loss and Degradation, in* Simmonds & Hutchinson, *supra* note 6, at 262, 267.

Although it is expected that mysticetes will suffer more than odonto-cetes from increases in ambient noise, because their hearing is more sensitive to the lower-frequency sounds generated by human activity, there are some indications that small cetaceans may also be at risk from the effects. A recent modeling exercise estimated that the noise generated by a Canadian Coast Guard icebreaker would be audible to beluga whales up to approximately 50 miles away. Some behavioral impacts would be expected within approximately that same range. Masking of beluga communication was predicted to occur within an approximately 9–44 mile range.[42] A previous study documented that beluga whales manifested noticeable avoidance behavior to icebreaker noise 30 miles away.[43]

Most dramatically, low-frequency active sonar (LFAS) was suggested as the likely cause of the atypical and fatal strandings of 12 Cuvier's beaked whales (*Ziphius cavirostris*) off Greece in 1998, an event that closely coincided with nearby testing of a NATO submarine-detection program.[44] The coincidence of whale strandings (including 18 beaked whales) and the presence of naval fleets has also been noted in three separate incidents around the Canary Islands during the 1980s.[45] More recently, U.S. Navy sonar used in an anti-submarine exercise appears to have been responsible for the strandings of some 17 cetaceans in the northern Bahamas over a two-day period in March of 2000.[46] At least six Cuvier's beaked whales and one Blainville's beaked whale (*Mesoplodon densirostris*) died, with autopsies showing injuries "consistent with an intense acoustic or pressure event."[47] Even sub-lethal

---

[42]   C. Erbe & D.M. Farmer, *Zones of Impact around Icebreakers Affecting Beluga Whales in the Beaufort Sea*, 108 J. ACOUSTIC SOC'Y AM. 1332 (2000).

[43]   Kerry J. Finley *et al.*, *Reactions of Belugas, Delphinapterus leucas, and Narwhals, Monodon monocerus, to Ice-Breaking Ships in the Canadian High Arctic*, 224 CANADIAN J. FISH. AQUATIC SCI. 97 (1990).

[44]   A. Frantzis, *Does Military Acoustic Testing Strand Whales?*, 392 NATURE 29 (1998).

[45]   M.P. Simmonds & L.F. Lopez-Jurado, *Whales and the Military*, 351 NATURE 448 (2000).

[46]   Anonymous, *Update of the Ongoing Investigation into the Stranding of Beaked Whales*, U.S. Navy Press Release, Nov. 15, 2000.

[47]   Anonymous, *Update on the Mass Stranding in the Bahamas*, 19/20 MMPA BULL. 3 (2000).

injury to odontocetes from intense sound could conceivably cause hearing impairment or complete hearing loss. Such an injury could reduce these animals' ability to detect prey, communicate, avoid predators and boats, or care for young animals.

An apparent increase in marine mammal *mass mortality* events (as distinguished from *mass strandings*, which appear to involve a social dynamic, and from mortalities due to *ice-entrapment*) over the last two decades is also of growing concern. While catastrophic mortality events involving marine organisms, including marine mammals, have been recorded with some frequency in the past, recent reviews have suggested that such events may now be occurring more frequently.[48] A range of marine mammal species from widely separated geographic areas have been affected; disease, algal biotoxins, or climatic factors are typically implicated.

A morbillivirus-related epizootic during 1987/88 reduced the inshore population of the U.S. east coast bottlenose dolphin (*Tursiops truncatus*) by over one-half.[49] A morbillivirus infection was also probably responsible for a die-off of this species in Texas waters in the beginning of 1994[50] and for the death of some 50 animals from a small resident population along Florida's east coast in 1982.[51] Unusually high numbers of dead bottlenose dolphins were also reported during early 1990 along

---

48   *See, for example*, Drew Harvell, *Emerging Marine Diseases—Climate Links and Anthropogenic Factors*, 285 SCIENCE 1505 (1999); Paul Epstein *et al.*, *Marine Ecosystems: Emerging Diseases as Indicators of Change*, HEALTH ECOLOGICAL AND ECONOMIC DIMENSIONS (HEED) OF GLOBAL CHANGE PROGRAM (1998); Ernest H. Williams & Lucy Bunkley-Williams, *Marine Major Ecological Disturbances of the Caribbean*, 2 INFECTIOUS DISEASE REV. 110 (2000).

49   Thomas P. Lipscomb *et al.*, *Morbilliviral Disease in Atlantic Bottlenose Dolphins from the 1987–1988 Epizootic*, 30 J. WILDLIFE DISEASE 567 (1994).

50   Amy Krafft *et al.*, *Postmortem Diagnosis of Morbillivirus Infection in Bottlenose Dolphins (Tursiops truncatus) in the Atlantic and Gulf of Mexico Epizootics by Polymerase Chain Reaction-Based Assay*, 31 J. WILDLIFE DISEASE 410 (1995).

51   Padraig J. Duignan *et al.*, *Morbillivirus Infection in Bottlenose Dolphins: Evidence for Recurrent Epizootics in the Western Atlantic and Gulf of Mexico*, 12 MAR. MAMMAL SCI. 499, 509 (1996).

the U.S. portion of the Gulf of Mexico. Adverse weather may have caused the deaths of some 23 animals in Matagorda Bay, Texas,[52] but the factor(s) for the bulk of the mortalities was never determined. Morbillivirus-related disease was ultimately responsible for the deaths of some 2,000 striped dolphins between 1990 and 1992 throughout the Mediterranean Sea[53] and probably responsible for the 47 common dolphins (*Delphinus delphis ponticus*) found dead along the northern Black Sea during an event in 1994.[54]

Morbillivirus-specific serum antibodies have now been found in various small cetacean populations, suggesting widespread infection. The fact that there have been no reported die-offs in these other populations suggests either herd immunity (*i.e.*, the disease is enzootic) or a disease that is less severe in some species. Alternatively, mass mortalities may indeed have occurred but were obscured by environmental conditions or simply went unnoticed because there were no human observers or recording equipment present. An overriding concern, however, is that, since viral outbreaks can cause heavy mortalities in a naïve population, some populations of small cetaceans—notably those with already reduced numbers—may be at serious risk from morbillivirus infection.

A series of mass mortalities affecting marine mammals and associated with algal toxins (*e.g.*, saxitoxin, brevetoxin, domoic acid) have occurred since the mid-1980s. Affected species include humpback whales off Massachusetts, manatee in Florida, California sea lions, and monk seals off Mauritania. It is suspected that deaths of bottlenose dolphins in the northeastern Gulf of Mexico in late 1999 were related to brevetoxin intoxication,[55] but evidence is circumstantial. Two mysterious

---

[52]   W.G. Miller, *An Investigation of Bottlenose Dolphin Tursiops truncatus Deaths East Matagorda Bay, Texas, January 1990*, 90 FISH B-NOAA 791 (1992).

[53]   William C. Burns, *The Agreement on the Conservation of Cetaceans of the Black Sea, Mediterranean Sea and Contiguous Atlantic Area (ACCOBAMS): A Regional Response to the Threats Facing Cetaceans*, 1(1) J. INT'L WILDLIFE L. & POL'Y 112, 115 (1998); A. Aguilar & A. Borrell, *The striped dolphin epizootic in the Mediterranean Sea*, 22 AMBIO 524 (1993).

[54]   A. Birkun *et al., Epizootic of Morbilliviral Disease in Common Dolphins (Delphinus delphis ponticus) from the Black Sea*, 144 VETERINARY REC. 85 (1999).

[55]   *See for example*, Victor Hull, *Red Tide Suspected in Panhandle Dolphin*

mass mortality events involving dolphins, sea lions, seabirds, and other species also occurred in the Gulf of California during the 1990s. These two events are considered by some to be related to algal poisoning,[56] although little evidence has been presented. The potential of the inter-action of marine mammals with algal toxins (via their food) may well be increasing in some areas in tandem with what is widely perceived as a general increase in the frequency, intensity, and geographic distribu-tion of harmful algal blooms, notably along developed coastlines, over recent years.[57] Nutrient enrichment,[58] alterations in freshwater influxes to coastal waters,[59] and ballast water introductions[60] are important human-related factors in some regions, while climate change may play an enhanced role in the future.[61]

4.   STATUS OF SOME SMALL CETACEAN POPULATIONS

The total number of small cetaceans worldwide is difficult to assess, and it is in many cases even more difficult to determine trends or ascer-

---

Deaths; More than 100 Bottlenose Dolphins Have Died in the Gulf since August, SARASOTA HERALD-TRIBUNE, Jan. 11, 2000, at 1; *Annual Report to Congress 1999,* Marine Mammal Commission 150 (2000).

[56]   J.L. Ochoa *et al., Toxic Events in the Northwest Pacific Coastline of Mexico during 1992–1995: Origin and Impact,* 352 HYDROBIOLOGIA 195 (1997).

[57]   *See, for example,* G.M. Hallegraeff, *A Review of Harmful Algal Blooms and Their Apparent Global Increase,* 32 PHYCOLOGIA 79 (1993); Ted J. Smayda, *Novel and Nuisance Phytoplankton Blooms in the Sea: Evidence for a Global Epidemic, in* TOXIC MARINE PHYTOPLANKTON 29 (Edna Graneli *et al.* eds., 1990).

[58]   Joanne M. Burkholder & Howard B. Glasgow, *Pfiesteria piscicida and other Pfiesteria-like Dinoflagellates: Behavior, Impacts, and Environmental Controls,* 42 LIMNOLOGY & OCEANOGRAPHY 1052 (1997); H.W. Paerl & D.R. Whitall, *Anthropogenically-derived Atmospheric Nitrogen Deposition, Marine Eutrophica-tion and Harmful Algal Bloom Expansion: Is There a Link?,* 28 AMBIO 307 (1999).

[59]   C. Humborg *et al., Effect of Danube River Dam on Black Sea Bioge-ochemistry and Ecosystem Structure,* 386 NATURE 385 (1997).

[60]   Gustav M. Hallegraeff, *Transport of Toxic Dinoflagellates via Ships' Ballast Water: Bioeconomic Risk Assessment and Efficacy of Possible Ballast Water Management Strategies,* 168 MARINE ECOLOGY PROGRESS SERIES 297 (1998).

[61]   Patricia A. Tester, *Climate Change and Marine Phytoplankton,* 2 ECOSYS-TEM HEALTH 191 (1996).

tain the status of a particular species or population. Of 68 species of small cetaceans (*i.e.*, all odontocetes except the sperm whale) listed in the IUCN Red Data Book on Cetaceans, published in 1991,[62] the status of fully 63 was classified as "insufficiently known." The other five were four species of river dolphin (the Ganges and Indus River dolphins listed as separate species, the former classified as vulnerable and the latter endangered); the baiji, listed as endangered; and the boto, classified as vulnerable; and the vaquita, listed as endangered. If there is any consensus regarding the status of small cetaceans, it is that the vaquita is probably the most endangered. This species is apparently restricted in range to the upper Gulf of California, and a recent survey estimated its population in this region at 567 animals.[63] Unknown numbers are caught accidentally each year in nets set by fishers targeting totoaba; even low levels of capture could be sufficient to push this species toward extinction.[64]

With respect to the other porpoise species, there is little information on status and abundance of the Burmeister's porpoise (*Phocoena spinipinnis*), spectacled porpoise (*Australophocoena dioptrica*), or finless porpoise (*Neophocoena phocaenoides*), although there is some concern that the finless porpoise population may have diminished throughout large parts of its range. The harbor porpoise and Dall's porpoise are generally considered abundant, but concern has been expressed over the status of some specific populations. This concern centers on, in the case of the harbor porpoise, incidental catches in the North Atlantic and, with respect to the Dall's porpoise, entanglement in drift-net fisheries in the North Pacific, along with direct hunting in Japanese coastal waters.

Of the river dolphin species, the most critically endangered is probably the baiji, as a result of a wide range of human activities including pollution, hunting, disturbance, conflicts with fisheries (especially entanglement in fishing gear), and habitat loss (particularly as a result of dam

---

[62]   MARGARET KLINOWSKA, DOLPHINS, PORPOISES AND WHALES OF THE WORLD: THE IUCN RED DATA BOOK (1991).

[63]   A.M. Jaramillo-Legoretta, L. Rojas-Bracho & T. Gerrodette, *A New Abundance Estimate for Vaquitas: First Step for Recovery*, 15(4) MAR. MAMMAL SCI. 957 (1999).

[64]   William F. Perrin, *Selected Examples of Small Cetaceans at Risk, in* Twiss and Reeves, *supra* note 19, at 296.

and barrage construction). Given the fact that it shares the Yangtze river drainage with approximately 10 percent of the human population, it has been written that the most remarkable thing about the baiji "is that it still exists at all."[65] The most optimistic estimates place its numbers in the low hundreds, and "it seems unlikely that it will survive far into the [21st] century."[66]

Of 29 recognized populations of beluga in the Arctic and sub-Arctic, 11 are considered depleted. One of these, in Cook Inlet, Alaska, is continuing to decline; one, in West Greenland, has been reduced by approximately 60 percent since 1981; and one, in Ungava Bay in the Canadian Arctic, is close to extirpation.[67]

Very little is known about the distribution and abundance of bottlenose and beaked whales, but the fact that some species, such as the Cuvier's beaked whale, strand with relative frequency suggests that they are more numerous than the paucity of sightings would indicate. The northern bottlenose whale was hunted intensively by Norwegian and, to a lesser extent, Scottish and Canadian whalers in the early part of the 20th century; as a result, it is possible that some stocks may be depleted.[68]

As mentioned earlier, the delphinidae contains more species than any other small cetacean family. Therefore it is not surprising that it contains the highest number of populations that may have been impacted by human activities.

Two populations, the northern form of the spotted dolphin (*Stenella attenuata*) and the eastern form of the spinner dolphin (*S. longirostris*), accounted for 80 per cent of the total mortality in the Eastern Tropical Pacific purse-seine fishery for yellowfin tuna. It is likely that these stocks were significantly reduced in the early years of the fishery, although more recent analysis suggests that, while probably depleted, the popu-

---

[65]   *Id.* at 298.

[66]   *Id.* at 300.

[67]   IWC 51/4, *supra* note 13, at 16.

[68]   STEPHEN LEATHERWOOD & RANDALL R. REEVES, THE SIERRA CLUB HANDBOOK OF WHALES AND DOLPHINS 115 (1983).

lations are now apparently stable.[69] On the other hand, the northern stock of common dolphins, also targeted by the fishery, has shown signs of significant recent decline.[70]

The IWC Scientific Committee has on several occasions expressed concern over the status of the striped dolphin in Japanese waters in the face of directed hunts there.[71] Dusky dolphin (*Lagenorhynchus obscurus*) numbers may be being impacted by directed takes off Peru.[72] Peale's dolphins (*Lagenorhynchus australis*) had become extremely rare in certain parts of the Magellan region by the late 1980s, presumably as a result of their being hunted for use as crab bait in Chile; however, indications suggest they are now returning to the region.[73]

Some populations of several species, including the northern right whale dolphin (*Lissodelphis borealis*), striped dolphin, common dolphin, and Pacific white-sided dolphin (*Lagenorhynchus obliquidens*), may have been significantly depleted by past mortality in high-seas driftnet fisheries.

The Irrawaddy dolphin (*Orcaella brevirostris*) may be in decline in parts of its range, possibly as a result of a wide combination of factors, including incidental mortality in fishing gear, shooting by soldiers and villagers, the use of explosives to catch fish, and general habitat deterioration as a result of the Vietnam war.[74]

As previously noted, species with limited range are especially vulnerable to human impacts. For example, the entanglement of Hector's

---

[69]   *Report of the sub-committee on small cetaceans*, 42 REP. INT'L WHALING COMMISSION, 172, 215 (1992).

[70]   *Report of the sub-committee on small cetaceans*. 43 REP. INT'L WHALING COMMISSION 130, 140 (1993).

[71]   *Report of the sub-committee on small cetaceans, supra* note 9, at 136.

[72]   *Report of the sub-committee on small cetaceans*, 47 REP. INT'L. WHALING. COMMISSION 169, 177 (1997).

[73]   *Id.* at 178.

[74]   *Report of the sub-committee on small cetaceans*, 44 REP. INT'L WHALING COMMISSION 108 (1994).

dolphins in fishing nets in New Zealand waters is apparently causing at least two of the species' three populations to decline.[75]

5.   MANAGEMENT: LEGISLATION, TREATIES, AND CONVENTIONS

Numerous legislative measures have been promulgated that, specifically or in part, address the threats facing small cetaceans; others are continually proposed. There is at present no overarching international convention or agreement that solely and specifically covers small cetacean species, although some observers have argued for one. Rather, the initiatives that address small cetacean species and issues can generally be divided into three principal types: *geographically based* international and regional treaties, which may consider small cetaceans as one aspect of a broader mandate; *issue-specific*, which may be national, regional, or global in scope, and which address topics such as overfishing; and *species-specific*, such as national sanctuaries and legislation to protect marine mammals.

5.1.   Geographically Based Initiatives

Cetaceans are mentioned explicitly in Article 65 of the United Nations Convention on the Law of the Sea (UNCLOS), which went into effect in 1994. This is the broadest of all treaties pertaining to marine issues. This article proclaims that states shall "cooperate with a view to the conservation of marine mammals and in the case of cetaceans shall in particular work through the appropriate international organizations for their conservation, management and study."

This point is affirmed in Chapter 17 of Agenda 21, adopted at the Rio Earth Summit in 1992; however, although Agenda 21 explicitly identified the International Whaling Commission as the appropriate authority for large whales, no such authority is identified for small cetaceans.[76]

---

[75]   K.K. Martien *et al.*, *A Sensitivity Analysis to Guide Research and Management for Hector's Dolphin*. 90(3) BIOLOGICAL CONSERVATION 182–91 (1999).

[76]   Cassandra Phillips, *Conservation in Action: Agreements, Regulations, Sanctuaries and Action Plans, in* Simmonds & Hutchinson, *supra* note 6, at 447, 451.

The 1979 Bonn Convention on the Conservation of Migratory Species of Wild Animals[77] is a "global convention that has also enabled subsidiary regional conservation agreements to be adopted."[78] Two of these agreements cover small cetaceans: the 1992 Agreement on the Conservation of Small Cetaceans of the Baltic and North Seas (ASCOBANS)[79] and the 1996 Agreement on the Conservation of Cetaceans in the Black, Mediterranean, and Contiguous Atlantic Area (ACCOBAMS).[80] Both these agreements commit the signatories to maintaining favorable conservation status for small—and, in the case of ACCOBAMS, all—cetacean species and populations in the region. Among other steps, parties are to work toward preventing the release of pollutants that could threaten cetaceans; reducing and eliminating incidental mortality in fishing gear; reducing the impact of human activities on cetaceans' food sources; preventing other significant environmental disturbances; and passing legislation to prohibit the intentional killing of small (or all) cetaceans.[81] Both agreements are examined in more detail elsewhere in this volume.

Also within European waters, the 1979 Berne Convention on Conservation of European Wildlife and Natural Habitats[82] charges signatories with conserving wild flora and fauna and their natural habitats, especially species whose conservation requires the cooperation of several states and particularly migratory species.[83] These migratory species, including all small cetaceans in the North and Baltic Seas, are listed in Appendix II of the Convention.[84]

---

[77]   19 I.L.M. 15 (1980).

[78]   Gregory A. Rose, *International Law and the Status of Cetaceans, in* Simmonds & Hutchinson, *supra* note 6, at 23, 27.

[79]   *Reprinted in* II THE MARINE MAMMAL COMMISSION COMPENDIUM OF SELECTED TREATIES, INTERNATIONAL AGREEMENTS, AND OTHER RELEVANT DOCUMENTS ON MARINE RESOURCES, WILDLIFE AND THE ENVIRONMENT 1612 (1994).

[80]   36 I.L.M. 777 (1997).

[81]   *Id.*

[82]   Available online at: <http://www.ecnc.nl/doc/europe/legislat/bernconv.html>.

[83]   *Id.* at 48.

[84]   Available online at: <http://www.ecnc.nl/doc/europe/legislat/bernapp2.html>.

In addition, member states of the European Union are bound by the Habitats Directive,[85] which requires members to establish measures to protect listed species, including all cetaceans, in their natural range.

Such broad-based regional and international agreements enable signatories to address a wide variety of issues—ranging from pollution control to habitat preservation and the prohibition of directed takes—within one treaty or directive. Although small cetaceans (or any other species) may be explicitly listed, they are not singled out for special attention. Rather, the focus is on maintaining the quality of the overall environment, with the goal that small cetaceans, along with all other species, will benefit accordingly.

5.2.   Issue-Specific Initiatives

These agreements and initiatives benefit small cetaceans by addressing specific issues which, directly or indirectly, may threaten them. So, in a sense, this category actually includes all broad environmental agreements on issues that may affect the marine environment, such as the 1987 Montreal Protocol on Substances that Deplete the Ozone Layer[86] and the 1998 Kyoto Protocol on Climate Change.[87] Of greater relevance to this discussion, however, are agreements and treaties that focus on matters of immediate import to small cetacean populations, even if protection of small cetaceans is not their sole objective. This particularly applies to restrictions on fishery practices, such as the 1989 Wellington Convention for the Prohibition of Fishing with Long Driftnets in the South Pacific Ocean[88] or the subsequent United Nations global moratorium on drift-net fishing on the high seas (see Section 3.2).

The Inter-American Tropical Tuna Commission (IATTC)[89] is the body responsible for regulating the purse-seine fishery of yellowfin tuna

---

[85]   Directive 92/43/EEC, 1992 O.J. (L 206) 7.

[86]   26 I.L.M. 1541 (1987), available online at <http://www.tufts.edu/departments/fletcher/multi/texts/BH906.txt>.

[87]   Kyoto Protocol to the United Nations Framework Convention on Climate Change, FCCC/CP/1997/L.7/Add. 1, Dec. 10, 1997.

[88]   Available online at <http://eelink.net/~asilwildlife/southpacific.html>.

in the Eastern Tropical Pacific (ETP). As mentioned previously, this fishery has been the single greatest source of human-caused mortality of small cetaceans (see Section 3.2). However, for many years, the IATTC refused to accept any responsibility for the dolphin kills or to acknowledge the need to take measures to reduce dolphin mortality. Concerted pressure from environmental NGOs and unilateral trade embargoes by the United States on nations whose subjects engage in tuna fishing operations and do not use "dolphin-safe" methods led the Commission's members to work toward an agreement to reduce dolphin mortality in their nets. The result was the 1992 Agreement to Reduce Dolphin Mortality in the Eastern Tropical Pacific Tuna Fishery. Under this agreement, annual national limits on dolphin mortality get progressively lower, with a dolphin mortality limit assigned to each registered vessel over 400 tons. Each vessel carries an observer and must cease fishing for the season when its dolphin mortality limit is reached.[90]

The largest issue-specific convention of particular relevance to small cetacean issues is the Convention on International Trade in Endangered Species (CITES).[91] As its name suggests, CITES regulates cross-border trade in species of flora and fauna and prohibits commercial trade in species listed on its Appendix I. For small cetaceans, these species are the baiji, Indian river dolphins, the *Hyperoodon* and *Berardius* species of beaked and bottlenosed whales, the *Sotalia* and *Sousa* species of delphinids, and two species of porpoise: the vaquita and the finless porpoise (*Neophocaena phocaenoides*). All other cetaceans are featured on Appendix II of the Convention, which allows limited trade but only under strict controls.

5.3.   Species-Specific Initiatives

Most agreements, initiatives, and laws created purely for the protection of cetaceans or marine mammals are strictly national in nature.

---

[89]   *See* the IATTC site online at <http://www.iattc.org/>.

[90]   *Id.*

[91]   12 I.L.M. 1088 (1973), available online at <http://sedac.ciesin.org/pidb/texts/cites.trade.endangered.species.1973.html>.

Countries adopt these agreements at the demand of their populace. These citizens have emotional and moral reasons (often at least as strong as environmental reasons) for their concern for marine mammals and their objection to their deaths.

These agreements take several forms. Several countries have declared specific sanctuaries and protected areas in their waters partly or wholly to protect local cetacean populations,[92] while many others have banned the hunting and killing of cetaceans, large and small, throughout their 200-mile Exclusive Economic Zones (EEZs).[93]

A number of nations have adopted national laws protecting marine mammals in general or cetaceans in particular, such as the 1978 Marine Mammal Protection Act of New Zealand and the 1980 Whale Protection Act of Australia.[94] The model remains the 1972 Marine Mammal Protection Act of the United States,[95] the seminal achievement of which probably remains the effective end of dolphin mortality by U.S.-flagged tuna seiners in the Eastern Tropical Pacific.

5.4.   Small Cetaceans and the International Whaling Commission

The IWC plays a significant role in the management and conservation of cetaceans, and many environmental organizations have advocated that it increase its efforts to protect small cetacean populations.

As noted at the beginning of this chapter, the very terminology and definition of "small cetacean" derives to a large extent from the exclusion of certain species from the IWC's initial Annex of Nomenclature. Some countries argue that any originally excluded species are ipso facto beyond the Commission's purview. These countries fall into two categories. One category includes those that actually hunt small cetaceans.

---

[92]   For a partial listing, *see* Cassandra Phillips, *supra* note 76, at 458–59.

[93]   *Id.* at 459.

[94]   Available online at <http://austlii.law.uts.edu.au/au/legis/cth/consol_act/wpa1980190.txt>.

[95]   16 U.S.C. § 1361, available online at <http://www.nmfs.noaa.gov/prot_res/laws/MMPA/MMPA.html>.

The other category includes countries that are reluctant to, as they see it, surrender *any* sovereignty over their EEZs to an international body, and this includes authority over the coast-dwelling small cetaceans within their EEZs.[96]

The legal rationale for excluding species purely because they were not initially listed is spurious at best. Former IWC Secretary Ray Gambell has observed that, essentially, it is up to the IWC itself to decide which species it is and is not competent to regulate.[97] Indeed, the Commission has already broadened its mandate once by adding a previously unlisted species: at the 29th Annual Meeting in 1977, a definition of the orca was added to the definitions of toothed whales under Section I of the Schedule to the ICRW. Then, at the 32nd Meeting, the orca was added to the list of species included in a provision imposing a moratorium on whaling with factory ships. This clearly demonstrates the capacity of the Commission to extend the scope of its regulations to species not included in the original Annex of Nomenclature.

However, aside from any issues of legal competence, it is not clear that the IWC is the *ideal* forum for addressing small cetacean issues. Environmentalists typically approach the Commission as if it were, to use the categories employed in this chapter, a *species-specific* body: in other words, an agreement convened to protect, conserve, and manage cetaceans. In fact, that is at best only partly true. It is really an *issue-specific* agreement: a fisheries convention designed to regulate an industry that happens to involve the killing of whales. The ICRW itself makes this clear in its preamble: "Recognizing . . . that increases in the size of whale stocks will permit increases in the *number of whales which may be captured*. . . . Having decided to conclude a convention for the proper conservation of whale stocks *and thus make possible the orderly development of the whaling industry*[.]"[98]

---

[96]   *See* Alexander Gillespie *Small Cetaceans, International Law and the International Wahling Commission*, 2(2) Melbourne J. Int'l L. 257–300 (2001). William C. Burns, *The International Whaling Commission and the Regulation of the Consumptive and Non-Consumptive Uses of Small Cetaceans: The Critical Agenda for the 1990s*, 13 WIS. INT'L L.J. 105, 128 (1994).

[97]   Kieran Mulvaney, *A Time of Transition*, 1 INT'L WHALE BULL. 3 (1987).

[98]   Preamble, International Convention for the Regulation of Whaling, Dec. 2, 1946 [emphasis added].

Even within that definition, there are some small cetacean issues that are clearly within the commission's remit, assuming its members can ever agree on competence: the hunt of the Baird's beaked whale, the ship-based hunts of pilot whales and Risso's dolphins in Japanese waters, and perhaps commercial-drive, hand-harpoon, and net hunts of small cetaceans, such as those conducted in the coastal waters of Japan, Peru, Chile, and Sri Lanka.

And, indeed, the IWC itself on many occasions has, if not exactly broadened its mandate, then certainly stretched it in ways that have benefited small cetaceans. In 1974, the IWC Scientific Committee established a Subcommittee on small cetaceans; this body was given the responsibility of establishing a clearer taxonomy of these species and identifying their conservation and research needs. In 1975, the Scientific Committee suggested that the Commission should consider managing the populations of some species of small cetaceans; one year later, the Subcommittee on Small Cetaceans recommended that the Commission establish a comprehensive framework to regulate the taking of small cetaceans. The Committee remains the primary scientific forum for, and source of, easily accessible information on the status of small cetaceans and the threats they face. Furthermore, on several occasions the Commission has expressed its opinion on small cetacean issues, sometimes to great effect. In 1990, for example, resolutions highly critical of the increasing size of Japan's Dall's porpoise hunt[99] werewwerewere arguably instrumental in persuading that country to reduce its annual kill significantly over the course of several years.[100]

However, the fact remains that most small cetacean kills take place in countries that are not members of the Commission and have no particular incentive to become members. Furthermore, legislation and critical resolutions at the national level may not by themselves be sufficient to bring such hunts under control. For example, dolphin hunts are apparently proceeding apace in Peru despite the existence of strict national laws prohibiting them. Additionally, many such hunts take place in less developed countries. These hunts are the result of necessity and des-

---

[99]   Resolution on the directed take of Dall's porpoises, available online at <http://eelink.net/~asilwildlife/42iwc.PDF>.

[100]   Mulvaney, *supra* note 8, at 101.

peration caused by complex socio-economic factors, including poverty, rapidly growing populations, and the unreasonable exploitation of marine resources by industrialized fishing fleets, frequently foreign-owned. Under such circumstances, protection of small cetaceans and other wildlife frequently takes a back seat to simple survival, an understandable situation that will take more than the disapproval of environmentalists and industrialized nations to resolve. And although the Commission and the Scientific Committee are well placed to advise on the impacts of commercial fisheries, chemical pollution, or climate change on cetaceans, it is powerless to address the causes.

6. CONCLUSION

At the dawn of the 21st century, knowledge of the current population status and the population trends of many small cetaceans remains woefully inadequate. It seems likely that most species are in a healthier state than many of their larger brethren, of which a number of populations remain highly endangered after decades and even centuries of commercial whaling.[101] Even so, several populations of small cetaceans are apparently greatly depleted, others are under threat, and the long-term survival of at least two species—the vaquita and the baiji—is uncertain. Most of the identified threats are the result of directed hunts and indirect takes in fisheries; however, less tangible are the current and potential impacts of broader environmental change.

Although many environmentalists have called for the International Whaling Commission to assume responsibility for management of small cetacean issues, or for the development of a new global convention dedicated to small cetaceans, such an approach can address only some of the problems facing small cetaceans worldwide. What is required instead is a multi-faceted approach that combines measures protecting small cetaceans' environment on a regional basis with measures tackling the numerous human activities that impact small cetaceans directly and indirectly (of which directed hunts or indirect takes are only a part). It also requires appreciating that the problems facing small cetaceans are not

---

[101] *See, for example*, Phil Clapham *et al., Baleen Whales: Conservation Issues and the Status of the Most Endangered Populations*, 29(1) MAMMAL REV. 35–60.

only matters of concern in their own right; they are symptomatic of broader environmental change and require understanding that complex social/cultural, political, and economic factors drive that change. By doing so, those concerned for the future of small cetaceans cannot only place those issues in context, but they also increase the chances of solving the threats some small cetacean populations now face.

Chapter 9
Small Cetaceans, International Law and the International Whaling Commission

Alexander Gillespie[1]

[Small cetaceans (small whales) represent something of an anomaly in international law, as it is not clear which international or regional organizations have primary competence over their management. It is my contention that the International Whaling Commission has a broad authority in relation to small cetaceans and clear primacy over all competing organizations on this question. This primacy is given further credence when considering small cetaceans which are either migratory or endangered.]

[I]t is essential to protect *all* species of whales from further overfishing.[2]

1. INTRODUCTION

Small cetaceans are currently struggling for adequate protection in international law. This is despite being on the agenda of a number of international fora since the early 1970s. In 1992 the primacy of the International Whaling Commission (IWC) with regard to cetaceans was clearly stated in the United Nations Environment Programme's Agenda 21.[3] However, in recent years a number of countries have argued that the IWC should not have jurisdiction over these creatures. The primary argument in support of this contention is that small cetaceans were not included in the

---

[1]   Lecturer, International Law, University of Waikato, Hamilton, New Zealand. This chapter is adapted from an article published in the Melbourne Journal of International Law.

[2]   International Convention for the Regulation of Whaling, opened for signature Dec. 2, 1946, 161 U.N.T.S. 72, sch I(1)(a), preamble (entered into force Nov. 10, 1948) (emphasis added).

[3]   United Nations Environment Programme, Agenda 21, <http://www.unep.org/Documents/Default.asp?DocumentID=52>, site visited on Sept. 9, 2001.

original nomenclature annexed to the International Convention for the Regulation of Whaling (ICRW),[4] the convention which established the IWC, and that to add them subsequently to the nomenclature would require the consent of all the signatories to the ICRW.

I believe this view is mistaken for a number of reasons. The first reason relates to the actual language of the ICRW and the simplistic view that the nomenclature was somehow a pivotal dividing mechanism within it. Secondly, there is a general misunderstanding of the way in which treaties evolve and change, and the mechanisms that allow entrenched majorities to modify annexes within the broad framework of their original treaties (that is, it is *not* an amendment of the treaty).

Closely aligned to the argument that the IWC does not have competence over small cetaceans is the assertion that coastal states have near absolute competence. This contention is typically bolstered by invocation of the United Nations Convention on the Law of the Sea (UNCLOS).[5] However, I contend that this view is also mistaken as UNCLOS does not accord coastal states complete sovereignty in their exclusive economic zones (EEZs), rather it affords them limited sovereignty. It is limited in the sense that nations can waive their rights under UNCLOS and freely consent to the authority of overlapping international organizations such as the IWC. Under UNCLOS, even if a nation does not wish to join the IWC, it is still necessary to cooperate with the relevant overlapping international organizations when dealing with critically endangered species, migratory species, and more specifically, cetaceans.

2.   DEFINING "SMALL CETACEANS"

Whales, dolphins, and porpoises are known collectively as whales—or to be more precise, as cetaceans. The word "cetacean" is derived from the Latin *cetus* (large sea animal) and the Greek *ketos* (sea monster). At least 79 species of cetaceans are currently recognized.[6] Broadly, the

---

4   *Id.*

5   Opened for signature Dec. 10, 1982, 1833 U.N.T.S. 3, 21 I.L.M. 1261 (entered into force Nov. 16, 1994).

6   MARK CARWARDINE, WHALES, DOLPHINS AND PORPOISES: THE VISUAL GUIDE TO ALL OF THE WORLD'S CETACEANS 6, 10 (1995).

order cetacean can be subdivided into large and small. However, trying to define exactly which species are contained within each category is a difficult task, as there are no accepted definitions that encompass both biological and political considerations.

Small cetaceans are a focus of concern in a number of international conventions, although the reason they are listed is not usually because they are "small." Rather, they are listed because they happen to possess certain characteristics that fall within various relevant categories: for example, they may migrate,[7] they may be endangered,[8] or they may inhabit a certain region.[9] However, such characteristics do not describe a small cetacean according to their physical characteristics.

The definition of a small cetacean is usually based on the false assumption that small cetaceans are limited in size. Dolphins (usually of the *Delphinidae* and *Ziphiidae* families) and porpoises (usually of the *Phocoenidae* family) are typically placed in this camp. However, this type of classification is simplistic and soon falls apart as consideration turns to the multitude of other species, which are clearly nowhere near the size of the larger whales, but which possess some of the characteristics of the smaller whales. That is, while it is true that some species of small cetaceans are no more than 1.2–1.5 meters in length,[10]

---

[7]   They are often listed in the Convention on the Conservation of Migratory Species of Wild Animals, opened for signature June 23, 1979, 1651 U.N.T.S. 333, 19 I.L.M. 15, art. I (entered into force Nov. 1, 1983) (CMS Convention): "Migratory species means . . . species [which] cross one or more jurisdictional boundaries." Hence, a number of small cetaceans are listed within the CMS Convention directly. UNCLOS also lists a number of migratory species, in which no distinction is made between large and small whales: UNCLOS, *supra* note 5, annex 1.

[8]   *See* Convention on International Trade in Endangered Species of Wild Fauna and Flora, opened for signature Mar. 3, 1973, 993 U.N.T.S. 243, 12 I.L.M. 1085 (entered into force July 1, 1975) (hereinafter CITES Convention).

[9]   *See* Agreement on the Conservation of Cetaceans of the Black Sea, Mediterranean Sea and Contiguous Atlantic Area, opened for signature Nov. 24, 1996, 36 I.L.M. 777 (entered into force June 1, 1997) (hereinafter ACCOBAMS) that covers all whales, including small cetaceans. For an overview of the agreement, *see also* William C.G. Burns, *The Agreement on the Conservation of Cetaceans of the Black Sea, Mediterranean Sea and Contiguous Atlantic Area*, 1(1) J. INT'L WILDLIFE L. & POL'Y 113–24 (1998).

[10]   These include the smallest cetaceans, such as the Hector's dolphin found

other small cetaceans may reach close to 10 or 13 meters in length.[11] In itself, this creates problems as it is generally accepted that the IWC has the authority to regulate minke whales which are deemed "large," but which are physically smaller (at 7–10 metres) than Baird's beaked whales, which are deemed "small."[12]

Due to such difficulties, marine biologists have sought more definitive biological yardsticks to establish the division between the various suborders of cetaceans—which typically correspond to the groupings of large and small. The first suborder is the *Odontoceti*, which includes individuals that have teeth (generally of only one kind) and an asymmetrical skull. This suborder is comprised of the families *Iniidae, Lipotidae, Pontoporiidae, Platanistidae, Delphinidae, Phocoenidae, Monodontidae, Ziphiidae* and *Physeteridae*.[13] The second suborder is the *Mysticeti*, which have plates of baleen[14] instead of teeth, and a symmetrical skull. The families *Eschrichtidae, Balaenopteridae, Balaenidae* and *Neobalaenidae* are included in this grouping.[15]

---

off New Zealand or the black dolphin found off Chile: CARWARDINE, *supra* note 6, at 31.

[11]   These include the killer whale (at 5.5–9.8 meters) and Baird's beaked whale (at 10.7–12.8 meters): CARWARDINE, *supra* note 6, at 34–5. In 1982 St. Lucia argued that the Commission should regulate catches of Baird's beaked whale "because the species is larger than the minke and killer whale whales [sic] already regulated," IWC, *Thirty-Third Report of the International Whaling Commission* 28 (1983).

[12]   In 1986 the Netherlands, UK, India and Sweden utilized this argument in suggesting that the IWC does have competence over small cetaceans, such as Baird's beaked whale "since . . . it is a larger animal than the minke whale regulated by the IWC," IWC, *Thirty-Sixth Report of the International Whaling Commission* 14 (1986).

[13]   CARWARDINE, *supra* note 6, at 12.

[14]   Baleen/baleen plates are "[c]omb-like plates hanging from the upper jaw of many large whales, used to strain small prey from sea water (also known as "whalebone")." *Id*. at 250.

[15]   *See also* RONALD NOWAK, II WALKER'S MAMMALS OF THE WORLD 969–70 (5th ed., 1991). *See also* LEONARD HARRISON-MATTHEWS, THE NATURAL HISTORY OF THE WHALE 23–48 (1978).

The IWC did not originally base its work on such biological distinctions. Indeed, the fact that there was no such dividing line between toothed and baleen whales in the original nomenclature that was annexed to the ICRW has been found to be the source of much controversy.[16] In fact, the nomenclature mixes toothed and baleen whales, as well as listing typically small cetaceans such as the northern and southern bottlenose dolphins of the *Hyperoodon ampullatus* and *Hyperoodon planifrons* species.[17] However, the Interpretation section of the Schedule to the ICRW (which is amended from time to time)[18] does work upon an assumption that baleen whales[19] and toothed whales[20] are different suborders.

Unfortunately, the debate over whether these categories translate into large and small cetaceans does not end there. The difficulty is that many of those species listed as having teeth are often very large in size and are typically thought of as large cetaceans. That is, species such as the sperm whale (and all genera within) possess teeth. Moreover, this species, which can mature to a length of 18 meters, has never been thought of as small in any sense of the word within the IWC.[21] With such considerations in mind, the Agreement on the Conservation of Small Cetacean of the Baltic and North Seas (ASCOBANS) defines

---

[16] ICRW, *supra* note 2, sch. I.

[17] Final Act of the International Whaling Conference, annex, reprinted in PATRICIA BIRNIE, INTERNATIONAL REGULATION OF WHALING: FROM CONSERVATION OF WHALING TO CONSERVATION OF WHALES AND REGULATION OF WHALE-WATCHING 701 (1985).

[18] ICRW, *supra* note 2, sch. I.

[19] "Baleen whale means any whale which has baleen or whale bone in the mouth, ie any whale other than a toothed whale" *id.*, sch. I(1)(A). *See also* CARWARDINE, *supra* note 6, at 250: Baleen whales are a "[s]ub-order of whales with baleen plates instead of teeth; scientific term Mysticeti, from the Greek *mystax*, meaning moustache, and *cetus*, meaning whale."

[20] "Toothed whale means any whale which has teeth in the jaws," ICRW, *supra* note 2, sch. I(1)(B).

[21] The sperm whale was included in the "large baleen whale" category when fishing quotas were initially set for this species. It was given earlier attention than the physically smaller minke whale, and is recognised as a toothed whale, *id.*

small cetaceans as: "Any species, subspecies or population of toothed whales *Odontoceti*, except the Sperm whale *Physeter macrocephalus*."[22]

In the Agreement on the Conservation of Cetaceans of the Black Sea, Mediterranean Sea and Contiguous Atlantic Area (ACCOBAMS) cetaceans are defined as animals, "including individuals, of those species, subspecies or populations of *Odontoceti* or *Mysticeti*."[23] In the 1998 Agreement on the International Dolphin Conservation Program, "dolphin" was defined as meaning "species of the family *Delphinidae* associated with the fishery for yellowfin tuna in the Agreement area."[24] Elsewhere, although the word dolphin appears in regional treaties, the word itself has remained undefined.[25]

Such types of definition have not appeared in the IWC, where, other than for administrative reasons,[26] the debate has moved away from biological distinctions towards political distinctions. The signatories have wrestled with jurisdictional questions as to whether the IWC has competence to manage stocks of small cetaceans. Within this broader debate, the answer to whether a species is small or not, and the interlinking question of whether that grants the IWC jurisdiction over them, will often depend on which side of the jurisdictional debate the protagonist falls.[27]

---

[22]   Opened for signature Mar. 17, 1992, 1772 U.N.T.S. 217, art. 1.2(a) (entered into force Mar. 29, 1994).

[23]   ACCOBAMS, *supra* note 9, art. 1(3)(a).

[24]   Agreement on the International Dolphin Conservation Program, opened for signature 15 May 1998, (1998) 37 I.L.M. 1246, art. I(2) (entered into force Feb. 15, 1999).

[25]   Agreement for the Establishment of the Intergovernmental Organization for Marketing Information and Cooperation Services for Fishery Products in Africa, opened for signature Dec. 13, 1991, 1777 U.N.T.S. 401, art. 2 (entered into force Dec. 23, 1993) explicitly excludes marine mammals "especially dolphins" (irrespective of whether they are endangered or not) from the provision of the services provided by the agreement.

[26]   The 28th Meeting of the IWC approved (for administrative reasons), a list of smaller cetaceans of the world: IWC, *Twenty-Eighth Report of the International Whaling Commission* 30 (1978).

[27]   It has been suggested that within the IWC, the classification of whether a species is large or small turns on the debate of whether or not it has traditionally been exploited by the commercial whaling industry. *See also* William C. Burns,

3.   THE SPECIES WITHOUT SUPERVISION

In the early 1970s, the Scientific Committee of the IWC began to examine the stocks of minke and other small whales that were essentially still unexploited,[28] but which certain nations were beginning to exploit.[29] At this interim stage the Scientific Committee requested information on the status and catches of small cetaceans by member countries. In addition, the Scientific Committee recommended that "[a] sub-committee on small cetaceans be set up to improve data collection on all world catches of these animals and to review species and stock identification and other problems."[30]

The intention of the Scientific Committee was to keep watch over small cetaceans and provide an early warning system, thus avoiding the IWC's earlier mistakes on larger species.[31] However, after reviewing the problem at hand, the Scientific Committee recommended that the IWC consider "initially the management of those small cetaceans which are taken in deliberate, direct fisheries."[32] In its report, the Sub-Committee on Small Cetaceans (SCSC) concluded that

---

The International Whaling Commission and the Regulation of the Consumptive and Non-Consumptive Uses of Small Cetaceans: The Critical Agenda for the 1990s, 13 WIS. INT'L L.J. 105, 106 (1994). *See also* James Scarff, *The International Management of Whales, Dolphins, and Porpoises: An Interdisciplinary Assessment* (Part One) 6 ECO. L.Q. 323 (1977).

28   International Commission on Whaling, *Twenty-Fourth Report of the International Whaling Commission* 29 (1974). Note that the International Commission on Whaling became the International Whaling Commission in 1978, and all subsequent references to the International Commission on Whaling or the IWC refer to the same body.

29   International Commission on Whaling, *Twenty-Second Report of the Commission* 32 (1972).

30   International Commission on Whaling, *Twenty-Fourth Report of the International Whaling Commission* 33 (1974). Publication of the initial report required special funding from the United States and Norway: IWC, *Twenty-Sixth Report of the Commission* 28 (1976).

31   International Commission on Whaling, *Twenty-Third Report of the Commission* 42 (1973).

32   IWC, *Twenty-Seventh Report of the Commission* 9 (1977).

there is an urgent need for an international body to effectively manage stocks of all cetaceans not covered by the present IWC Schedule. This body should concern itself with all types of exploitation of cetaceans, both incidental and deliberate. All nations involved in such exploitation of small as well as large cetaceans should be included in such a body. . . . The Sub-Committee therefore recommends that the present Convention for the regulation of whaling [that is, the *ICRW*] should be revised so that [it] covers all cetaceans and all forms of exploitation.[33]

In a resolution on small type coastal whaling the following year, it was noted that

the Commission is at the present time the sole international authority exclusively concerned with the regulation of major species of cetaceans [and] the Commission has under study proposals for the revision of the International Whaling Convention to include all species of Cetacea.[34]

Over the following years, the IWC urged that attention should remain focused on this issue.[35] However, direct involvement with the substance of the suggestion was avoided as a series of working groups examined the problem both indirectly, as part of a complete reworking of the ICRW,[36] and directly, with the creation of working groups and steering committees to address the specific problem presented by small cetaceans. These groups repeatedly tried to "broaden the discussion"[37] and "go deeper into the problem of small cetaceans."[38] In doing so, attempts were made to find common ground which did "not seek in any way to prejudice different members' positions."[39] Accordingly, non-controver-

---

[33] *Id*. at 480. *See also id*. at 49.

[34] IWC, *Reporting Requirements for Small Type Whaling, Twenty-Eighth Report, supra* note 26.

[35] *Id*.

[36] BIRNIE, *International Regulation of Whaling, supra* note 17, at 494–96.

[37] IWC, *Forty-First Report of the International Whaling Commission* 16 (1991).

[38] IWC, *Forty-Second Report of the International Whaling Commission* 17 (1992).

[39] IWC, "Resolution on Small Cetaceans," *Forty-First Report, supra* note 37, 48: "[The IWC is] aware that there exist differences in views between member

sial and cooperative solutions[40] to the problems surrounding small cetaceans meant that the signatories sought solutions while simultaneously "setting aside without precedent the different legal views over competence and sovereign rights."[41]

Despite these attempts, controversy broke out periodically during the 1980s and more regularly during the 1990s. The controversies arose as the IWC did three things. First, they invited governments with adversely impacted populations of small cetaceans "to seek advice from the IWC on ways in which those impacts may be assessed, and to this end to share catch statistics and data."[42] Second, they urged

> Parties to undertake relevant research and to continue to provide information on directed and incidental catches of small cetaceans to assist the Scientific Committee in assessing the status of, and threats to, small cetacean populations.[43]

Finally, when specific populations of small cetaceans were being pushed to the limits of sustainable catch levels, by both direct and indirect means, the IWC issued direct resolutions to specific governments, such as Chile and Argentina,[44] Japan,[45]

---

states on the regulatory competence of the IWC with regard to small cetaceans, and noting that this resolution does not seek in any way to prejudice different members' positions." *See also* IWC, *Forty-Third Report of the International Whaling Commission* 32, 36–37 (1993).

[40]   IWC, "Resolution on Addressing Small Cetaceans in the IWC" *in Forty-Fourth Report of the International Whaling Commission* (1994) 31–32. *See also* IWC, *Forty-Third Report of the International Whaling Commission* 36 (1993).

[41]   IWC, *Forty-Third Report of the International Whaling Commission* 36 (1993).

[42]   *Id.* at 51: "Resolution on Small Cetaceans."

[43]   IWC Resolution 1997-8, "Resolution on Small Cetaceans" *in Forty-Eighth Report of the International Whaling Commission* 49 (1998).

[44]   IWC, *Thirty-Fifth Report of the International Whaling Commission* 27 (1985). This resolution called on the governments of Chile and Argentina to "investigate the levels of direct takes" and, due to the problem of Chilean black dolphins being used as bait in crab traps, to "initiate research to identify alternative sources of bait and provide information on such bait to fishermen."

[45]   IWC, "Resolution on the Directed Takes of Striped Dolphins in Drive

Canada,[46] and Mexico,[47] to try to rectify the problems at hand. Other resolutions overlapping with the issue of the humane killing of small cetaceans were also issued.[48] This approach was reinforced in 1999 with

---

Fisheries" *in Forty-Third Report, supra* note 41, at 51–52. This resolution invited the Japanese government to "consider the advice from the Scientific Committee on this species" which suggested that this population was overexploited and "to take appropriate action as soon as possible that will allow recovery of the population." *See also* IWC Resolution 1999–9, "Resolution on Dall's Porpoise" *in Annual Report of the International Whaling Commission 1999* 55–6 (1999).

[46]   IWC, "Resolution in the Directed Takes of White Whales and Narwhals" *in Forty-Third Report, supra* note 41, at 52. This resolution, though it noted the differences of opinion on this issue, still requested the Canadian government to cooperate with the IWC on this species, and to make sure that the take would be sustainable. A similar resolution directed at Canada from the 50th Meeting "expressed the Commission's concern that directed takes of white whales might not be sustainable, and invited all states having white whales in their waters to conduct further research on white whales" in accordance with "a precautionary approach," IWC, "Resolution on Directed Takes of White Whales" *in Annual Report of the International Whaling Commission 1998* 46 (1998). *See also* IWC Resolution 1999–7, "Resolution on Small Populations of Highly Endangered Whales" *in Annual Report of the International Whaling Commission 1999* 55 (1999). Again, this resolution was specifically forwarded to Canada.

[47]   IWC Resolution 1994–3, "Resolution on Biosphere Reserve of the Upper Gulf of California and the Colorado River Delta" *in Forty-Fifth Report of the International Whaling Commission* 42 (1995). In fact, apart from noting the status of this species, the Resolution congratulated, commended and complimented Mexico on its decision to declare a "Biosphere Reserve," and to take other steps to protect the vaquita. The commendation of Mexico was later repeated and was matched with congratulations to the People's Republic of China for their attempts to conserve the baiji: IWC Resolution 1996–4, "Resolution on Small Cetaceans" *in Forty-Seventh Report of the International Whaling Commission* 49 (1997).

[48]   In 1980 it was agreed that due to often inhumane practices, considerations of humane killing should be extended to small whales: "[E]very attempt should be made to investigate ways and means to shorten time-to-death of killing small whales such as minke whales," IWC, "Recommendations Adopted by the International Whaling Commission at its 31st Annual Meeting Concerning the Humane Killing of Whales," *Thirtieth Report of the International Whaling Commission* 36 (1980). *See also* issues relating to the take of Dall's porpoise (which also had humane killing considerations): IWC, *Forty-First Report, supra* note 37, at 41–42. However, it was the killing of pilot whales which led to direct resolutions from the 43rd and 44th Meetings: *See also* IWC, "Resolution on the Killing of Pilot Whales

the Resolution on Small Populations of Highly Endangered Whales.[49] This resolution welcomed the focus of the Scientific Committee upon the status and trends of small populations of highly endangered whales, and encouraged member and non-member governments to participate in this work. Moreover it called

> upon all governments whose nationals have in recent years taken whales from any of these populations of highly endangered whales to refrain from authorising any further takes until the Scientific Committee concludes that adequate scientific advice is available to demonstrate that such takes will not cause a continued threat to the survival or recovery of these populations.[50]

The result of these resolutions was that the countries that had traditionally rejected the IWC's competence over small cetaceans reiterated their views with vigour.[51] As such, by the end of the 1990s the need for an effective international body to manage small cetaceans was as acute as it had been when the issue was first raised 25 years earlier. With this predicament, it is no surprise that the IWC has likewise continued to reiterate the "urgent need for further international cooperation to ensure the conservation of small cetaceans."[52]

(*Globicephala melas*)" in *Forty-Third Report, supra* note 41, at 52; IWC, "Resolution on Pilot Whales," *Forty-Fourth Report of the International Whaling Commission* 31 (1994). For further information on this hunt, *see also* Environmental Investigation Agency, *Don't Buy the Faroese Pilot Whale Slaughter* (1994).

[49] IWC Resolution 1999–7, "Resolution on Small Populations of Highly Endangered Whales" *in Annual Report of the International Whaling Commission 1999* 55 (1999).

[50] *Id.*

[51] According to Mexico, such resolutions "not only [exceed] the IWC mandate but especially [purport] to dictate behaviour to sovereign governments on these matters," IWC, *Forty-Third Report, supra* note 41, at 37. In the following years, Mexico, Japan, Norway, Austria and Denmark all took exception to various resolutions on small cetaceans due to a belief that they did not reflect the cooperative approach of the past, IWC, *Annual Report of the International Whaling Commission 1998* 40–41, 49 (1998). *See also* IWC, *Annual Report 1999, supra* note 49, at 55.

[52] IWC, "Resolution on Small Cetaceans" *in Forty-First Report, supra* note 37, at 48. The resolution from the 44th Meeting noted "the need for international cooperation to address problems relating to small cetaceans and to facilitate the

4.   THE RESPONSIBILITIES OF THE SCIENTIFIC COMMITTEE

In the early 1970s the IWC decided it was necessary to monitor the catches of species of whales that had previously been largely unexploited.[53] To assist the achievement of this goal, member states were asked, for the first time, to provide information concerning their take (both deliberate and incidental) of small cetaceans.[54] The 1976 "Resolution on Reporting Requirements for Small Type Whaling" noted that

> existing international commissions and organisations concerned with marine resources do not, at the present time, provide a central agency for the collection of scientific information on captures of small cetaceans. . . . [T]he Commission has had brought to its attention the need for such an agency, and the need to commence the collection of such information on an urgent basis.[55]

To help collate and synthesise information with regard to species review, stock identification and other problems worldwide, the SCSC

---

conservation and restoration of depleted or threatened stocks," IWC, "Resolution on Addressing Small Cetaceans in the IWC" in *Forty-Fourth Report of the International Whaling Commission* 31-2 (1994). The 45th Meeting added the importance of "cooperation to conserve and restore threatened and depleted stocks" of small cetaceans, IWC Resolution 1994–2, "Resolution on Small Cetaceans" in *Forty-Fifth Report of the International Whaling Commission* 41 (1995). Finally, the 47th Meeting largely repeated the call at the beginning of the decade for the "continuing urgent need for cooperation to conserve and restore depleted stocks of small cetaceans," IWC Resolution 1996–4, "Resolution on Small Cetaceans" in *Forty-Seventh Report of the International Whaling Commission* 49 (1997).

[53]   International Commission on Whaling, *Twenty-Second Report, supra* note 29. In this context, the term "unexploited" means that these species of whales were not previously subject to IWC quotas. Further, if they were not also subject to other previous conservation measures that predated the ICRW, like the blue and right whales, then it was generally presupposed that they were not endangered, and therefore could be hunted.

[54]   International Commission on Whaling, *Twenty-Third Report, supra* note 31.

[55]   IWC, "Reporting Requirements for Small Type Whaling," *Twenty-Eighth, supra* note 26.

was established.[56] The SCSC later[57] (through the Scientific Committee) made recommendations for future action with regard to small cetaceans.[58] It recommended that member nations report information regarding the direct and indirect take of small cetaceans to the IWC.[59] By the mid-1970s, although the Scientific Committee was *not* in a position to classify stocks of small cetaceans in the same way as the classification of large whales, it was nevertheless determined not to repeat the mistakes that had plagued them with regard to larger whales. Accordingly, it emphasized the need to "provide an early warning system for signs of depletion" of small cetaceans.[60] Until the larger issue of the place of small cetaceans within the IWC regime was concluded, the IWC agreed (and reiterated throughout the 1980s[61] and 1990s[62]) that the harvesting of *all* cetaceans is subject to consideration by the Scientific Committee.

---

[56]   International Commission on Whaling, *Twenty-Third Report, supra* note 31, at 26. For the proposal for the establishment of the SCSC, See also International Commission on Whaling, *Twenty-Fourth Report of the International Whaling Commission* 33 (1974).

[57]   Following the publication of their report on small cetaceans, the Scientific Committee recommended that the IWC should seek funds to have this report published: International Commission on Whaling, *Twenty-Sixth Report of the Commission* 28 (1976). This request was supported by the IWC. Norway, Canada and the U.S. all made contributions to this report.

[58]   IWC, *Twenty-Seventh Report, supra* note 32, at 12.

[59]   IWC, "Reporting Requirements for Small Type Whaling," *Twenty-Eighth Report, supra* note 26.

[60]   BIRNIE, INTERNATIONAL REGULATION OF WHALING, *supra* note 17, at 425.

[61]   It requested that "member nations collect and submit full statistics on small cetacean catches as previously requested," IWC, *Thirty-Second Report of the International Whaling Commission* 27 (1982). At the 35th Meeting, with regard to overall statistics related to direct catches, the Scientific Committee "requested re-emphasis of the Commission's agreement in 1976 to collect and report" to it that statistics for small cetaceans directly taken as well as by live captures be included in Annual Progress Reports, IWC, *Thirty-Fifth Report, supra* note 44.

[62]   This work was commended by the IWC at the 42nd and 43rd Meeting. The IWC urged the Russian government to continue to "[advise] on ways in which those threats [may] be eliminated or minimised," IWC Resolution 1994–2, "Resolution on Small Cetaceans" *in Forty-Fifth Report of the International Whaling Commission*

Despite the objections,[63] the IWC stipulated that the Scientific Committee should "provide such scientific advice as may be warranted to Contracting Governments, coastal States and other interested governments and inter-governmental organisations as appropriate."[64] This aim was reiterated in the 1990s, with it being affirmed that the SCSC should direct its attention to the "severe reduction of certain stocks of small cetaceans through directed exploitation and incidental catches in fishing operations."[65] In the pursuit of this objective, the Scientific

---

(1995) 41. *See also* IWC, "Resolution on Small Cetaceans" *in Forty-Second Report, supra* note 38, at 48; IWC, "Resolution on Small Cetaceans" *in Forty-Third Report, supra* note 41, at 51.

63   As the Scientific Committee attempted to carry out its mandate, the issue of jurisdiction arose with regard to its advice. As such, countries like Mexico, while recognizing the general scientific value of the information requested by the Scientific Committee, expressed reservations about making this advice available to the Commission on species that were not listed in the ICRW, IWC, *Thirty-Fifth Report, supra* note 44, at 19. Likewise, it was argued that "no Schedule amendments [should] be made upon the recommendation of the Scientific Committee" and that the Scientific Committee should not adopt recommendations or decisions on management issues, IWC, *Thirty-First Report of the International Whaling Commission* 24 (1981); IWC, *Forty-Seventh Report of the International Whaling Commission* 22–23 (1997). Conversely, Japan has broadly accepted the role and value of the Scientific Committee with regard to the examination of issues relating to small cetaceans: *See also* IWC/46/SM1, *Japan's View on Addressing Small Cetaceans within the IWC*, (Paper presented at the 45th Meeting of the IWC, Puerto Vallarta, May 1994), at [a], [c].

64   IWC, *Resolution Concerning Extension of the Commission's Responsibility for Small Cetaceans*, in *Thirty-First Report of the International Whaling Commission* 31 (1981). This point followed the Canadian recommendation that "a working process [be adopted] whereby the Scientific Committee would consider all cetaceans and be able to make scientific advice available to Contracting Parties, coastal states and other interested Governments and [inter-governmental] organisations." *Id.* at 24.

65   IWC, "Resolution on Small Cetaceans" *in Forty-First Report, supra* note 37, at 48: This resolution requested that the Scientific Committee
> commence a process of drawing together all available relevant information on the present status of the stocks of small cetaceans which are subjected to significant directed and incidental takes, on the impact of those takes on those stocks, and providing an assessment of the present threats to the stocks concerned.

Committee instigated a series of regional and global reviews[66] by which they could begin to ascertain which species of small cetaceans were severely threatened.[67]

5.   THE IWC'S COMPETENCY TO MANAGE SMALL CETACEANS

Although member states were initially ambivalent on the issue of IWC management of small cetaceans,[68] by the 32nd Meeting the lines of debate were clearly established, having been triggered by attempts to list some small cetaceans as protected species.[69] The essence of the debate, as it unfolded at the 32nd meeting was in three parts and has remained largely the same ever since.[70] The issues are whether the ICRW confers competence upon the IWC to manage small cetaceans, the possible conflict with coastal states' rights and the role of regional organisations in small cetacean management.

See also IWC, *Forty-Second Report, supra* note 38, at 35–36.

[66]   These included studies of Latin America, Africa, the Indian Ocean and the Red Sea, with special reference to the Middle East: *See also* IWC, *Forty-Fourth Report of the International Whaling Commission* 25 (1994); IWC, *Annual Report of the International Whaling Commission 1998* 33–34 (1998).

[67]   For the SCSC, the priority areas were endangered species, species under direct or indirect threat, and global and regional reviews: IWC, *Forty-Fifth Report of the International Whaling Commission* 20 (1995). *See also* IWC Resolution 1999–7, "Resolution on Small Populations of Highly Endangered Whales" *in Annual Report 1999, supra* note 49, at 55.

[68]   A number of "delegations were unsure of their position on this matter, but it was thought important that the recommendations on the small cetaceans should not be ignored," IWC, *Thirtieth Report of the International Whaling Commission* 30 (1980). This uncertainty continued for a few countries such as Norway, which initially reserved its position in this debate: IWC, *Thirty-Second Report of the International Whaling Commission* 27 (1982). However, other states were clear in their belief that small cetaceans were not within the jurisdiction or competence of the IWC, including Argentina, Brazil, Chile, Denmark, France, Japan, Mexico, Peru and the USSR: IWC, *Thirty-Fourth Report of the International Whaling Commission* 24 (1984).

[69]   *See also* IWC, *Thirty-First Report, supra* note 64, 23–24; IWC, *Thirty-Fourth Report of the International Whaling Commission* 16 (1981).

[70]   IWC, *Thirty-First Report, supra* note 64, at 24.

5.1.   Interpreting the Language of the ICRW and Related Documents

The answer to the question of whether the IWC is the competent body to manage small cetaceans is, in part, found within the ICRW and its associated documents: the Schedule and the nomenclature.[71] The specific issue is whether the text of these documents grants the IWC management rights over small cetaceans as well as large ones. The instrument to guide this analysis is the Vienna Convention on the Law of Treaties (VCLT),[72] which sets forth specific rules for treaty interpretation:

(1)   A treaty shall be interpreted in good faith in accordance with *the ordinary meaning* to be given to the terms of the treaty in their context and in the light of its object and purpose.

(2)   The context for the purpose of the interpretation of a treaty shall comprise in addition to the text, including *its preamble and annexes:*

    (a)   any agreement relating to the treaty which *was made between all* the parties in connection with the conclusion of the treaty.

    (b)   any instrument which was made by one or more parties in connection with the conclusion of the treaty and *accepted* by the other parties as an instrument related to the treaty.

(3)   There shall also be taken into account, together with the context:

    (a)   any *subsequent agreement between the parties* regarding the interpretation of the treaty or the application of its provisions;

    (b)   any *subsequent practice* in the application of the treaty which establishes the agreement of the parties regarding its interpretation.[73]

---

[71]   ICRW, *supra* note 2, sch. I.

[72]   Opened for signature May 23, 1969, 1155 U.N.T.S. 331, 8 I.L.M. 679 (entered into force Jan. 27, 1980).

[73]   *Id.* art. 31 (emphases added).

5.1.1. The Language of the Convention

The primary rule applying to the interpretation of international documents is that "the ordinary meaning" shall be "given to the terms of the treaty." In scientific terms, whales covers both large and small cetaceans. In this regard, no distinction between the two suborders of the genus whale can be made.

This failure to make a distinction between the suborders is also reinforced by the language of the ICRW, which talks of whales generally.[74] The objectives of the ICRW include "the interest of the nations of the world in safeguarding for future generations the great natural resources represented by the *whale stocks*."[75] In addition, Article V explains that

[t]he Commission may amend from time to time the provisions of the Schedule by adopting regulations with respect to the conservation and utilisation of *whale resources*, fixing protected and unprotected *species*.[76]

Article VI adds: "The Commision may . . . make recommendations . . . on any matters *which relate to whales or whaling* and to the objectives and purposes of this Convention."[77] The exception to such generic language continues in the preamble, which stipulates that "it is essential to protect *all* species of whales from further overfishing."[78] Clearly, the word "all" would give weight to the argument that the use of the generic term "whale" in the ICRW was designed to cover *all* cetaceans, both large and small.

The ordinary reading of such language has led several countries, including Switzerland, to suggest that

---

[74]   IWC, *Thirty-First Report, supra* note 64: "[T]he Convention itself does not define the species covered by the term whale and Contracting governments are not of one view on such a definition as regards the Convention."

[75]   ICRW, *supra* note 2, preamble (emphasis added).

[76]   Emphasis added.

[77]   Emphasis added.

[78]   Emphasis added. Note, however, that the "history of whaling," as noted in the preamble, was about the hunting of large cetaceans. *See also* WILLIAM BURKE, THE NEW INTERNATIONAL LAW OF FISHERIES 293 (1994).

no distinction has been made between large and small whales. The ordinary meaning of the term "whale" is therefore legally extended to any kind of whale without any restriction due to size or classification in the whale family.[79]

This approach is in accordance with the view that the focus of the ICRW is to afford protection to whale stocks generally.[80] Indeed, the ICRW does not deal with individual species. Rather, the question of the management of individual species is dealt with in the Schedule, which is amendable at each IWC meeting.[81]

The original 1931 Convention for the Regulation of Whaling supports the contention that the ICRW was designed to accommodate small as well as large cetaceans.[82] This document took a different approach to the ICRW in that it stipulated: "the present Convention applies only to baleens or whalebone whales."[83] Likewise, the 1937 International Agreement for the Regulation of Whaling[84] was directly linked to baleen whales, which were clearly defined as "any whale other than a toothed whale."[85] As such, the 1931 and 1937 conventions were clearly limited to large whales. Conversely, while the 1946 ICRW took a somewhat generic approach with its textual language (apart from the phrase "all whales"), it adopted a different approach with its annexed nomencla-

---

[79]   Antoine Goetschel, "A Legal Analysis of IWC Competence to Manage Small Cetaceans" (Paper presented by Switzerland at the 51st Meeting of the IWC, Grenada, May 1999) at [7]. The comments were later reported in IWC, *Annual Report 1999, supra* note 49, at 42.

[80]   James Cameron, *Legal Opinion in* THE GLOBAL WAR AGAINST SMALL CETACEANS 6–7 (Duncan Currey *et al.*, eds. 2d ed. 1991).

[81]   This argument was suggested by New Zealand. *See also* IWC, *Forty-First Report, supra* note 37.

[82]   Opened for signature Sept. 24, 1931, 155 L.N.T.S. 349 (entered into force Jan. 16, 1935).

[83]   *Id.* art. 2. "Whalebone" is another word for baleen plates—hence this refers to baleen whales, or "large cetaceans."

[84]   Opened for signature June 8, 1937, 190 L.N.T.S. 79, art. 9 (entered into force July 1, 1937).

[85]   *Id.* at art. 18.

ture, which included *Hyperoodon ampullatus* and *Hyperoodon plani-frons* (the northern and southern bottlenose). Both of these species are beaked whales and have teeth;[86] thus they may be classified as small cetaceans. Hence, the *ICRW* was clearly intended to be different to its predecessors because it did not limit its coverage to baleen or whale-bone whales.

5.1.2. The Schedule and Subsequent Practice

When interpreting a treaty it is permissible to examine the annexes of the treaty. However, the ICRW does not have any document denominated as an annex. Rather, it has a Schedule—which for all effective purposes effectuates many of the same purposes traditionally associated with treaty annexes. The difference between the two terms is that the term "annex" means to tie or bind to, and in essence "expresses the idea of joining a smaller or subordinate thing with another, larger, or of higher importance."[87] The Schedule is annexed to the principal instrument and "exhibits in detail the matters mentioned or referred to in the principal document."[88] In legal terms, it often comes in the form of an appendix "arranged under headings prescribed by official authority."[89]

Such an approach, where the text of the treaty and the annex/schedule are closely linked, was clearly recognised in the *ICRW*. Article I of the *ICRW* explains:

This Convention includes the Schedule attached thereto which forms *an integral part* thereof. All references to "Convention" shall be understood as including the said Schedule either in its present form or as amended in accordance with the Provisions of Article V.[90]

---

[86]   This point was raised by Switzerland in 1999: "The mentioned species are of the toothed small cetacean family," Goetschel, *supra* note 79.

[87]   Joseph Nolan *et al., Black's Law Dictionary* (6th ed., 1990).

[88]   *Id.*

[89]   2 THE NEW SHORTER OXFORD ENGLISH DICTIONARY ON HISTORICAL PRINCIPLES 2710 (Leslie Brown ed., 1993).

[90]   Emphasis added.

The Rules of Procedure for the ICRW explain that "a three-fourths majority of those casting an affirmative or negative vote shall be required for action in pursuance of Article V of the Convention."[91]

This approach by the ICRW is in accord with many international documents dealing with environmental problems, whereby the annex forms an integral part of the treaty. This practice can be seen in areas as diverse as wildlife treaties,[92] conventions dealing with ecosystems or other important areas,[93] and international and regional fisheries treaties,[94]

---

[91]   IWC, "Rules of Procedure and Financial Regulations," *Annual Report of the International Whaling Commission 1998* 85, Rule E(3) (1998).

[92]   The Convention Relative to the Preservation of Fauna and Flora in their Natural State, opened for signature Nov. 8, 1933, 172 L.N.T.S. 241, annex (entered into force Jan. 14, 1936) listed species needing immediate protection due to "special urgency and importance." The African Convention on the Conservation of Nature and Natural Resources, opened for signature Sept. 15, 1968, 1001 U.N.T.S. 3, art. VIII (entered into force June 16, 1969) created a series of annexes with lists of protected species. The Convention for the Conservation of Antarctic Seals, opened for signature June 1, 1972, 1080 U.N.T.S. 175, 11 I.L.M. 251, art. 3 (entered into force Mar. 11, 1978) contains annexes on all matters relating to the take of seals around the Antarctic. The CITES Convention, *supra* note 8, art. II, and the Convention on the Conservation of Migratory Species of Wild Animals, opened for signature June 23, 1979, 1651 U.N.T.S. 333, 19 I.L.M. 15, art. II (entered into force Nov. 1, 1983) have annexes of protected species, as does the Convention on the Conservation of European Wildlife and Natural Habitats, opened for signature Sept. 19, 1979, 1284 U.N.T.S. 209, art. 4 (entered into force June 1, 1982).

[93]   The Convention of Wetlands of International Importance, Especially as Waterfowl Habitat, opened for signature Feb. 2, 1971, 996 U.N.T.S. 245, 11 I.L.M. 963, art. 2 (entered into force Dec. 21, 1975) created an annexed 'list' of wetlands. The Convention for the Protection of the World Cultural and Natural Heritage, opened for signature June 1, 1972, 1037 U.N.T.S. 151, 11 I.L.M. 1358, art. 11 (entered into force Dec. 17, 1975) contains a list of property forming part of the cultural and natural heritage.

[94]   Annexes are also an integral part of the Agreement for the Implementation of the Provisions of the United Nations Convention of the Law of the Sea of 10 December 1982, Relating to the Conservation and Management of Straddling Fish Stocks and Highly Migratory Fish Stocks, opened for signature Dec. 4, 1995, 34 I.L.M. 1542, art. 48.1 (not yet in force) (Agreement on Straddling and Highly Migratory Fish Stocks); International Convention for the High Seas Fisheries of the North Pacific Ocean, opened for signature May 9, 1952, 205 U.N.T.S. 65, art. V.1 (no longer in force); Convention on Future Multilateral Cooperation in the

through to treaties dealing with such diverse topics as the trade in toxic waste,[95] dumping of waste at sea,[96] ozone depletion[97] and climate change.[98]

The similarity between schedules and annexes extends to the fact that they both do the same thing; they focus on the subject areas that are of particular concern to regime members. These may be anything from the specific endangered species to the gases that need to be restricted to prevent ozone layer deterioration. The essence of such lists is that they evolve to meet the needs of the signatories as time progresses. Thus, all of the above examples of conventions with annexes allow amendments to the annex by the signatories provided that a majority of the signatories approve (typically three-quarters).[99] This practice is exactly the same

---

Northwest Atlantic Fisheries, opened for signature Oct. 28, 1978, 1135 U.N.T.S. 369, art. XX (entered into force Jan. 1, 1979).

[95]   Basel Convention on the Control of Transboundary Movements of Hazardous Wastes and Their Disposal, opened for signature Mar. 22, 1989, 1673 U.N.T.S. 57, 28 I.L.M. 657, art. 18 (entered into force May 5, 1992) (Basel Convention) contains annexes which "form an integral part of this Convention."

[96]   1996 Protocol to the Convention on the Prevention of Marine Pollution by Dumping of Wastes and Other Matter, opened for signature Nov. 7, 1996, 36 I.L.M. 1, art. 20 (not yet in force).

[97]   Vienna Convention for the Protection of the Ozone Layer, opened for signature Mar. 22, 1985, 1513 U.N.T.S. 293, 26 I.L.M. 1529, art. 10.1 (entered into force Sept. 22, 1988) created a system of annexes which were deemed 'an integral part of this convention.'

[98]   In the United Nations Framework Convention on Climate Change, opened for signature May 9, 1992, 1771 U.N.T.S. 107, 31 I.L.M. 849, art. 16.1 (entered into force Mar. 21, 1994), the annexes to the Convention were deemed 'an integral part thereof.' In the *Kyoto Protocol to the United Nations Framework Convention on Climate Change*, opened for signature Dec. 11, 1997, 37 I.L.M. 22, art. 21 (not yet in force) (Kyoto Protocol) the annexes were again deemed "an integral part" of the Protocol.

[99]   *E.g.*, the African Convention on the Conservation of Nature and Natural Resources, *supra* note 92, art. VIII(2) provides that "additional species shall be placed in [the annexes] by the state concerned." This convention also allows for amendment of the annexes (art. XVI(3)) and revision of the Convention by "Meetings of the Parties" (art. XXIV). The Convention for the Conservation of Antarctic Seals, opened for signature June 1, 1972, 1080 U.N.T.S. 175, 11 I.L.M.

as the ICRW. Moreover this approach is well established through the axiomatic principle of international law that countries have the right to object (and not be bound) by majority decisions. This right is also commonly acknowledged in the treaties themselves and is recognized in Article V(3) of the ICRW.

Once it is recognized that the schedule/annex of an international agreement is typically what the signatories negotiate and vote upon, it then becomes possible to examine some of the other rules of interpreting international documents. These are typically based on the

251, art. 9 (entered into force Mar. 11, 1978) contains annexes, which may be amended by "any contracting party" and a two-thirds majority of signatories who agree with it. It is, of course, not binding on those who object: art. 9(4). CITES Convention, *supra* note 8, art. XV allows for amendments to its appendices, if these are accepted by a two-thirds majority. The CMS Convention, *supra* note 7, art. XI also allows for amendments to its appendices if these are accepted by a two-thirds majority. The Convention on the Conservation of European Wildlife and Natural Habitats, *supra* note 92, art. 14(1), "keep[s] under review the provisions of the Convention, including its Appendices, and examine[s] any modifications necessary." Modifications are achieved by a three-quarters majority: *See also* art. 16. The Convention of Wetland of International Importance, Especially as Waterfowl Habitat, opened for signature Feb. 2, 1971, 996 U.N.T.S. 245, 11 I.L.M. 963, art. 2 (entered into force Dec. 21, 1975) created an annexed list of wetlands. The parties who possess the wetlands in their sovereign territory may amend, delete or add to this list as they please, art. 5. In a similar vein, the Convention for the Protection of the World Cultural and Natural Heritage, opened for signature June 1, 1972, 1037 U.N.T.S. 151, 11 I.L.M. 1358, art. 11 (entered into force Dec. 17, 1975) contains a list of "world heritage," which is submitted by sovereign governments (to see also if it qualifies for World Heritage Status). With the Agreement on Straddling and Highly Migratory Fish Stocks, *supra* note 95, art. 48(2) (not yet in force), "the Annexes form an integral part of [the] Agreement." The annexes "may be revised from time to time by State Parties" by either consensus or majority, art. 45(2). The International Convention for the Conservation of Atlantic Tunas, opened for signature May 14, 1966, 673 U.N.T.S. 63, art. XIII (entered into force Mar. 21, 1969) can be amended by three-quarters majority. UNCLOS, *supra* note 5, art. 312(2) allows amendment of its annexes by either consensus or majority. The International Convention for the High Seas Fisheries of the North Pacific Ocean, opened for signature May 9, 1952, 205 U.N.T.S. 65, art. VII (no longer in force) contains a series of annexes, which the International North Pacific Fisheries Commission has the power to alter. Note the importance of consent in order to be bound by such alterations. Likewise, the North-East Atlantic Fisheries Convention, opened for signature Jan. 24, 1959, 486 U.N.T.S. 157, art. 5(4) (no longer in force) contained

ideal of subsequent practice (or subsequent agreement) between the parties regarding the interpretation of the treaty or the application of its provisions.[100]

In terms of subsequent practice, the question that needs to be and was asked[101] is whether the IWC did anything subsequent to its formation

---

annexes which set out fishing boundaries which may "be subject to alterations." Note, however, the importance of consensus for the alteration of boundaries in this convention. The Convention on Future Multilateral Cooperation in the Northwest Atlantic Fisheries, opened for signature Oct. 28, 1978, 1135 U.N.T.S. 369, art. XX(2) (entered into force Jan. 1, 1979) contained annexes that may be altered "by a two-thirds majority vote of all Contracting Parties." The International Tropical Timber Agreement, opened for signature Nov. 18, 1983, 1393 U.N.T.S. 119, art. 38 (provisionally entered into force Apr. 1, 1985) could originally only be amended if an 85% majority was achieved. The International Tropical Timber Agreement, opened for signature Apr. 1, 1994, 33 I.L.M. 1014, art. 42 (entered into force Jan. 1, 1997) changed this to a two-thirds majority. The Basel Convention, *supra* note 95, contains annexes that can be amended as needed by a three-quarters majority, art. 17(3). The Vienna Convention for the Protection of the Ozone Layer, *supra* note 97, arts. 6(e)–(g), 10 created a system which could be amended as needed, by a three-quarters majority, art. 9(3). The annexes of the United Nations Framework Convention on Climate Change, *supra* note 98, art. 15(3) can be amended by a three-quarters majority, as can those of Kyoto Protocol, *supra* note 98. Of course, these are non-binding on those who disagree, art. 20(5). The 1996 Protocol to the Convention on the Prevention of Marine Pollution by Dumping of Wastes and Other Matter, *supra* note 96, arts. 21, 22 can be amended by a two-thirds majority. Amendments of annexes and appendices to the Convention for the Protection of the Marine Environment of the North-East Atlantic, opened for signature Sept. 22, 1992, 32 I.L.M. 1069, (entered into force Mar. 25, 1998), require a three-quarters majority. Amendments to the Convention on the Prior Informed Consent Procedure for Certain Hazardous Chemicals and Pesticides in International Trade, opened for signature Sept. 10, 1998, 38 I.L.M. 1, arts. 21(3), 22(3)(a) (not yet in force) also require a three-quarters majority. Note, however, that amendments to annex III (the list of chemicals needing prior informed consent) require consensus, art. 22(5)(b). To amend the Agreement for the Establishment of a General Fisheries Council for the Mediterranean, opened for signature Sept. 24, 1949, 126 U.N.T.S. 237, art. VII (entered into force Feb. 20, 1952) a two-thirds majority is required. Finally, ACCOBAMS, *supra* note 9, art. X, annex, can be amended by a two-thirds majority.

[100] *See also* VCLT, *supra* note 72, at art. 31(3)(b).

[101] The articulation of the argument that competence can be obtained by subsequent practice appeared at the 34th Meeting. The broader question of competency

to suggest that it has competence over small cetaceans? The place to look for evidence is in the Schedule, which provides direct evidence that the ICRW's jurisdiction has been extended by the parties to cover species that were not considered to be within its original purview. This is evidenced by the nomenclature to the agreement, which listed the multiple names of a number of whales for which the IWC regularly set quotas.

An important species missing from this original list, which nevertheless falls within the large whale category, is the minke whale (contained in the general family of rorqual whales). Despite the fact that the minke is now the smallest and most abundant of all rorqual whales, it was *not* listed in the original nomenclature. Despite this exclusion, by the third meeting of the IWC catches of minke (with no minimum size restrictions) were being authorized.[102] Moreover, this interest in taking minkes increased rapidly after the 1970s, as restrictions on the takings of other whales (from stocks that were clearly plummeting) took effect.[103]

A similar subsequent practice, but with regards to small cetaceans, is evinced in both the general interpretation section of the Schedule for baleen and toothed whales,[104] and in its general section (which defines small type whaling). Small type coastal whaling, which the IWC collects information on, but does not regulate, is defined as a "catching operation" which focuses upon "minke, bottlenose, beaked, pilot or killer whales."[105] The evidence of subsequent practice, which incorporated some small cetaceans, can be found in the fact that the killer whale was not included in the original chart of nomenclature. It was subse-

---

with regard to small cetaceans in general was deemed to be outside the Steering Committee's mandate, IWC, *Thirty-Fourth Report of the International Whaling Commission* 16 (1984).

[102]   International Commission on Whaling, *Third Report of the Commission* 6, 16 (1952).

[103]   For a discussion of the beginning of this interest in the 1970s, *see also* International Commission on Whaling, *Twenty-Third Report of the Commission*, *supra* note 31, at 9–10.

[104]   ICRW, *supra* note 2, sch. I(A), (B).

[105]   *Id.*, sch. I(C).

quently defined in Schedule I at the IWC's 29th Meeting in 1977.[106] It was not until the 32nd Meeting in 1980 that the killer whale was added to the list of species included in a provision to the moratorium.

Likewise, the Baird's beaked whale was not listed in the original nomenclature.[107] However, it was later included in the definition of bottlenose whale in Schedule I.[108] This inclusion took place at the IWC's 29th Meeting in 1977 and was not objected to by any government. There is also a heading "bottlenose" in table 3 of the Schedule, which deals with their capture. For many countries, such subsequent practices suggest that "the application of small cetaceans to the IWC has been concordant, common and consistent through the years."[109] To many this practice by the IWC "shows a willingness to exercise its competence to conserve small cetaceans."[110]

However, not all countries are willing to accept this view, especially with regard to the extension of competence to cover small cetaceans. That is, although the inclusion of the minke has not been challenged,, the latter examples of small cetaceans have been. Accordingly, it has been suggested that

> the inclusion of Baird's beaked whale in the definition of bottlenose whale in section I does not confer upon the Commission the competence to classify and set catch limits for it. . . . [S]uch action would be outside the scope of the Convention.[111]

Moreover with regard to the killer whale example:

> the inclusion of killer whales in the factory ship moratorium was a special measure taken in response to a particular situation. . . .

---

[106] *See also* IWC, *Twenty-Eighth Report, supra* note 26, at 22.

[107] For the origins of this debate in the IWC, *see also* IWC, *Thirty-Third Report of the International Whaling Commission* 28 (1983).

[108] ICRW, *supra* note 2, sch I.

[109] Goetschel, *supra* note 79.

[110] Cameron, *supra* note 80, at 6–7.

[111] IWC/35/15, "Report of the Steering Committee on Regulation of Baird's Beaked Whale" (Paper presented at the 35th Meeting of the IWC, Brighton, July 1983), at [7.1.1].

[K]iller whales in other situations [are] not regulated by the Commission.[112]

5.1.3. The Scope and Purposes of Nomenclatures

At the core of the aforementioned objections is the fact that the two small cetaceans now listed in the Schedule were not listed on the original nomenclature that was annexed to the ICRW.

The word *"nomen"* is Latin and in the civil law it was recognized, when referring to a particular group, as meaning "the name showing to what gens or tribe . . . they belonged . . . as distinguished from . . . [an] individual name."[113] Thus the *nomen* became recognised as "the name or style of a class or genus of persons or objects."[114] Working from this base definition, a nomenclature is

[a] set of names used, or intended to be used, to designate things, classes, places etc; a system of technical terms used in a science or other discipline. . . . Zoological nomenclature is the application of distinctive names to each of the groups recognized in the . . . classification.[115]

Attempts have been made to standardize nomenclature within the disciplines of biology, ecology, botany and zoology since the end of the 19th century. To this end, international codes have been devised and continually updated as new editions are necessitated by new discoveries and refinements pertaining to the knowledge of species.[116]

---

[112]   *Id.* at [7.1.3].

[113]   *Black's Law Dictionary, supra* note 87.

[114]   *Id.*

[115]   Brown, *supra* note 89, at 1932.

[116]   *The International Code of Zoological Nomenclature: Adopted by the International Union of Biological Sciences* (4th ed., 1999) has one fundamental aim, which is to provide the maximum universality and continuity in the scientific names of animals compatible with the freedom of scientists to classify all animals according to taxonomic judgment. The code is available from the International Commission on Zoological Nomenclature.

The idea that the nomenclature changes, and can accordingly only ever be a guide to the interpretation of the vernacular names of cetaceans (and hence, is not a tool that can circumscribe the ultimate purview of regimes), is well recognized in other wildlife treaties. Accordingly, ACCOBAMS clearly explains that the annex listing the cetaceans covered is only indicative of species names and classifications.[117] Likewise, in the 1996 Inter-American Convention for the Protection and Conservation of Sea Turtles, the list of sea turtles that the Convention covers are listed in Annex I,[118] but it is specifically noted that "[d]ue to the wide variety of common names, even within the same State, this list should not be considered exhaustive."[119]

The nomenclature guide is continually evolving. For example, in 1999 the Convention on Migratory Species (CMS) passed a recommendation on standardized nomenclature for the CMS Convention Appendices.[120] This noted that "biological nomenclature is dynamic,"[121] and that "the taxonomy used in the Appendices to the Convention will be most useful to the Parties if standardised by nomenclatural references."[122] The CMS then recommended that a series of standard

---

[117] ACCOBAMS, *supra* note 9, annex 1.

[118] Inter-American Convention for the Protection and Conservation of Sea Turtles, opened for signature Dec. 1, 1996, annex I (not yet in force) <http://www.seaturtle.org/iac/convention.shtml>.

[119] *Id.*

[120] CMS, Recommendation 6.1, "Standardised Nomenclature for the CMS Appendices" *in Proceedings of the Sixth Conference of the Parties* (1999) vol 1, 75. Note that the CMS as an organization should be distinguished from the *CMS Convention*. The CMS is administered by the CMS Secretariat, whose responsibilities include overseeing the development of the Convention in accordance with the resolutions of its member states, collectively known as the Conference of Parties (COP). *See also* United Nations Environment Programme, "Introduction to the Convention on Migratory Species" <http://www.wcmc.org.uk/cms/>, site visited on Sept. 21, 2001: "A Secretariat under the auspices of the United Nations Environment Programme ('UNEP') provides administrative support to the Convention. The decision-making organ of the Convention is the Conference of the Parties."

[121] CMS, Recommendation 6.1, *supra* note 120.

[122] *Id.*

references "be recognised and used as the bases on which the *CMS Convention* Appendices and amendments thereto, are prepared."[123]

Likewise, following problems with the identification of species through nomenclature at CITES,[124] at the tenth Conference of Parties of CITES it was similarly noted that "biological nomenclature is dynamic."[125] Moreover within the realm of CITES, "the names of the genera and species of several families are in need of standardization."[126] Without such a process, and due to discrepancies between the parties with regard to nomenclature, the effectiveness of the CITES Convention may have been damaged, as the signatories would otherwise be hemmed in by outdated scientific opinion.[127] Soon after, at the 11th Conference of Parties in 2000, the Nomenclature Committee was (re)established as the driving force to achieve a standardized nomenclature.[128] Its responsibilities include reviewing some of the species that are already listed, or about to be listed on, the CITES Convention Appendices to ensure consistency with "the correct use of zoological and botanical nomenclature."[129] Despite these clear indications, the idea that nomenclature is something that evolves and cannot (and should not) be held back by

---

[123]   *Id.*, "[T]he working group had been guided by the desire to keep to a minimum the number of consequential changes that would be required and the need for consistency with nomenclature of other organisations, in particular, CITES." *Id.* at 114.

[124]   *See also* CYRILLE DE KLEMM, GUIDELINES FOR LEGISLATION TO IMPLEMENT CITES 14–15 (1993). Note the distinction between CITES the international organisation, and the CITES Convention, *supra* note 8.

[125]   CITES, "Standard Nomenclature" (11th Conference of Parties) Doc 11.22 <http://www.cites.org/eng/resols/11/11_22.shtml>, site visited on Sept. 21, 2001.

[126]   *Id.*

[127]   CITES, "How to Improve the Effectiveness of the Convention" (10th Conference of Parties) Doc 10.22.

[128]   CITES, "Standard Nomenclature," *supra* note 125.

[129]   CITES, "Establishment of Committees: Establishment of the Nomenclature Committee of the Conference of Parties" (11th Conference of Parties) Doc. 11.1, annex 3 <http://www.cites.org/eng/resols/11/11_1.shtml>, site visited on Sept. 21, 2001. *See also* CITES, "Report of the Nomenclature Committee" (11th Conference of Parties) Doc. 11.11.4.1.

earlier (and often scientifically flawed) assumptions, is not well received within the IWC.

The history of the nomenclature of the ICRW can be found in Article IV of the Final Act of International Agreements for the Regulation of Whales[130] from the 1946 International Whaling Conference. This article recommended "[t]hat the chart of Nomenclature of whales annexed to this Final Act be accepted as a guide by the Governments represented at the conference."[131] The "Nomenclature of Whales," which was then admitted as an annex, listed the scientific names for 17 different species of whales.

The Japanese government has suggested that if a whale species was not on this original list, the ICW cannot claim jurisdiction over it unless all of the members agree. According to the wording of the VCLT, the justification for this assertion is that the nomenclature was an "*agreement relating to the treaty which was made between all the parties in connection with the conclusion of the treaty.*"[132] Accordingly, any attempt by the IWC to exercise management over stocks that were not listed in the original nomenclature would be *ultra vires* unless *all* members subsequently agreed to change the nomenclature:[133]

> [A]s the chart was adopted unanimously, the only way in which the Commission's competence to adopt regulatory measures can be extended to additional species is if there is unanimous agreement among Contracting Governments to do so.[134]

Therefore, as Japan argued, an agreement on whether to admit other species or not was a "matter for discussion between Contracting

---

[130]   Final Act of the International Whaling Conference (1946) art. IV, *reprinted in* BIRNIE, INTERNATIONAL REGULATION OF WHALING, *supra* note 17, at 695.

[131]   *Id.*

[132]   VCLT, *supra* note 72, at art. 31.2(a) (emphases added).

[133]   IWC, *Thirty-Fourth Report, supra* note 101, 16. *See also* IWC, *Thirty-Seventh Report of the International Whaling Commission* 15 (1987).

[134]   IWC/35/15, "Report of the Steering Committee on Regulation of Baird's Beaked Whale," *supra* note 111, at [7.1.3].

Governments."[135] Consequently, "since it is a matter of differing views between Contracting Governments . . . [it] cannot be dealt with by the Commission."[136] The implications of this view are that if there is no *agreement* between the parties on whether to regulate a species or not, it would not be regulated. This would lead to an outcome—as Denmark has suggested—of IWC jurisdiction being restricted to "*only* species named in the [original] Annex."[137] Any other approach, according to the former USSR, would be "legally unjustified."[138]

The question that this assertion raises is: was the nomenclature an agreement that was intended to limit the Commission in its jurisdiction? To answer this, a number of points need to be highlighted. First, the Final Act of the International Whaling Conference expressly recommended "[t]hat the chart of nomenclature of whales . . . be accepted as a guide."[139] The word guide does not suggest that the nomenclature was to constitute an exhaustive list.[140] Second, it appears that the nomenclature was designed as a guide to help achieve consistency with regard to the vernacular names of the whales which were, at that time, the subject of particular attention.[141] This explanation is supported by the title and layout of the chart.[142] As such, the nomenclature was only a guide to name usage. As New Zealand suggested, the nomenclature was "for

---

135　Japan also stated that it would "take no more than forty Baird's beaked whales in the next year, with catch control under its national regulations," IWC, *Thirty-Fifth Report, supra* note 44, at 13.

136　IWC, *Thirty-Sixth Report, supra* note 12.

137　IWC, *Thirty-Fifth Report, supra* note 44, at 13 (emphasis added). This view was followed by Mexico, Brazil, Peru and Argentina and was subsequently repeated in IWC, *Fortieth Report of the International Whaling Commission* 22 (1990).

138　IWC, *Thirty-First Report, supra* note 64, at 21.

139　Final Act of the International Whaling Conference, *supra* note 130, at 695.

140　IWC/35/15, "Report of the Steering Committee on Regulation of Baird's Beaked Whale," *supra* note 111, at [7.1.3].

141　IWC, *Thirty-Fourth Report, supra* note 101, at 16.

142　IWC/35/15, "Report of the Steering Committee on Regulation of Baird's Beaked Whale," *supra* note 111, at [7.1.3].

information only" or "illustrative, not exclusive."[143] Finally, the question of the overall status of the nomenclature within the ICRW needs to be examined. As it is set out, the nomenclature was annexed to the Final Act of the International Whaling Conference.[144] It was not annexed to the ICRW. By contrast, when the drafters of the ICRW attached the Schedule to the ICRW, the Schedule was deemed "an integral part thereof."[145] Conversely, in keeping with the assumption that the nomenclature was only to be a guide, the ICRW is completely silent as to its enforceability.[146] Hence it was never deemed an "integral part" of the Convention.

Those who are seeking to suggest that the ICRW cannot extend its mandate have confused the nomenclature and the Schedule. The nomenclatures that are attached to treaties are typically about vernacular names. By contrast, schedules or annexes attached to conventions typically reflect the decision by the parties as to the scope of their purview. There is no evidence to suggest, either within the history of the ICRW or comparable international regimes, that nomenclatures are, or were, ever intended to be ultimately the defining mechanisms in international treaties. Conversely, the defining mechanisms in treaties are those that expand or restrict the mandates of the majority of the signatories. Indeed, because such decisions are so important, the principle of

---

[143]   IWC, *Fortieth Report of the International Whaling Commission* 22 (1990). This idea has already been partly accepted. As was explained in the debate about the regulation of Baird's beaked whale,

> the reference in the Final Act to the acceptance of the chart as a guide only means that the names therein are to be taken as a guide, whereas the list of species which the Conference regarded as whales is exhaustive and the Commission's competence to adopt regulatory measures is restricted to those species listed.

IWC/35/15, "Report of the Steering Committee on Regulation of Baird's Beaked Whale," *supra* note 111, at [7.1.3].

[144]   Final Act of the International Whaling Conference, *supra* note 130, at 695.

[145]   *ICRW, supra* note 2, art. I(1). Article I goes on to explain that "all references to the Convention shall be understood as including the said Schedule."

[146]   IWC, *Thirty-Fourth Report, supra* note 101, at 16. *See also* IWC, *Fortieth Report of the International Whaling Commission* 22 (1990).

consensus[147] (but more commonly large majorities, as opposed to simple majorities) is required, along with the ability of dissenting signatories to object and not be bound by the weighted majority. Without this ability, which is clearly established in international environmental law (from climate change to toxic waste), neither the IWC nor the multiple other international organizations with similar purposes would allow for the positions of the vast majority of their signatories to *evolve* (while retaining mechanisms that allow dissent to the majorities) as new situations necessitated new responses.[148]

With such considerations in mind, it has been repeatedly suggested by various countries that the nomenclature cannot be used as a justification to limit the Commission's competence in this area.[149] Rather, the

---

[147]   *See also, e.g.*, the Convention on the Protection of the Marine Environment of the Baltic Sea Area, opened for signature Apr. 9, 1992, 32 I.L.M. 1101, art. 19(5) (entered into force Jan. 2, 2000) which requires decisions to be unanimous. The Convention on the Protection of the Black Sea Against Pollution, opened for signature Apr. 21, 1992, 1764 U.N.T.S. 3, 32 I.L.M. 1101, art. 21 (entered into force Jan. 15, 1994) stipulates that amendments must be on the basis of consensus. Under the Convention for the Conservation of Antarctic Marine Living Resources, opened for signature May 20, 1980, 1329 U.N.T.S. 47, 19 I.L.M. 84, art. XII (entered into force Apr. 7, 1982) decisions of the Commission for the Conservation of Antarctic Marine Living Resources on matters of substance are taken by consensus. Decisions on other matters are taken by a simple majority. Amendments to annexes of the Draft Protocol on the Convention of Long-Range Transboundary Air Pollution on Persistent Organic Pollutants, 37 I.L.M. 505, art. 14 (entered into force Mar. 16, 1983) require consensus. The Agreement on the International Dolphin Conservation Program, *supra* note 24, arts IX, XXX requires all decisions to be made by consensus. The Inter-American Convention for the Protection and Conservation of Sea Turtles, *supra* note 118, also requires all decisions to be made by consensus: art. V(5). Under the Convention on the Conservation and Management of Pollock Resources of the Bering Sea, opened for signature June 16, 1994, 34 I.L.M. 67, art. V(2) (entered into force Dec. 8, 1995) "matters of substance shall be taken by consensus." However, if consensus cannot be reached, alternative procedures for decision making exist. *See also* annex, pts. 1, 2.

[148]   IWC, *Thirty-Fourth Report, supra* note 101, 16.

[149]   IWC, *Thirty-Fifth Report, supra* note 44, 13. *See also* IWC, *Forty-Third Report, supra* note 41, 31; Goetschel, *supra* note 79, at [4]; IWC/35/15, "Report of the Steering Committee on Regulation of Baird's Beaked Whale," *supra* note 111, at [7.1.3].

answer as to whether a species is within the competence of the Commission hinges on whether it is placed in the Schedule by a three-quarters majority of the voting members of the IWC.[150]

5.2. Coastal States, UNCLOS Negotiations and EEZs

Developments in the law of the sea following the first and second Geneva conferences have been of decisive importance in the evolution of a coastal state's fishing rights and the acceptance of the EEZ concept. Although there were few claims to exclusive fishery zones before 1958 (notably made in the late 1940s and 1950s by some Latin American states), the exclusive fishery zone is actually a product of the failure of the Geneva conferences to settle the question of the territorial sea breadth, and to confer on the coastal state any special rights to exclusive access to fisheries beyond its 12-mile territorial sea. Accordingly, states decided to act unilaterally for both the purpose of conservation and economic gain.[151] The period of 1960–74 was characterized by a wave of unilateral claims to exclusive fishing zones and a considerable number of bilateral and regional agreements recognising these claims—despite the rulings of the ICJ to the contrary.[152]

By the time UNCLOS was concluded, the 200-mile EEZ had evolved into customary international law. Article 55 of UNCLOS defines the EEZ as "an area beyond and adjacent to the territorial sea . . . under which the rights and jurisdiction of the coastal [s]tate and the rights and freedoms of other States are governed by the relevant provisions of this Convention."[153] Within the EEZ the coastal state has "sovereign rights

---

[150]   *See also* IWC/35/15, "Report of the Steering Committee on Regulation of Baird's Beaked Whale," *supra* note 111. Competence is due "since it is included in the Schedule definitions," IWC, *Thirty-Sixth Report, supra* note 12. *See also* Goetschel, *supra* note 79, at [2].

[151]   *See also* Ramprakash Anand, *The Politics of a New Legal Order for Fisheries*, 11 OCEAN DEV. & INT'L L. 263, 268–70 (1982); Ann Hollick, *The Origins of 200-Mile Offshore Zones*, 71 AM. J. INT'L L. 494, 500 (1977).

[152]   *See also, e.g., Fisheries Jurisdiction (UK v Iceland) (Merits)* [1974] ICJ Rep. 3.

[153]   UNCLOS, *supra* note 5.

for the purpose of . . . exploiting, conserving and managing natural resources."[154] Moreover "the coastal State shall determine the allowable catch of the living resources in its exclusive economic zone."[155] Coastal States therefore have "full powers to determine the allowable catch of the living resources in the EEZ."[156]

5.2.1. Coastal States, EEZs and the IWC

Given the highly coastal nature of small cetaceans, it became apparent in the mid-1970s that parallel debates in the law of the sea negotiations and the IWC regarding the rights of states within their respective 200-mile EEZs, would render the issue of IWC competence "even more sensitive and difficult."[157] Moreover, due to states' mixed agendas, the IWC tended to defer to the then ongoing UNCLOS negotiations. Thus the Resolution from the 31st meeting acknowledged the ongoing overlap with these negotiations[158] and that contracting governments and other interested parties might have to consider the question of possible amendments to, or the renegotiation of, the Whaling Convention in a manner that would reflect a consideration of the developments in the law of the sea.[159] Once the UNCLOS negotiations were concluded, and the sovereign rights of coastal states were established, the IWC consistently expressed its awareness of these rights both generally[160] and

---

[154]   *Id.* at art. 56(1)(a).

[155]   *Id.* at art. 61(1). *See also* art. 62(1).

[156]   *See also* JOSE YTURRIAGA, THE INTERNATIONAL REGIME OF FISHERIES: FROM UNCLOS 1982 TO THE PRESENTIAL SEA 115 (1998).

[157]   BIRNIE, INTERNATIONAL REGULATION OF WHALING, *supra* note 17, at 470.

[158]   IWC, *Thirty-First Report, supra* note 64, at 31: "[T]he rights and responsibilities of Contracting Governments with respect to the conservation, management and study of cetaceans are matters under consideration of the UN Conference on the Law of the Sea."

[159]   *Id.*

[160]   The Resolution from the 41st Meeting noted, "the sovereign rights of coastal states" as set out in UNLCOS, *supra* note 5, at art. 65. Likewise, the Resolution on small cetaceans from the 43rd Meeting was "conscious of the sovereign rights of

specifically,[161] when dealing with individual countries.

This awareness can be juxtaposed with a growing number of countries (Mexico, Chile, Uruguay, Brazil, Argentina, Peru, Spain, Japan, the former USSR, and Costa Rica) that have all recorded their reservations on the Commission's competence in relation to small cetaceans and within coastal waters.[162] These countries broadly argued that "IWC membership is small compared to the 139 coastal States and it would be wrong for this body to take such responsibility."[163] To rectify this problem, the IWC attempted to promote "a cooperative dialogue with relevant States"[164] that are not members of the IWC, but have small cetaceans in their waters. To assist this dialogue, a funding mechanism to facilitate the participation of coastal states on relevant small cetacean issues was established in the mid-1990s.[165]

Despite these initiatives, a number of coastal states have refused outright to engage the IWC on this matter. This position was exemplified

---

coastal states," IWC, *Forty-Third Report, supra* note 41, at 32, 36–37. The Resolution from the 45th Meeting was also "[conscious] of the sovereign rights of coastal states, as set out in the United Nations Convention on the Law of the Sea," IWC, *Forty-Fifth Report of the International Whaling Commission* 41 (1995).

[161]   *See also* IWC Resolution 1994–3, "Resolution on Biosphere Reserve of the Upper Gulf of California and the Colorado River Delta" *in Forty-Fifth Report of the International Whaling Commission* 42 (1995). This Resolution noted the highly endangered status of this species and was "conscious of the sovereign rights of Mexico within its coastal waters."

[162]   IWC, *Thirty-Third Report, supra* note 107, at 26–27, 30. IWC, *Fortieth Report of the International Whaling Commission* (1990) 22.

[163]   IWC, *Forty-Second Report, supra* note 38, at 16.

[164]   IWC, *Forty-Fifth Report, supra* note 161, at 20. *See also* IWC Resolution 1994–2, "Resolution on Small Cetaceans" *in Forty-Fifth Report, supra* note 161, at 41: "[The IWC] recommends that efforts be made to continue to consider the problems facing small cetacean stocks, including . . . engaging . . . the coastal and range States concerned . . . [and] assessing the condition of stocks."

[165]   IWC, "Resolution on Addressing Small Cetaceans in the IWC" *in Forty-Fourth Report of the International Whaling Commission* 31–32 (1994). IWC, *Forty-Fifth Report, supra* note 161, at 20, 41. *See also* IWC, *Forty-Sixth Report of the International Whaling Commission* 38–9 (1996); IWC, "Voluntary Fund for Small Cetaceans," *Forty-Seventh Report of the International Whaling Commission* 54 (1997).

at the 46th Meeting by a Resolution (from which the countries seeking IWC small cetacean competence abstained) introduced by St. Vincent and the Grenadines to show the importance of sovereignty over coastal matters.[166] The Resolution noted:

> The governments of St Vincent and the Grenadines, St Lucia, Dominica and Grenada . . . do not accept the competence of the Commission in the management of small cetaceans and related research and . . . these governments may not therefore permit IWC research on small cetaceans in their territorial seas and Exclusive Economic Zones.[167]

5.2.2. Toward a Correct Understanding of the Implications of UNCLOS for the IWC

The thrust of these arguments seems to indicate that neither the IWC, nor any other body, has the authority to fetter the discretion of a coastal state with regards to sovereign decisions relating to the exploitation (or conservation) of their EEZ. A rider to this claim has been the implicit assertion that the IWC is somehow *ultra vires* even to be suggesting that it may have a mandate to assist in the management of small cetaceans in sovereign waters. However, these contentions are incorrect for a number of reasons. Moreover, in a number of cases clear limitations upon the authority and unfettered freedom of the coastal state to act have been imposed by UNCLOS.[168] That is, UNCLOS does not accord to coastal states complete sovereignty in the territorial seas that abut their terra firma. Rather, it affords them limited sovereignty, comprised of rights and jurisdiction over the listed economic resources of the EEZ.

5.2.3. Consenting to the IWC

The aforementioned "Resolution on Small Cetaceans" from the 46th Meeting is correct, in the sense that countries (as sovereign states) can

---

[166]   IWC, *Forty-Sixth Report of the International Whaling Commission* 20–21 (1996).

[167]   IWC Resolution 1995–4, "Resolution on Small Cetaceans" *in id.* 44.

[168]   *See also* UNCLOS, *supra* note 5, pts. V, XI.

quite legitimately object to the authority of the IWC (or any other international body for that matter) over the whales in their waters. However, the converse is also true in that parties to the ICRW (like any other international treaty) can agree to confer competence on the IWC regarding the exercise of their rights.[169] As such, UNCLOS and the development of the EEZ did not make the IWC redundant with regard to the coastal zone management of cetaceans (unless the signatories choose to exclude the IWC from these areas).

Moreover, throughout the history of the international management of cetaceans, whales in territorial waters have always been a direct concern. Indeed, from the outset the 1931 Convention for the Regulation of Whaling stated that "[t]he geographical limits within which the Articles of this Convention are to be applied shall include all the waters of the world, including both the high seas and territorial and national seas."[170] Six years later, the 1937 International Agreement for the Regulation of Whaling stated the coverage slightly differently: "The present agreement applies to factory ships and whale catchers and to land stations . . . and to all waters in which whaling is prosecuted."[171] It was this latter definition, which was followed in the 1946 ICRW, where no restriction upon the geographical limits of the IWC's competence over cetaceans was set forth. Accordingly, the *ICRW* was to apply

> to factory ships, land stations, and whale catchers under the jurisdiction of Contracting Governments and all waters where whaling is prosecuted by such factory ships, land stations and whale catchers.[172]

In accordance with this broad coverage, the IWC sought to ensure that full compliance with the ICRW by its signatories in all the areas in which whaling has been practiced. The inspection and observation regimes of the IWC, which were originally only designed to cover pelagic whaling (in non-territorial waters), were extended to cover land

---

[169]   Burke, *supra* note 78, at 296.

[170]   Convention for the Regulation of Whaling, *supra* note 82, at art. 9.

[171]   International Agreement for the Regulation of Whaling, *supra* note 84, at art. 2.

[172]   ICRW, *supra* note 2, at art. I(2).

stations in the 1960s,[173] and aboriginal whaling by the late 1970s.[174] In the last two examples, not only was the IWC exercising authority over territorial areas, it was also ensuring compliance of the signatories in these places, with what is in effect a very intrusive regime. Of course, the question that needs to be asked is did UNCLOS eclipse the possibility of such regimes? Clearly this is not the case—Article 311(2) of UNCLOS specifically addresses the problem of prior agreements and declares that

> the Convention shall not alter the rights and obligations of States Parties which arise from other agreements compatible with this convention and which do not affect the enjoyment by other States Parties of their rights or the performance of their obligations under this Convention.

Accordingly, two questions arise:

(1) Did the signatories agree to cede their own sovereignty on such issues to the IWC (which is quite permissible);[175] and.

(2) Is the *ICRW* compatible with *UNCLOS*?

5.2.4. Article 65

It has been my contention elsewhere that Article 65 of UNCLOS, and the supporting principles of primacy in international law,[176] make the

---

[173]   International Commission on Whaling, *Seventeenth Report of the Commission* 17, 19–22 (1967). International Commission on Whaling, *Eighteenth Report of the Commission* 18, 23 (1968). International Commission on Whaling, *Nineteenth Report of the Commission* 15, 18, 21–22 (1969) .

[174]   IWC, *Twenty-Eighth Report of the International Whaling Commission* 24 (1976).

[175]   *See also* BURKE, *supra* note 78, at 268, 291: "The argument that whales are not subject to the *ICRW* is difficult to follow. . . . [I]t stems from the belief that the *UNCLOS* over-rides the *ICRW* and reinstalls coastal state authority over whales." This position is not consistent with Article 65, which indicates the clear expectation regarding cetaceans that states shall "work through the appropriate international organizations for their conservation, management and study."

[176]   The principle of primacy suggests that an international organization, through either specific provisions in their founding treaties or through the general princi-

IWC the central and uppermost international authority for cetaceans.[177] With regard to the different question of the overlap of EEZs and the management of cetaceans, it is important to note that Article 65, in making no distinction between large and small marine mammals within EEZs, stipulates that

> [s]tates shall cooperate with a view to the conservation of marine mammals and in the case of cetaceans shall in particular work through the appropriate international organisations for their conservation, management and study.

When dealing with stocks on the high seas, Article 120 notes that "[a]rticle 65 also applies to the conservation and management of marine mammals in the high seas." The question of marine mammals was dealt with, once more, in chapter 17 of Agenda 21. Paragraph 17.47 of Agenda 21 begins by repeating Article 65 of UNCLOS, however it goes further than *UNCLOS* in paragraph 17.61. After expressing the hope that countries that are not members of appropriate organisations for the management of high seas fisheries should be encouraged to join these, it explains that:

States recognise:

(a)   The responsibility of the International Whaling Commission for the Conservation and management of whale stocks and the regulation of whaling pursuant to the 1946 International Convention for the Regulation of Whaling.

(b)   *The work of the International Whaling Commission Scientific Committee in carrying out studies of large whales in particular, as well as of other cetaceans.*

(c)   The work of other organizations, such as the Inter-American Tropical Tuna Commission and the Agreement on Small Cetaceans in the Baltic and North Sea under the Bonn Convention, in the

---

ples of treaty law, can gain precedence over another upon certain questions at international law.

[177]   Alexander Gillespie, *Forum Shopping In International Environmental Law: Between the IWC, CITES and The Management Of Cetaceans*, 33(1) OCEAN DEV. & INT'L L. 17–56 (2002).

conservation, management and study of cetaceans and other marine mammals.[178]

The same section is later repeated under the section on "marine living resources *under national jurisdiction.*"[179] Paragraph 17.63 ends with the plea that "[s]tates should cooperate for the conservation, management and study of cetaceans."

Therefore Agenda 21 is very important (with respect to small cetaceans) for three reasons. First, the IWC was clearly listed as the "appropriate body" for whale stocks.[180] Second, this recognition was not just limited to large cetaceans, but included other cetaceans as well, although the exact responsibility of the IWC for other cetaceans is far from clear.[181] Nevertheless, this sentence was clearly linked to the importance of the Scientific Committee of the IWC.[182] Finally, the work of other organizations that deal with small cetaceans was noted.[183]

5.2.5.   Article 64

In addition to Article 65, there are two further areas under which the jurisdiction of coastal states may be circumscribed under UNCLOS. These two areas are particularly important for states that are signatories to UNCLOS, but not the IWC. The first area in which there is "a certain limitation of a coastal State's sovereign rights in its zone"[184] is that of highly migratory species.[185] Annex I of UNCLOS lists the highly migratory species in question. In addition to listing the large cetacean

---

178   Agenda 21, *supra* note 3, at [17.62] (emphasis added).

179   *See also id.* at [17.75]–[17.96] (emphasis added).

180   *Id.* at [17.90](a).

181   *Id.* at [17.90](b). *See also* Patricia Birnie, *Small Cetaceans and the International Whaling Commission*, 10 GEO. INT'L ENVTL. L. REV. 1, 20 (1997).

182   Agenda 21, *supra* note 3, at [17.90](a), (b).

183   *Id.* at [17.90](c).

184   *See also* YTURRIAGA, *supra* note 156, 129.

185   *Id.* at 127–30. *See also* DOUGLAS JOHNSTON, THE INTERNATIONAL LAW OF FISHERIES: A FRAMEWORK FOR POLICY-ORIENTATED INQUIRIES LXXI (1987).

families, annex I also includes the families of *Monodontidae* (narwal and beluga); *Ziphiidae* (including 20 small to medium sized beaked whale species) and *Delphinidae* (with 26 species of dolphins).[186] Accordingly, Article 64 also covers cetaceans, and makes no distinction between large and small whales.

However, an important distinction is made in Articles 63 and 64. Article 63, which deals with stocks occurring within the EEZs of two or more coastal states and/or within an EEZ and in an area beyond and adjacent to it, specifies that

> these States shall seek, either directly or through appropriate *subregional or regional* organisations, to agree upon the measures necessary to coordinate and ensure the conservation and development of such stocks.[187]

Conversely, under Article 64 "highly migratory species":

> The coastal State and other States whose nationals fish in the region for the highly migratory species listed in Annex 1 shall cooperate directly or through *appropriate international organisations* with a view to ensuring conservation and promoting the objective of optimum utilisation of such species throughout the region. . . . In regions for which *no* appropriate international organisation exists, the coastal State and other States whose nationals harvest these species in the region shall cooperate to establish such an organisation and participate in its work.[188]

The importance of Article 64 is twofold. First, it makes no distinction between the listing of the two suborders in the same appendix. This would appear to legitimize the assumption that they should be dealt with together. Second, countries which share these species are mandated (note the operative words, as with Article 65, are "*shall* cooperate" "through appropriate international organisations").[189] The use of the word international is crucial here, in contrast with the regional vision of Article 63.

---

[186]   For a discussion of the various species involved. *See also* CARWARDINE, *supra* note 6, at 99–223.

[187]   Emphasis added.

[188]   Emphasis added.

[189]   Emphasis added.

5.2.6.   The Bottom Line on the High Seas and Within EEZs: Extinction

UNCLOS gives the coastal state considerable discretion in determining resource exploitation levels within its EEZ. As such, UNCLOS does not provide detailed management and conservation schemes. Rather, it provides a framework in which the coastal state retains an independent discretion to formulate its own management plans, within certain broad conservation constraints and standards.[190] This power is so wide that Article 297(3)(a) specifically excludes the coastal states' discretion to determine the total allowable catch from the application of the compulsory dispute settlement procedure provided for in section 2 of part XV of UNCLOS. Due to coastal states' freedom in managing their own catch limits within their EEZ's, it has been contended that "little if anything in these management and conservation requirements seems sufficiently precise and mandatory to constitute an effective restriction on the coastal State's sovereign rights to exploitation."[191] The worst-case result of such freedom may be that a state chooses, either directly or through negligence, to fish a species to extinction. For example, with regard to small cetaceans and Japan, it has been contended that

> with virtually no government control on the hunt of dolphins, porpoises and small whales, the increased efficiency and hunting effort [has] resulted in extensive over-catching, causing the numbers of several species or populations to plummet. . . . [T]he exploitation of small cetaceans in Japan's waters [has] resulted in a domino effect, overhunting one species after another.[192]

The question that this scenario raises is whether or not such exploitation is permissible under UNCLOS? It would appear that no state may drive a marine species towards extinction. This point may be inferred

---

190   UNCLOS, *supra* note 5, pts V, XI. *See also generally* Mohamed Dahmani, The Fisheries Regime of the Exclusive Economic Zone 42–50 (1987).

191   *See also* Johnston, *supra* note 185, LIX, LXVII–LXIX. *See also* Burke, *supra* note 78, at 266.

192   *See also* Environmental Investigation Agency, *Towards Extinction: The Exploitation of Small Cetaceans in Japan* 2 (2000.

from the principles surrounding both the management of the high seas and EEZs.

With regard to the high seas, it has been evident since the 1958 Convention on Fishing and Conservation of the Living Resources of the High Seas that "the development of modern techniques for the exploitation of the living resources of the sea . . . has exposed some of these resources to the danger of being over-exploited."[193] Therefore, "[a]ll States have the duty to adopt, or to cooperate with other States in adopting, such measures for their respective nationals as may be necessary for the conservation of the living resources of the high seas."[194] This principle was later reiterated by the International Court of Justice:

> It is one of the advances of maritime international law resulting from the intensification of fishing, that the former laissez-faire treatment of the living resources of the sea in the high seas has been replaced by a recognition of a duty to have due regard to the rights of other States and the needs of conservation for the benefit of all.[195]

Very similar rules were later expressed in UNCLOS in parts V and VII. Part V dealt primarily with species in the EEZ (which I shall come to shortly) and part VII focused upon the "Conservation and Management of Marine Living Resources of the High Seas." In this latter section, the cooperation of states in the "Conservation and Management of Living Resources,"[196] and the "Conservation of the Living Resources of the High Seas,"[197] was highlighted.

The necessity of protecting and restoring marine species was reiterated in Agenda 21,[198] as well as in the United Nations Food and

---

[193]   Convention on Fishing and Conservation of the Living Resources of the High Seas, opened for signature Apr. 29, 1958, 559 U.N.T.S. 285, preamble (entered into force Mar. 20, 1966).

[194]   *Id.* art. 1(2).

[195]   *Fisheries Jurisdiction (UK v Iceland) (Merits)* [1974] ICJ Rep. 3, 31.

[196]   UNCLOS, *supra* note 5, at art. 118.

[197]   *Id.* at art. 119.

[198]   Agenda 21, *supra* note 3, at [17.46] (b), (e), [17.74] (c), (e).

Agriculture Organization's Code of Conduct for Responsible Fisheries (FAO Code of Conduct)[199] and in the 1995 Agreement on Straddling and Highly Migratory Fish Stocks.[200] Indeed, the first general principle of the FAO Code of Conduct is that "[t]he right to fish carries with it the obligation to do so in a responsible manner so as to ensure effective conservation and management of marine living resources."[201] The Agreement on Straddling and Highly Migratory Fish Stocks added to this by securing a convention that has the objective of the long term conservation of the stocks in question.[202]

Finally, the rule not to drive species on the high seas towards extinction was strengthened in discussion before the International Tribunal for the Law of the Sea in the *Southern Bluefin Tuna Cases*.[203] Specifically, New Zealand and Australia argued that Japan had "breached its obligations under Articles 64 [with regard to highly migratory species] and 116 to 119 [with regard to conservation and management of the living resources of the high seas] of *UNCLOS* in relation to the conservation and management of the [southern bluefin tuna]."[204] More specifically, they contended that Japan had failed "to adopt necessary conservation measures for its nationals fishing in the high seas so as to maintain the [Southern Bluefin Tuna] stock to levels which can produce the maximum sustainable yields, as required by article 119 and contrary to the obligation in article 117 to take necessary conservation measures for its nationals."[205] While the Tribunal found against New Zealand and Australia at the jurisdiction stage of these cases, it effectively left open

---

[199]   (Adopted Oct. 31, 1995), at art. 6.1.

[200]   Agreement on Straddling and Highly Migratory Fish Stocks, *supra* note 94.

[201]   FAO Code of Conduct, at art. 6.1. Thereafter, the importance of making sure that "depleted stocks are allowed to recover or, where appropriate, are actively restored" is emphasized. *See also* arts. 7.2.2(e), 7.6.10.

[202]   Agreement on Straddling and Highly Migratory Fish Stocks, above 94, art. 2. *See also* preamble (1), art. 5(a).

[203]   *New Zealand v Japan; Australia v Japan (Provisional Measures)* (1999) 38 I.L.M. 1624 (Order of Aug. 27, 1999).

[204]   *Id.* at [28(1)].

[205]   *Id.* at [28(1)(a)].

the possibility of finding Japan in breach of Articles 64 and 116–119, which would further support the protection of these species from extinction.[206]

With regard to EEZs, states are also obliged to protect marine species from extinction.[207] The primary provision relating to this obligation is in Article 61(2):[208]

The coastal State, taking into account the best scientific evidence available to it, shall ensure through proper conservation and management measures that the maintenance of the living resources in the exclusive economic zone is not endangered by over-exploitation.

Article 61(4) adds:

[T]he coastal State shall take into consideration the effects on species associated with or dependent on harvested species with a view to maintaining or restoring populations of such associated or dependent species above levels at which their reproduction may become seriously threatened.

To help achieve these goals, "[a]s appropriate, the coastal State and competent international organisations, whether sub-regional, regional, or global, shall co-operate to this end."[209] Moreover,

available scientific information, catch and fishing effort statistics, and other data relevant to the conservation of fish stocks shall be contributed and exchanged on a regular basis through competent *international* organizations, whether sub-regional, regional or global.[210]

---

[206]   *Id.*

[207]   *See also* UNCLOS, *supra* note 5, at art. 61(2). For a discussion of this obligation, *see also* YTURRIAGA, *supra* note 156, at 15, 118; DAHMANI, *supra* note 190, at 43, 49.

[208]   *See also* UNCLOS, *supra* note 5, at art. 194(5), which requires parties to take measures necessary to protect the habitats of depleted, threatened or endangered species.

[209]   *Id.* at art. 61(2).

[210]   *Id.* at art. 61(5) (emphasis added).

As these last provisions show, when dealing with the bottom line of the extinction of a species, a role is contemplated for the most competent international organisations. In dealing with such questions with regard to cetaceans, that body would have to be the IWC.

5.3.   Regional Organizations

For the countries that have objected to the universal managerial competency of the IWC, a common argument has been that the only organisations competent to manage small cetaceans (apart from the coastal states themselves) are regional organizations.[211] As Japan explained:

> [The] IWC should recognize that small cetacean [sic] migrating and distributing within [the] 200 nautical mile zone of coastal States [are] subject to the management [of] the regional organisation or [the] coastal States concerned.[212]

The emphasis on the importance of regional cooperation for such matters is well established with regard to marine pollution (and the UNEP's Regional Seas Programme)[213] and in international fisheries law and policy. With regard to the latter, a number of documents such as Agenda 21,[214] the Agreement on Straddling and Highly Migratory Fish Stocks,[215] the FAO Code of Conduct,[216] and the Kyoto Plan of Action,[217] recognize the importance of regional cooperation.

---

[211]   *See also* IWC, *Forty-Fifth Report, supra* note 161, at 20; IWC, *Forty-Third Report, supra* note 41, at 31. *See also* BURKE, *supra* note 78, at 264–65, 293–94.

[212]   IWC/46/SM1, "Japan's View on Addressing Small Cetaceans within the IWC" (Paper presented at the 45th Meeting of the IWC, Puerto Vallarta, May 1994), at [e].

[213]   *See also* Terttu Melvasalo, *Cleaning the Seas*, 9 OUR PLANET 1–5 (1998). Regional cooperation is in accordance with UNCLOS, *supra* note 5, at art. 197.

[214]   Agenda 21, *supra* note 3, at [17.10].

[215]   *Supra* note 200, at arts. 7, 9.

[216]   FAO Code of Conduct, *supra* note 199, at arts. 6.12, 10.3.

[217]   The "Plan of Action" encourages states to "enhance subregional and regional cooperation and establish, where it is considered appropriate, subregional and regional fishery conservation and management organizations," United Nations Food

The assertion that regional organizations have a "crucial role to play with respect to small cetaceans"[218] is one on which most sides of the debate agree. Some regional agreements already encompassed small cetaceans by the early 1980s.[219] By the late 1980s the IWC was seeking to work with related regional organizations on issues affecting small cetceans—the Inter-American Tropical Tuna Commission[220] and UNEP[221] both received direct attention. In the 1990s the focus on organisations with an overlapping interest in small cetaceans moved indirectly to CITES,[222] and directly to the CMS.

With particular regard to the CMS and the question of small cetaceans, an invitation to the CMS Secretariat "to exchange information with the Secretary of the IWC" was made.[223] The engagement with the

---

and Agriculture Organization, "Kyoto Plan of Action" (Paper presented at the International Conference on the Sustainable Contribution of Fisheries to Food Security, Kyoto, December 1995), at [2] <http://www.fao.org/fi/agreem/kyoto/kyoe. asp>, visited on Sept. 21, 2001.

[218]   IWC, "Resolution on Addressing Small Cetaceans in the IWC," *supra* note 40.

[219]   *See also, e.g.*, Convention on the Conservation of European Wildlife and Natural Habitats, *supra* note 92, art. 4. The Convention requires its parties to ensure the special protection of all species listed in appendix II, which includes all species of small cetaceans normally found in European waters. For a discussion of the limitations of this convention with regard to small cetaceans, *see also* Robin Churchill, *Sustaining Small Cetaceans: A Preliminary Evaluation of the ASCOBANS & ACCOBAMS Agreements, in* INTERNATIONAL LAW AND SUSTAINABLE DEVELOPMENT: PAST ACHIEVEMENTS AND FUTURE CHALLENGES 225, 231–33 (Alan Boyle & David Freestone, eds. 1999). The European Union also has regulations which deal with the protection of small cetaceans. *Id.* at 240–43.

[220]   IWC, *Thirty-Ninth Report of the International Whaling Commission* 24–45 (1989).

[221]   *Id.* at 29.

[222]   CITES Convention, *supra* note 8. In 1993 it was noted that "CITES had requested comments on five species of small cetaceans, which it proposed be transferred onto its appendix I: the Irrawaddy, rough-tooth and Risso's dolphins, short-finned pilot and dwarf sperm whales," IWC, *Forty-Fourth Report of the International Whaling Commission* 25 (1994).

[223]   IWC, "Resolution on Small Cetaceans" *in Forty-Third Report, supra* note 41, at 51.

CMS was undertaken because this organisation was seen as important and largely complementary to the work of the IWC.[224]

The relationship between the CMS and the IWC took a further step forward in 2000, when a memorandum of understanding was signed by the two organizations, despite objections from Japan. This memorandum seeks to "establish a framework of information and consultation between CMS and IWC." In particular, the two bodies will "to the extent possible, coordinate their programme of activities to ensure that their implementation is complementary and mutually supportive."[224] In 2001 this approach was strengthened by the suggestion that the IWC, under this memorandum with the CMS, seek to "pursue complementary and mutually supportive actions in respect of small cetaceans."[225]

With respect to such considerations, the IWC has encouraged cooperation between ASCOBANS work, the range states[226] of the North Atlantic harbor porpoise, and the IWC.[227] This cooperation was furthered in 1998 when the Scientific Committee recommended the establishment of a joint working group with ASCOBANS to consider scientific matters relating to the status of harbor porpoises in the eastern North Atlantic.[228]

Despite growing cooperation with organizations like the CMS, the IWC has been selective about which organizations it will work with in regard to small cetaceans. Accordingly, there have been repeated suggestions by those who oppose the competence of the IWC with regard

---

[224]   *See also* Gillespie, *supra* note 177.

[225]   IWC, "Resolution on Small Cetaceans" *in Annual Report of the International Whaling Commission 2001* 47 (2001).

[226]   *See also* CMS Convention, *supra* note 7, at art. I(1)(h): "'Range State' in relation to a particular migratory species means any State . . . that exercises jurisdiction over any part of the range of that migratory species, or a State, flag vessels of which are engaged outside national jurisdictional limits in taking that migratory species."

[227]   IWC, "Resolution on Harbour Porpoise in the North Atlantic and the Baltic Sea" *in Forty-Fourth Report of the International Whaling Commission* 34–35 (1994).

[228]   IWC, *Annual Report 1998, supra* note 91, at 34.

to small cetaceans, that a more appropriate regional organization to work with would be the North Atlantic Marine Mammal Commission (NAMMCO).[229] In comparison with other regional bodies dealing with small cetaceans, NAMMCO has received little recognition from the IWC as it competes with the IWC for primacy in a number of areas.[230] Nevertheless, it is useful to note that, with regard to the above quoted documents and the drift to regionalism, an equally strong current can be found with regards to international bodies (as opposed to just regional ones) when it comes to management of the oceans. Thus Agenda 21 stresses the importance of "international and regional cooperation and coordination."[231] With particular regard to coastal states and the high seas it was noted that "[e]ffective cooperation within existing subregional, regional or global fisheries bodies should be encouraged."[232] The FAO Code of Conduct also took this approach by suggesting that "[s]tates should, within their respective competences and in accordance with international law, cooperate at subregional, regional and global levels through fisheries management organisations."[233]

Moreover, the FAO Code of Conduct emphasized that the organization in question must have "the competence to establish conservation and management measures."[234] In a similar way, the Agreement on Straddling and Highly Migratory Fish Stocks also highlights the point that the appropriate management organizations must have competence.[235] In addition, it is important to note that this Convention did not

---

[229]   IWC, *Forty-First Report, supra* note 37, at 39–40. *See also* IWC, *Forty-Seventh Report of the International Whaling Commission* 22 (1997).

[230]   I have written at length elsewhere on the primacy of the IWC in international law with regard to the management of cetaceans and thus the necessity to treat organizations such as NAMMCO as possessing limited integrity. Accordingly, I do not intend to reiterate my earlier arguments: *See also* Gillespie, *supra* note 177.

[231]   Agenda 21, *supra* note 3, at [17.10].

[232]   *Id.* at [17.60]. *See also* [17.45].

[233]   FAO Code of Conduct, *supra* note 199, at art. 6.12.

[234]   *Id.* at art. 7.1.4.

[235]   Agreement on Straddling and Highly Migratory Fish Stocks, *supra* note 94, at art. 8(6).

eclipse the principles espoused by UNCLOS. Rather, it noted the importance of taking "into account previously agreed measures established and applied in accordance with the Convention."[236] Hence parts V and VII of UNCLOS (with the general conservation requirements within the EEZs and the high seas) and the special status of marine mammals (Articles 64, 65 and 120) remain standing.

These principles, when taken in conjunction with the rules of international organizations, should make it apparent that the IWC, as an international body, may have a strong role to play in the management of small cetaceans. Moreover, although regional organizations also have an important part to play in the management of small cetaceans, it is necessary that these are complementary to the IWC and not inconsistent with its work.

5.3.1.   The Limits of Regional Organizations and the Antarctic Killer Whale

The case of Antarctica is a particularly useful example of such regional complementarity, as cetaceans are intricately connected with the marine environment of the South Pole.[237] One of the cetacean species which often resides at the South Pole (and in numerous other locations) is the killer whale, which is commonly recognized as a small cetacean. The Antarctic killer whale was the subject of controversy in 1980 when the USSR objected to attempts by the IWC to set catch quotas for this species in the Antarctic area:

> [K]iller whales belong to small cetaceans which are not regulated by the Commission within the framework of the existing Convention. Thus [the USSR believed] that the inclusion of killer

---

[236]   *Id.* at art. 7(2)(c).

[237]   *See also* IWC, "Resolution to Consider the Implications for Whales of Management Regimes for Other Marine Resources" *in Thirtieth Report of the International Whaling Commission* 34 (1980). "[Recognising] that certain marine resources in the Southern Ocean, especially krill, are food species of whales, and that exploitation of these resources may affect the demography of whale stocks to an extent that is as yet largely unknown."

whales in the number of species regulated by the Commission is legally unjustified.[238]

The correlate to this position to the above quotation is that such small cetaceans should be dealt with by appropriate regional organizations. However, regional organizations (for example, NAMMCO) are not always appropriate or they may not even exist, let alone claim authority for the species in question. The killer whale, which resides in Antarctic waters, cannot be claimed as a species which resides in either sovereign waters (as Antarctica is governed by the signatories to the Antarctic Treaty and is not subject to sovereignty claims)[239] or within the jurisdiction of strong regional or international organisations which have an interest in it. Rather, the governing international legislation for Antarctica has been explicit in displaying deference to the greater body—the IWC—in dealing with whaling matters.[240] This deference was carefully spelt out in the 1980 Convention for the Conservation of Antarctic Marine Living Resources:

> Nothing in this Convention shall derogate from the rights and obligations of Contracting Parties under the International Convention for the Regulation of Whaling.[241]

Moreover article XXIII (3) states:

> The Commission and the Scientific Committee [of the Commission for the Conservation of Antarctic Marine Living Resources] shall seek to develop cooperative working relationships, as appropriate, with inter-governmental and non-governmental organisations which could contribute to their work, including . . . the International Whaling Commission.[242]

---

[238]   IWC, *Thirty-First Report, supra* note 64, at 21.

[239]   Antarctica Treaty, opened for signature Dec. 1, 1959, 402 U.N.T.S. 71, art. IV(2) (entered into force June 23, 1961).

[240]   Agreed Measures for the Conservation of Antarctic Fauna and Flora, opened for signature June 2, 1964, UKTS 1978 No 45, art. II(a) (entered into force Nov. 12, 1982).

[241]   Convention for the Conservation of Antarctic Marine Living Resources, opened for signature May 20, 1980, 1329 U.N.T.S. 47, 19 I.L.M. 84, art. VI (entered into force Apr. 7, 1982).

[242]   Art. XXIII(3) operates together with art. IX(5) to ensure that "[t]he

From this background, the Commission for the Conservation of Antarctic Marine Living Resources (CCAMLR) and the IWC have established a strong (but informal) coordinated working relationship[243] in areas of mutual interest.[244] This relationship has flourished since the mid-1980s.[245]

The result of these provisions is that it is unclear which organization should manage these species if, as Russia earlier claimed, the killer whale in Antarctica cannot be governed by the IWC because it is a small

---

Commission shall take full account of any relevant measures or regulations established or recommended by . . . existing fisheries Commissions responsible for species which may enter the area to which this Convention applies [including the IWC] in order that there shall be no inconsistency between the rights and obligations of a Contracting Party under such regulations or measures and conservation measures which may be adopted by the Commission."

[243] In 1982 the secretariats of the IWC and CITES had discussions about possible cooperative arrangements and a possible agreement: IWC, *Thirty-Third Report, supra* note 107, at 37. With regard to other organizations, the IWC policy was "to welcome and extend every facility to organisations such as [the CCAMLR], and [hope] for similar treatment in return," IWC, *Thirty-Fourth Report, supra* note 101, at 15, 28–29. In 1984 the CCAMLR "confirmed that it did not wish to formalise working relationships with other organisations, but [would] continue to communicate between Secretariats," IWC, *Thirty-Fifth Report, supra* note 44, at 25.

[244] *See also* IWC, "Resolution on Cooperation and Coordination between the International Whaling Commission and the Proposed Commission for the Conservation of Antarctic Marine Living Resources" *in Thirty-First Report, supra* note 64, at 30. Argentina and Chile reserved their positions on this resolution. *Id.* at 20.

[245] In 1986 the Scientific Committee responded to requests for assistance by CCAMLR, in relation to questions on the suitability of whales as indicator species for krill availability, IWC, *Thirty-Seventh Report of the International Whaling Commission* 24 (1987). The Scientific Committee continued its cooperation with the Scientific Committee of the CCAMLR over this issue. *See also* IWC, *Forty-First Report, supra* note 37, at 46. In 1992 joint workshops between CCAMLR and the IWC were being considered, IWC, *Forty-Third Report, supra* note 41, at 30. In 1998 the Commission noted formation of a small liaison group between the Scientific Committee's of the IWC and the CCAMLR. The IWC Scientific Committee noted the "great importance it attached to cooperation with CCAMLR and . . . endorsed the formation of the liaison group," IWC, *Annual Report 1998, supra* note 91, at 32.

cetacean. Clearly, the species needs to be managed by some international organization and cannot under any circumstance be considered
"free for the taking." In this instance, the appropriate organization is
probably the CCAMLR, as it is primarily concerned with managing the
marine living resources of the Antarctic.[246] However, in accordance with
the rules of primacy and the necessity to complement (and not compete
against) the IWC, the CCAMLR has clearly specified that any question
of cetacean management is best dealt with by the IWC.

6.   THE CONVENTION ON MIGRATORY SPECIES

The only other international instrument that has specifically addressed
the protection of small cetaceans is the CMS Convention. As I have
explained elsewhere,[247] the CMS Convention (and more specifically, the
CMS Secretariat) has a close working relationship with the IWC.

The interest of the CMS in cetaceans began at the first Conference of
Parties of the CMS in 1985, when a number of large cetaceans were
listed in appendix I of the CMS Convention.[248] In addition, at the same
meeting, it was proposed that the Indus River dolphin be listed in appendix I.[249] At the same time, the CMS Scientific Committee advised that
since a number of small cetaceans were clearly threatened,[250] and
although significant scientific information on many of them was lacking,[251] many of these should be considered for inclusion in appendix II

---

[246]   *See also* Convention for the Conservation of Antarctic Marine Living
Resources, *supra* note 241.

[247]   *See also* Gillespie, *supra* note 177.

[248]   CMS Convention, *supra* note 7, at appendix I. These were the blue, humpback, bowhead, southern and northern right whales, CMS, *Proceedings of the First
Meeting of the Parties* 10 (1985).

[249]   CMS, 2 *Proceedings of the First Meeting of the Conference of the Parties*
23 (1987).

[250]   *Id.* at 8–9, "Working Group on Marine Mammals." A working group on
marine migratory animals drafted a paper on the biological elements for agreement
on certain small cetaceans. This paper listed the harbor porpoise (North and Baltic
Sea) as well as Commerson's dolphin as examples of threatened species.

[251]   CMS, Resolution 3.3, "Small Cetaceans" *in Proceedings of the Third*

of the CMS Convention at the next Conference of Parties.[252] These suggestions helped lead to Resolution 1.7.[253] Although this resolution was not passed unanimously,[254] a working group on small cetaceans[255] was established. This group was required to work "in conjunction with . . .

Meeting of the Conference of Parties 20 (1991): "The Scientific Council's report on the global review of the conservation of small cetaceans provides a detailed basis for the conservation measures to be included in agreements for the species and populations identified for listing in Appendix II." Scientific knowledge is also deficient with regard certain important behaviour like migration. In turn, this makes "the nature and scope of international conservation programmes difficult to determine, and making regional and international co-operation difficult to achieve." *See also* CMS, Recommendation 4.2, "Research on Migration of Small Cetaceans" *in Proceedings of the Fourth Meeting of the Conference of the Parties* (1994). Therefore, it has been recommended that the parties to the CMS Convention, and the CMS itself, carry out specific studies to investigate various issues. In 1994 the Scientific Council of the CMS discussed the status of small cetaceans in other regions such as Latin America and South East Asia, where despite a common "paucity of data," "it was clear from the limited amount of information available that there were many problems facing the small cetaceans of the area." *Id.* at 17, 121. In 1996 the Scientific Council began studies of cetacean-fishery interactions in the southwestern Sulu Sea and northeastern Malaysia: *Small Cetaceans/ Migratory Mammals* (1996). 5 CMS BULL. 6. In 1997 studies were conducted into cetacean distribution and cetacean-fisheries interactions in coastal waters around West Africa. The results from these studies revealed that at least 24 species of whales and dolphins inhabited these waters, of which at least one (the Atlantic humpback) whose "long-term survival" would be under "serious threat," largely due to problems of by-catch: Koen van Waerebeek, *Update on Conservation Activities: A Survey on the Conservation Status of Cetaceans in Senegal, The Gambia and Guinea* (2000). 10 CMS BULL. 11. A dedicated workshop to the La Plata dolphin, which inhabits the coastal waters of Argentina, Brazil and Uruguay, was also held in 1997, discussing the risks posed to its survival through incidental catch: *From the Regions: Americas and the Caribbean*, 7 CMS BULL. 4, 7 (1997).

[252]   CMS, Resolution 1.7, "Small Cetaceans" *in Proceedings of the First Meeting, supra* note 249, at 51.

[253]   *Id.*

[254]   Denmark reserved its position on this resolution, as it had not consulted either Greenland or the Faroe Islands, *id.* at 30.

[255]   *Id.*

[256]   *Id.*

appropriate national and international [organisations]."[256] The appropriate organization in question was clearly the IWC, and the IWC Observer at the meeting responded that this would be ready to cooperate with the group to be established under the resolution.[257]

At the second CMS Conference of Parties in 1988, despite objections against the listing of certain species from Norway[258] and Denmark,[259] the majority of the CMS parties followed the advice of the working group[260] and added seven species of small cetaceans to appendix II of the CMS Convention.[261] This list was extended substantially in 1991 at the third Conference of Parties,[262] where once more Norway and Greenland spoke against the inclusion of certain species in the list.[263] By 1994, 27 species of small cetaceans were included in appendix II of

---

[257]   *Id.* at 30.

[258]   *See also* CMS, *Proceedings of the Second Meeting of the Conference of the Parties* (1988) 16. Norway, while not wishing to break the consensus, noted that had these been voted on separately, they would have abstained on the white-beaked, and white-sided dolphins.

[259]   Denmark did not believe that the pilot whale should be included in the appendices, *id.* at 36.

[260]   After issuing its report, the working group was disbanded, and folded back into the Scientific Committee of the CMS, *id.* at 14–15.

[261]   *Id.* at 35–36. These were the bottlenose dolphin, the common dolphin, Risso's dolphin, the pilot whale, the harbor porpoise, and the white-beaked, and the white-sided dolphins.

[262]   CMS, *Proceedings of the Third Meeting of the Conference of the Parties* 10-11 (1991). In 1991 the Scientific Committee of the CMS recommended that a number of other small cetaceans be added to appendix II. These were the Indus and Ganges river dolphins, Franciscana, the boto river dolphin, the narwhal, the Indo-Pacific humpbacked dolphin, the harbor porpoise (Black Sea and North Atlantic populations), the finless porpoise, the Atlantic humpback dolphin, tucuxi, Peale's dolphin, the bottlenose dolphin (Black Sea population), the long snouted spinner dolphin (eastern tropical pacific), the striped dolphin (eastern tropical Pacific and western Mediterranean), the common dolphin, Baird's beaked whale, Commerson's dolphin, the irrawaddy dolphin, the killer whale (North Atlantic and north-eastern Pacific), the Northern bottlenose and Heaviside's dolphin.

[263]   *Id.* at 10. Norway spoke against inclusion of the killer whale in this list, as did Greenland against the inclusion of the narwhal, which they stated was already the subject of an agreement between Greenland and Canada.

the CMS Convention.[264] At the seventh Conference of the Parties earlier this year, an additional small cetacean, the pygmy right whale, was added to Appendix II.[265]

With regard to the question of whether or not the issues raised by the CMS Conferences of Parties were overlapping with debates in other fora, such as the IWC, it was concluded that the two governing conventions were dealing with substantively different topics. Thus

[t]here was general agreement among the representatives of the Parties to the convention that coverage of any given species in another Convention was not per se, an argument against coverage in the Bonn convention. The International Whaling Convention, for example, was mainly concerned with matters such as catch levels rather than habitat protection.[266]

This view was reiterated in discussions in the Scientific Committee at the Fourth CMS Conference of Parties in 1994,[267] where it was sug-

---

[264]   *See also* CMS, Recommendation 4.2, "Research on Migration of Small Cetaceans," *Proceedings of the Fourth Meeting of the Conference of the Parties* (1994).

[265]   UNEP, *"Jaws" Win Tough New Protection from Human Predators*, <http://www.unep.org/Documents/Default.asp?ArticleID=3135&DocumentID=264>, Sept. 25, 2002, site visited on Oct. 5, 2002.

[266]   CMS, *Proceedings of the Second Meeting, supra* note 258, at 35.

[267]   *See also* CMS, *Proceedings of the Fourth Meeting, supra* note 264. This relationship between the IWC and the CMS was the subject of discussion at the Fourth Meeting of the Conference of Parties of the CMS in 1994. India expressed concerns that efforts should be made to avoid duplication between the CMS and the IWC where small cetaceans were concerned. In response, the Coordinator of the CMS Secretariat pointed out that closer contacts with the secretariats of other conventions were favored, but that they were constrained in their efforts to collect and exchange data on small cetaceans due to their limited resources. However, agreement had been reached with the Secretary of the IWC on the need for closer contact and a regular exchange of information. Such contact had been strengthened by the election of the Vice-Chairman of the CMS Standing Committee as Chairman of the IWC. In addition, it was noted that at the 1994 meeting, the CMS Scientific Committee had recommended holding consultations with the IWC on the question of small cetaceans. It noted that although there were differences of opinion within the IWC on its competency to deal with the matter of small cetaceans, there had

gested that "there [were] significant prospects for complementarity"[268] between the CMS and the IWC. This was because "the [CMS] . . . focus [is] on the migratory aspects of small species while the IWC Scientific Committee was concerned with its habitat and population."[269] This complementarity on small cetaceans was soon recognised by the IWC, which in 1993 acknowledged the relevance of the CMS agreements with regard to work on small cetaceans.[270]

With such delineated functions in mind, the CMS passed a resolution on small cetaceans in 1998, which noted "the need to look at the conservation of migratory small cetacean species globally."[271] Moreover it suggested that one of the best ways to achieve this was through regional agreements. As such, the signatories were asked to further consider the species listed in the appendices as candidates.[272] This suggestion was directly linked to Article IV(4) of the CMS Convention, which provides:

> Parties are encouraged to take action with a view to concluding AGREEMENTS for any population or any geographically separate part of the population of any species or lower taxon of wild

---

recently been some detailed studies in the area, including one done for the UN Conference on Environment and Development. As such, the Scientific Committee of the CMS was in agreement that any conflict of interest or duplication between the CMS and the IWC was unlikely, as the CMS would focus upon the migratory aspects of small cetaceans, while the IWC's Scientific Committee would remain concerned with their habitat and population. It further noted that there were significant prospects for the two bodies to complement each other: "Report of the Scientific Committee" *in* CMS, *Proceedings of the Fourth Meeting, supra* note 264, UNEP/CMS/Conf.4.16 (1994), at ch. II [42]–[43].

[268] CMS, "Report of Scientific Committee" *in Proceedings of the Fourth Meeting, supra* note 264, at 52.

[269] *Id.*

[270] 4 CMS BULL. 8–9 (1993).

[271] CMS, Resolution 2.3, "Small Cetaceans" *in Proceedings of the Second Meeting, supra* note 258, at 21.

[272] *Id.* Note that there was disagreement in the CMS to whether such agreements, which would cover small cetaceans, should be restricted to just small cetaceans, and just to set geographic areas. *Id.* at 37.

animals, members of which periodically cross one or more national jurisdictional boundaries.[273]

With particular regard to marine species, it was suggested that additional matters to be considered in regional agreements were:

(1)   That range states should include not only those bordering on international waters, but also those whose vessels operate in those waters.

(2)   Conservation and management plans need to extend into international waters.

(3)   Hindrances to migration need to take into account boat traffic and noise pollution.

(4)   Harmful substances for marine species, including ghost nets and other non-degradable debris.[274]

From these considerations, two largely independent agreements which cover small cetaceans have been concluded in the Baltic and North Seas, and the Black and Mediterranean Seas.[275] It is hoped that such agreements will also occur in Latin America.[276]

6.1.   ASCOBANS

ASCOBANS had its genesis in 1985 when an "Agreement on Small Cetaceans in the North Sea" was outlined at the First CMS Conference of Parties.[277] Nevertheless it took a further seven years until it was

---

[273]   *See also* art. V, which sets out the guidelines for any such agreement.

[274]   CMS, "Working Group on Marine Mammals" *in Proceedings of the First Meeting, supra* note 249.

[275]   For a useful discussion on the stand-alone aspects of ASCOBANS, *see also* Hugo Nijkamp & Andre Nollkaemper, *The Protection of Small Cetaceans in the Face of Uncertainty: An Analysis of the ASCOBANS Agreement*, 9 GEO. INT'L ENVTL L. REV. 281, 288–89 (1997).

[276]   CMS, Recommendation 6.2, "Co-Operative Actions for Appendix II Species" *in Proceedings of the Sixth Conference of the Parties* (1999) vol 1, 111. This recommendation drew special attention to the dolphins of South America which, although already listed on appendix II, were not subject to cooperative action.

[277]   CMS, "Working Group on Marine Mammals," *Proceedings of the First Meeting, supra* note 249.

opened for signature in March 1992[278] and finally came into force in 1994. ASCOBANS applies to the "marine environment" of the whole of the Baltic Sea (including the Gulfs of Bothnia and Finland), the Kattegat, Skagerrak, North Sea and English Channels.[279] ASCOBANS begins by recognizing that "small cetaceans are and should remain an integral part of marine ecosystems."[280] However, "the population of harbour porpoises of the Baltic Sea has drastically decreased"[281] and "by-catches, habitat deterioration and disturbance may adversely effect these populations."[282] As such, the parties were "concerned about the status of small cetaceans"[283] in this area, and recognized "that their vulnerable and largely unclear status merits immediate attention in order to improve it."[284] Accordingly, the parties undertook "to cooperate closely in order to achieve and maintain a favourable conservation status for small cetaceans"[285] by attempting to fulfil the objectives prescribed in the Conservation and Management Plan that was set out in the annex.[286]

---

[278]   1 CMS BULL. 6 (1992): The final Act was signed in mid-1991 by Belgium, Denmark, Finland, Germany, Netherlands, Sweden, the UK and the European Economic Community. Soon after, Ireland expressed interest in extending the agreement to cover the Irish Sea; *see also* 4 CMS BULL. 6 (1993). The final impetus for this agreement came from the Memorandum of Understanding on Small Cetaceans in the North Sea, adopted at the Third Ministerial Conference on the North Sea, held in March 1990, as cited in (ed), THE NORTH SEA: BASIC LEGAL DOCUMENTS ON REGIONAL ENVIRONMENTAL CO-OPERATION 276 (David Freestone ed., 1991).

[279]   ASCOBANS, *supra* note 22, at art. 1(2)(b). This was clarified by ASCOBANS Secretariat, Resolution 6, "Resolution on the Clarification of the Definition of the Area of the Agreement" adopted at the first Meeting of the Parties of *ASCOBANS* <http://www.ascobans.org/>files/1994-1.pdf>, site visited on Sept. 21, 2001.

[280]   ASCOBANS, *supra* note 22, preamble.

[281]   *Id.*

[282]   *Id.*

[283]   *Id.*

[284]   *Id.*

[285]   *Id.* at art. 2.1.

[286]   *Id.* at annex.

The Plan, which prohibits "the intentional taking and killing of small cetaceans,"[287] also requires the signatories to work towards:

(a)   the prevention of the release of substances which are a potential threat to the health of the animals.

(b)   the development . . . of modifications to fishing gear and fishing practices in order to reduce by-catches.

(c)   the effective regulation, to reduce the impact on the animals, of activities which seriously affect their food resources.

(d)   the prevention of other significant disturbance, especially of an acoustic nature.[288]

At the time that ASCOBANS was introduced to the CMS, there was some general concern that "the agreement just signed was a weak one, in that it lacked teeth"[289] and that the items in the Conservation and Management Plan "did not seem sufficiently focused."[290] Moreover, although seven parties have signed the agreement,[291] a number of other range states have not committed themselves (but "remain receptive"),[292] whereas some, such as Norway, will do no more than cooperate at a scientific level "owing to [their] desire to maintain a consistent national policy."[293] Norway's interests coincide with other regional organizations

---

[287]   *Id.* at [4].

[288]   *Id.* at [1]. These goals reflected the earlier recognition by the CMS that "pollution, accidental and deliberate catches, habitat changes, and depletion of food supplies" were the primary sources of destruction to small cetaceans, *See also* CMS, Resolution 2.3, "Small Cetaceans" *in Proceedings of the Second Meeting, supra* note 258, 21. *See also* CMS, "Working Group on Marine Mammals," *Proceedings of the First Meeting, supra* note 249.

[289]   Comments from Australia and the Environmental Investigation Agency, *in* CMS, *Proceedings of the Third Meeting, supra* note 262, at 14.

[290]   *Id.*

[291]   The 1996 Progress Report on ASCOBANS noted that the number of parties to the agreement was seven (Belgium, Denmark, Germany, Netherlands, Sweden, UK and France) with the EU expected to join soon after, *see also* "Progress Report on the Agreement on the Conservation of Small Cetaceans of the Baltic and North Seas" *as cited in* 2 CMS BULL. 2–3 (1996).

[292]   *Id.* These included Estonia, Finland, Latvia, Lithuania and Russia.

[293]   *Id.* Norway had "cooperated actively with ASCOBANS scientific endeav-

(such as NAMMCO) which, although overlapping in geographical range with ASCOBANS, have differing objectives.[294] That is, Norway believes in the sustainable use of cetaceans and also wishes to be able to kill cetaceans while carrying out research.[295] Accordingly, Norway would not sign ASCOBANS.[296] In addition, the five Baltic States (Estonia, Finland, Latvia, Lithuania and Russia) have said that they do not intend to ratify ASCOBANS for the time being for a number of reasons, ranging from financial concerns to other priorities in wildlife conservation. In response to such concerns, the Netherlands, speaking on behalf of the EU, responded that although some states were dissatisfied with the provisions, "[t]hey should not overlook the fact that hitherto there had been no international agreement whatever."[297] Sweden added that "[e]ven a weak text, however, could be strongly implemented where the will to do [so] existed or could be aroused."[298]

ASCOBANS has developed its own Action Plans aimed at studying the effects of pollution on cetaceans (in conjunction with the IWC)[299]

---

ours and has participated in Advisory Committee meetings with the intention to continue, independently of its signatory status."

[294]   *Id*. It has been noted that surveys of the geographical range of the small cetaceans covered by ASCOBANS included areas "operated under the umbrella of NAMMCO."

[295]   *See also* CMS, *Proceedings of the First Meeting, supra* note 249, at 2. *See also* CMS, *Proceedings of the Second Meeting, supra* note 258, at 2, 25.

[296]   *CMS Agreement Update: Small Cetaceans of the Baltic and North Seas (ASCOBANS)*, 7 CMS BULL. 2 (1997).

[297]   CMS, *Proceedings of the Third Meeting, supra* note 262, 14.

[298]   *Id*. 15.

[299]   *News from CMS 19th Standing Committee*, 9 CMS BULL. 4 (1999). In 1999, with regard to the study of the adverse effects of marine pollution on cetaceans, the Advisory Committee of ASCOBANS recommended continuing "the positive cooperation" with other international organisations, such as the IWC. For a discussion of the other international organizations involved, *see also* Robin Churchill, *Sustaining Small Cetaceans: A Preliminary Evaluation of the ASCOBANS & ACCOBAMS Agreements, in* INTERNATIONAL LAW AND SUSTAINABLE DEVELOPMENT: PAST ACHIEVEMENTS AND FUTURE CHALLENGES 225, 239–43 (Alan Boyle & David Freestone eds., 1999).

and reducing the offending substances;[300] reducing disturbances upon cetaceans (including regulating whale-watching, seismic testing and military activities); establishing protected areas for cetaceans;[301] and controlling the problem of by-catch.

6.2. ACCOBAMS

The ASCOBANS treaty is accompanied in the CMS by ACCOBAMS.[302] ACCOBAMS applies to all the maritime waters of the Black Sea and the Mediterranean and their gulfs and seas.[303] The desirability of such an agreement was driven by the precarious status of some of the cetacean populations, and the serious threats they face in this region.[304] Accordingly, in February 1991 the secretariats of the Barcelona, Bern and Bonn Conventions, as well as the World Conservation Union and Greenpeace International, met to discuss the possibility of creating a new legal agreement to deal with such problems.[305] Soon after, the third CMS Conference of Parties adopted a resolution that urged

> Parties and non-parties to the Convention that are Range States for the species and populations of small cetaceans listed by the Conference of the Parties in Appendix II of the Convention, to give priority to concluding agreements for their conservation.[306]

---

[300]   *CMS Agreement Update: Advisory Committee*, 7 CMS BULL. 2 (1997).

[301]   *Agreement Secretariats: ASCOBANS*, 10 CMS BULL. 16 (2000). A working group on Protected Areas was formed in 2000.

[302]   ACCOBAMS, *supra* note 9.

[303]   *Id*. at art. I(1)(a). It also covers the internal waters connected to or interconnecting these maritime waters, and of the Atlantic area contiguous to the Mediterranean Sea west of the Straits of Gibraltar and bounded by the line joining Cape St Vincent (Portugal) and Casablanca (Morocco).

[304]   2 CMS BULL. 4 (1992): The CMS Secretariat attempted to draw attention to the need for "urgent action with respect to Black Sea dolphins." Attention was also brought to the mass strandings of harbor porpoises and common dolphins, which were due to multiple factors of causation (from parasitic infection to loss of habitat).

[305]   1 CMS BULL. 7 (1992).

[306]   CMS, Resolution 3.3, "Small Cetaceans" *in Proceedings of the Third Meeting, supra* note 262, at 50.

With particular regard to the Mediterranean and the Black Seas, the CMS

> urges Range States to collaborate, under the sponsorship of a Party
> Range State, with a view to concluding under the Convention an
> Agreement for the conservation of small cetaceans of the Mediter-
> ranean and Black Seas.[307]

The negotiating phase for this treaty, which was not concluded until late 1996,[308] was drawn out as a result of a number of factors. First, it sought to "bind the countries of two sub-regions to work together on a subject of common concern."[309] Second, it is also open to "membership of non-coastal States ('third countries') whose vessels are engaged in activities that may affect cetaceans."[310] The third factor was that it was extended to include all cetaceans frequenting the Mediterranean and Black Seas (small and large cetaceans) since their conservation require- ments are similar.[311] Finally, unlike with ASCOBANS, a number of important range states have not signed the ACCOBAMS; as it stands, only 11 of the 28 range states have signed.[312] Although a number of the non-signatories may sign at some point in the future, four (Bulgaria, Russia, Turkey and the Ukraine) have declared that although they sup- port the general principles of the agreement, they have reserved their position until further questions regarding the Black Sea fisheries are fully resolved.[313] This refusal to sign came despite the preamble of *ACCOBAMS*, which, unlike *ASCOBANS*, recognized

---

[307]   *Id.* at 20.

[308]   CMS, *Proceedings of the Fourth Meeting, supra* note 264, at 15, 122. *See also* 5 CMS BULL. 3 (1996); 6 CMS BULL. 1 (1997). Nevertheless by 1998 13 states were signatories to ACCOBAMS.

[309]   CMS, *Proceedings of the Fifth Meeting of the Conference of the Parties* 21 (1997).

[310]   *Id.*

[311]   5 CMS BULL. 3 (1996).

[312]   *See also ACCOBAMS: Agreement Summary Sheet*, <http://www.wcmc. org.uk/cms/acc_summ.htm>, site visited on Sept. 21, 2001.

[313]   For a useful discussion of this refusal, *see also* Churchill, *supra* note 219, at 231–33, 246–47.

the importance of integrating actions to conserve cetaceans with activities related to the socio-economic development of the Parties concerned by this Agreement, including maritime activities such as fishing and the free circulation of vessels in accordance with international law.[314]

ACCOBAMS, like ASCOBANS before it, emphasizes that "cetaceans are an integral part of the marine ecosystem which must be conserved for the benefit of present and future generations, and that their conservation is a common concern."[315] Moreover the signatories are aware that

the conservation status of cetaceans can be adversely affected by factors such as degradation and disturbance of their habitats, pollution, reduction of food resources, use and abandonment of non-selective fishing gear, and by deliberate and incidental catches.[316]

In an attempt to control some of these activities, the signatories agreed to

take co-ordinated measures to achieve and maintain a favourable conservation status for cetaceans. To this end, Parties shall prohibit and take all necessary measures to eliminate . . . any deliberate taking of cetaceans and shall co-operate to create and maintain a network of specially protected areas to conserve cetaceans.[317]

This objective was assisted by the adoption of the annexed Conservation Plan[318] which aimed to reduce pollution, reduce by-catch, reduce indirect interactions with fisheries, reduce disturbance (for example, seismic surveys and whale-watching), establish protected areas and monitor cetacean populations in the area.

---

[314]   ACCOBAMS, *supra* note 9, preamble.

[315]   *Id.*

[316]   *Id.*

[317]   *Id.* at art. II(1).

[318]   *Id.* at annex 2.

7.   CONCLUSION

A concern for small cetaceans has been evident within the IWC since the early 1970s. Since that time, the IWC has slowly brought the focus of its Scientific Committee to the status and trends of a number of these species. From the information collected, it has issued resolutions calling for restraint, often directed to specific countries, where the species are clearly at risk.

Despite this slowly evolving approach, a number of countries have continually objected to the practice of overfishing small cetaceans, suggesting that any questions of small cetaceans are not within the competence of the IWC. This is due to the fact that the power to make such moves was not conferred by the ICRW, since small cetaceans were not singled out for coverage, nor were they listed in the Nomenclature.

This approach is misplaced for a number of reasons. First, the language of the ICRW would appear to be broad and encompass all types of whales and whaling. This is especially so, considering that the earlier international conventions clearly excluded small types of whales from coverage, yet the later ones did not. Second, the idea that the nomenclature to the ICRW was the cut-off line for what species were to be under the purview of the *ICRW*—or the core of some form of agreement which delineated authority—is simplistic. Not only was the nomenclature specifically set out only as a guide in the ICRW, it has subsequently been shown in other conventions to be in itself something that needs to evolve to reflect changing scientific knowledge. Third, international law and multi-lateral environmental agreements are evolving. Thus, subsequent practices exist that are reinforced through mechanisms within conventions that allow entrenched majorities to change the direction of treaties. Typically, a three-quarters majority acting through a specified annex, schedule or carefully detailed provision has this power. The IWC has a clear and direct history of using subsequent practice to encompass species of whales and methods of capture and lethal utilization which were not specifically listed or utilized in earlier times.

It is with respect to this last point that those who object to the authority of the IWC to manage small cetaceans are particularly mistaken.

That is, what they are trying to argue is that the ICRW cannot evolve without the full and unanimous consent of all the signatories. Moreover, this argument is closely tied to the role of the Convention's nomenclature. The ICRW was clearly designed to evolve through subsequent practice and the operation of a three-quarters majority (but dissenting nations still have the clear right to object and not be bound). As it would not be possible to argue against this clearly established practice, these objecting States have chosen to confuse the issue with the spurious argument of nomenclature.

A related argument is the assertion that coastal states have competence over small cetaceans. States usually invoke UNCLOS to support this proposition. However, this assertion is mistaken as UNCLOS does not accord coastal states complete sovereignty in territorial seas. It is limited in the sense that nations can waive their rights under UNCLOS and freely consent to the authority of overlapping international organizations such as the IWC. For example, signatories to the complementary Convention for the Conservation of Antarctic Marine Living Resources clearly deferred authority to the IWC. The need to cooperate with the relevant overlapping international organizations under UNCLOS is strengthened when dealing with critically endangered species, migratory species and cetaceans in general. The primacy of the IWC in these discussions was affirmed in Agenda 21 in 1992.

This is not to suggest that decentralised approaches to the management of small cetaceans are not important. Rather, the point is that regional organizations need to be complementary to the primary international organization, namely the IWC, in this area. The regional ASCOBANS and ACCOBAMS agreements are prime examples of this.

In conclusion, the IWC has a broad authority in relation to small cetaceans and a clear primacy over regional organisations on this question. This authority becomes stronger when the small cetaceans migrate between countries or are endangered. This authority is in accordance with state practice in implementing the ICRW, subsequent state practice and the language of the Convention.

Chapter 10
The Agreement on the Conservation of Small Cetaceans of the Baltic and North Seas

Robin R. Churchill[1]

1. INTRODUCTION

The broad aim of the Agreement on the Conservation of Small Cetaceans of the Baltic and North Seas (ASCOBANS),[2] which has been in force since 1994, is to try to halt the substantial decline in the populations of the small cetaceans of the Baltic and North Seas that has occurred over the past few decades.

This chapter begins by looking at the causes of this population decline. It then proceeds to describe the origins, content and operation of ASCOBANS. This is followed by an examination of the relationship of ASCOBANS to the considerable number of other international agreements and organizations whose work is relevant to the conservation of small cetaceans in the Baltic and North Seas. The chapter ends with an attempt to evaluate the effectiveness of ASCOBANS.

---

[1] Professor of Law, Cardiff University, United Kingdom. This chapter draws to a considerable extent on an earlier work by the author: Robin Churchill, *Sustaining Small Cetaceans: A Preliminary Evaluation of the Ascobans and Accobams Agreements, in* INTERNATIONAL LAW & SUSTAINABLE DEVELOPMENT: PAST ACHIEVEMENTS & FUTURE CHALLENGES 225 (Alan Boyle & David Freestone eds., 1999). The writing of both this chapter and the earlier work would not have been possible without the assistance of the ASCOBANS Secretariat in providing me with ASCOBANS reports. I would therefore like to take this opportunity to express my gratitude to the members of the Secretariat (past and present) concerned.

[2] 1772 U.N.T.S. 217.

2.   THE SMALL CETACEANS OF THE BALTIC AND NORTH SEAS AND THE CAUSES OF THEIR POPULATION DECLINE

As is the case with small cetaceans[3] in most regions, there is considerable uncertainty over the population levels of small cetaceans in the waters to which ASCOBANS applies, as well as their patterns of migration and the threat to their continued well being. The first truly comprehensive survey of small cetaceans in the North Sea, the Small Cetacean Abundance in the North Sea (SCANS) project, was not carried out until 1994. The survey found nine species of small cetaceans in the North Sea and adjacent areas (including the western Baltic): harbor porpoise, bottlenose dolphin, white-beaked dolphin, white-sided dolphin, common dolphin, striped dolphin, long-finned pilot whale, Risso's dolphin and killer whale, of which the last five were thought to be extremely rare.[4]

There has been no directed hunting or killing of small cetaceans in the Baltic and North Seas for many years. Nevertheless, their populations are in long-term decline. Populations of harbor porpoises are thought to have been in significant decline in the North Sea, English Channel and Baltic since 1980 and the bottlenose dolphin may have declined drastically in

---

[3]   "Small cetacean" is not a scientific term. It is generally taken as referring to cetacean species which have not been traditionally exploited by the commercial whaling industry and which are under 30 feet in length. Of 75 cetacean species, about 65 are small cetaceans. Of these, two (minke and pygmy right whales) belong to the order *mysticeti* (baleen whales); the remainder (which include many species of dolphin, porpoise and beaked whales) belong to the order *odontoceti* (toothed whales). Small cetaceans are generally more coastal and less migratory than large cetaceans and have a more variable reproductive cycle. *See* further Cynthia E. Carlson, *The International Regulation of Small Cetaceans*, 21 S.D. L. REV. 557, 580–82 (1984); G. Rose & S. Crane, *The Evolution of International Whaling, in* GREENING INTERNATIONAL LAW 159, 175 (P. Sands ed., 1993).

[4]   Hugo Nijkamp & Andre Nollkaemper, *The Protection of Small Cetaceans in the Face of Uncertainty: An Analysis of the Ascobans Agreement*, 9 GEO. INT'L ENVTL. L. REV. 281, 282–84 (1997), who quote from DISTRIBUTION AND ABUNDANCE OF THE HARBOUR PORPOISE AND OTHER CETACEANS IN THE NORTH SEA AND ADJACENT WATERS (P. S. Hammond *et al.*, eds., 1995).

the North Sea since 1960.[5] Other species probably also have declining populations, but much less is known about them. The actual or presumed causes of these population declines are several. First, small cetaceans are caught incidentally in various kinds of fishing nets, particularly drift nets floating near the surface, gill nets attached to the seabed and large pelagic trawls.[6] In the central and southern North Sea, bottom-set gill nets are esti-mated to kill about 4,500 harbor porpoises a year out of a total popula-tion of about 170,000 (some 2.6 percent), while in the Skagerrak the by-catch level is even higher, probably in excess of 4 percent.[7] By com-parison, the Scientific Committee of the International Whaling Commission advises that an annual by-catch of 2 percent may cause pop-ulations to decline.[8] A second, but not well-understood, threat to the pop-ulations of small cetaceans comes from pollution of the marine environment. Various pollutants, notably organochlorines (particularly PCBs), polybrominated compounds, organotin compounds and heavy metals, are thought to inhibit the reproduction of marine mammals and weaken their immune systems.[9] Third, the migratory patterns and possi-bly the population levels of small cetaceans may be affected by distur-bances to their habitat. Such disturbances include noise from high-speed ferries and the offshore oil and gas industry (especially seismic testing, which may damage the hearing of small cetaceans), increased shipping traffic, and whale watching. Finally, it is possible that in some areas (espe-cially the North Sea) population levels have been affected by a decrease in food supplies because of intensive fishing.[10] Of these various causes of population decline, by-catch is certainly the most significant.

---

[5]   S. GUBBAY, A COASTAL DIRECTORY FOR MARINE NATURE CONSERVATION 207 (1988); S. Northridge, *Ecological Effects of Man's Activities on Mammals, in* NORTH SEA FORUM REPORT 97, 98 (1987); preamble to ASCOBANS.

[6]   ASCOBANS, *Report of the 7th Meeting of the Advisory Committee*, at 32 (2000) (hereafter *Report of AC7*).

[7]   *See* <http://www.ascobans.org/page 13.html> (Nov. 29, 2000).

[8]   ASCOBANS, *Report of the Second Meeting of Parties to Ascobans*, at 59 (1997) (hereafter *Report of Second Mop*).

[9]   *Id.* at 11 and 70; ASCOBANS, *Report of the 6th Advisory Committee Meeting*, at 7 (1999) (hereafter *Report of AC6*); *Report of AC7, supra* note 6, at 11, 38–39.

[10]   Patricia Birnie, *Problems concerning Conservation of Wildlife including*

3.   THE ORIGINS AND NEGOTIATION OF ASCOBANS

ASCOBANS was developed under the auspices of the Bonn Convention on the Conservation of Migratory Species of Wild Animals, 1979.[11] The Convention addresses threats to endangered migratory species listed in Appendix I (which does not include any small cetaceans), in relation to which it directly requires its parties to take protective measures, and those migratory species with an "unfavourable conservation status" listed in Appendix II. The latter includes over 30 species of small cetaceans, including nearly all of those found in the Baltic and North Seas. In relation to species listed in Appendix II, Article IV(3) provides that range states "shall endeavour to conclude AGREEMENTS where these would benefit the species." Article IV(4) authorizes the creation of stand-alone agreements, providing that states parties "are encouraged to take action with a view to concluding agreements for any population . . . of any species . . . , members of which periodically cross one or more national jurisdictional boundaries." On a strict reading of the Convention, agreements under Article IV(4) are not limited to Appendix II species.

It was decided to proceed under the umbrella of the Bonn Convention when concerns about the decline in the populations of small cetaceans in the Baltic and North Seas prompted the states concerned to consider international legislative action in the mid-1980s.

The negotiation of the text that ultimately became ASCOBANS was, nevertheless, a protracted and somewhat tortuous affair. Discussion of a possible agreement for the conservation of small cetaceans of the Baltic and North Seas took place at the First Conference of the Bonn Convention Parties in 1985. A Small Cetacean Working Group was subsequently established, and it elaborated a draft agreement during 1986–1987. The draft was discussed in the Bonn Convention's Scientific Council in 1988 but there were irreconcilable differences over its contents and scope. The Second Conference of the Bonn Convention

---

Marine Mammals in the North Sea, in THE NORTH SEA: PERSPECTIVES ON REGIONAL ENVIRONMENTAL CO-OPERATION 252, 254 (David Freestone & Ton IJlstra eds., 1990).

[11]   1651 U.N.T.S. 333.

Parties, held in 1988, disbanded the Working Group and asked the Scientific Council to give priority to a global review of small cetaceans.[12] A Memorandum of Understanding on Small Cetaceans in the North Sea, adopted at the Third Ministerial Conference on the North Sea held in March 1990,[13] provided new impetus for a regional agreement. In the Memorandum, the ministers noted their concern over the population status of small cetaceans in the North Sea and the "close link" between the small cetacean populations of the North and Baltic Seas. They also undertook to cooperate with Baltic Sea states on elaborating an agreement under the Bonn Convention, and, pending the conclusion of such an agreement, to apply the interim conservation measures contained in Annex I to the Memorandum. Many of these measures are similar to those that were ultimately set forth in ASCOBANS. Following the Memorandum, several rounds of negotiations took place resulting in the conclusion and approval of ASCOBANS' Agreement at the Third Conference of the Bonn Convention Parties in September 1991 and its formal opening for signature on March 17, 1992. The Agreement came into force on March 29, 1994, after having received the requisite six ratifications.

ASCOBANS is an agreement under Article IV(4) of the Bonn Convention, rather than an Agreement under Article IV(3).[14] It is therefore a stand-alone Agreement separate from the Bonn Convention, although obviously quite closely connected to it.[15] There are likely to be several reasons why ASCOBANS was concluded under Article IV(4) rather than under Article IV(3). First, as pointed out earlier, Article IV(3) Agreements, unlike Article IV(4) Agreements, are limited to the species listed in Appendix II. However, not all the small cetaceans of the Baltic and North Seas were listed in Appendix II at the time of the negotiation of ASCOBANS. Second, Article V of the Bonn Convention requires

---

[12]   Birnie, *supra* note 10, at 287.

[13]   The Memorandum is reproduced in THE NORTH SEA: BASIC LEGAL DOCUMENTS ON REGIONAL ENVIRONMENTAL CO-OPERATION 276 (David Freestone & Ton IJlstra eds., 1991).

[14]   ASCOBANS art. 8(1).

[15]   For the benefits of agreements being concluded within the Bonn Convention framework rather than being completely separate, *see* Nijkamp & Nollkaemper, *supra* note 4, at 289.

Article IV(3) Agreements to describe the range and migratory route of the species concerned. This information, however, was not available for the small cetaceans of the Baltic and North Seas at the time ASCOBANS was being negotiated. Third, states have more flexibility when concluding Article IV(4) Agreements: Article V lays down guidelines for Article IV(3) Agreements, whereas the Bonn Convention provides no guidelines in the case of Article IV(4) Agreements.[16]

4.   THE SCOPE OF ASCOBANS *RATIONE LOCI* AND *MATERIAE*

Ratione loci, ASCOBANS applies to the "marine environment" of the whole of the Baltic Sea (including the Gulfs of Bothnia and Finland), the Kattegat, Skagerrak, North Sea and English Channel.[17] This area falls entirely within the zones of national jurisdiction—territorial sea, fishing zones and EEZ, as well as internal waters to the extent that such waters are regarded as being part of the "marine environment"—of the 14 coastal states of the region—Norway, Sweden, Finland, Russia, Estonia, Latvia, Lithuania, Poland, Germany, Denmark, Netherlands, Belgium, France and the United Kingdom. This area does not include the full migratory range of all the stocks of small cetaceans found in the North and Baltic Seas. Thus, there has been some discussion about extending the area of application of ASCOBANS. Negotiations with Ireland have been taking place for some years to extend the area of application of ASCOBANS to include the waters west of the United Kingdom and around the whole of Ireland, but, so far, these negotiations appear to have made little progress.[18] There has also been discussion

---

[16]   *Id.* at 288–89. But note that the Bonn Convention Conference of the Parties has adopted guidelines for Article IV(4) Agreements: *see* Clare Shine, *Selected Agreements concluded pursuant to the Convention on the Conservation of Migratory Species of Wild Animals, in* COMMITMENT AND COMPLIANCE: THE ROLE OF NON-BINDING NORMS IN THE INTERNATIONAL LEGAL SYSTEM 196, at 202–204 (Dinah Shelton ed., 2000).

[17]   The area is defined precisely in art. 1(2)(b) of ASCOBANS, as clarified by Resolution 6 adopted at the first meeting of the parties: ASCOBANS, *Report from the First Meeting of the Parties*, at 38 (1994) (hereafter *Report of First MoP*).

[18]   *Id.* at 7; ASCOBANS, *Report of the Third Meeting of the Advisory*

about extending ASCOBANS' area southwards to include the 200-mile zones of Portugal and Spain and to meet up with the waters covered by ACCOBAMS,[19] but here, too, little progress has been made.[20] The possibility of extending the area of application of ASCOBANS northwards would seem remote, given the whaling policies of the Faroes, Iceland and Norway (all of which favor the utilization of small cetaceans, which, as will be seen, is incompatible with ASCOBANS) as well as the existence of the North Atlantic Marine Mammal Commission.

Ratione materiae, ASCOBANS applies to "all small cetaceans found within the area of the agreement."[21] "Small cetaceans" are defined as "any species, subspecies or population of toothed whales *Odontoceti* except the sperm whale."[22] The sperm whale is in fact the only member of the order *Odontoceti* that is not considered a small cetacean. The definition excludes the minke whale (which is harvested by Norway, although outside the North Sea) as it belongs to the order *Mysticeti* (baleen whales). The somewhat imprecise nature of the definition of small cetacean is doubtless explained by the uncertainty at the time ASCOBANS was being negotiated, as to which species of small cetacean were found in the North and Baltic Seas.

5. PARTICIPATION IN ASCOBANS

ASCOBANS is open to participation by any range state. The latter is defined in Article 1.2(f), in a manner similar to that of Article I(1)(h) of the Bonn Convention, as any state

---

Committee, at 7 (1996) (hereafter *Report of AC3*); ASCOBANS, *Report of the Fourth Meeting of the Advisory Committee*, at 16 (1997) (hereafter *Report of AC4*); ASCOBANS *Report of the Fifth Meeting of the Advisory Committee*, at 13 (1998) (hereafter *Report of AC5*); *Report of AC6*, *supra* note 9, at 13.

[19] Agreement on the Conservation of Cetaceans of the Black Sea, Mediterranean Sea and Contiguous Atlantic Area, Nov. 24, 1996, 36 I.L.M. 777 (1997).

[20] *Report of AC4*, *supra* note 18, at 16; *Report of AC5*, at 13; *Report of AC7*, at 16; ASCOBANS *Proceedings of the Third Meeting of Parties to ASCOBANS*, at 4, 7, 35 (2000).

[21] Art. 1.1.

[22] Art. 1.2(a).

that exercises jurisdiction over any part of the range of a species covered by this agreement, or a State whose flag vessels, outside national jurisdictional limits but within the area of the agreement, are engaged in operations adversely affecting small cetaceans.

The first category of range state includes the 14 coastal states of ASCOBANS' area listed above, and could conceivably include other coastal states beyond that area. The second category is entirely theoretical as there are no parts of the ASCOBANS' area beyond the limits of national jurisdiction.

Of the 14 range states mentioned, eight have so far become parties to ASCOBANS—Belgium, Denmark, Finland, Germany, Netherlands, Poland, Sweden and the United Kingdom. Of the other six states, none of which has signed ASCOBANS, France has been indicating for some time that it intends to ratify ASCOBANS; Norway has said that it will not ratify because it believes in the sustainable use of cetaceans and also wishes to be able to kill cetaceans for research (neither of which is possible under ASCOBANS); and the four eastern Baltic states (Estonia, Latvia, Lithuania and Russia) have said that they do not intend to ratify for the time being for a mixture of reasons, including the small numbers of small cetaceans found in their waters, other priorities in wildlife conservation and financial problems.[23]

ASCOBANS is also open to participation by any regional economic integration organization, defined as "an organisation constituted by sovereign States which has competence in respect of the negotiation, conclusion and application of international agreements in matters covered by this agreement."[24] In practice, the only such organization is the European Community (EC). Although it has signed ASCOBANS, the EC has not ratified it. On several occasions, the EC Commission has stated that the EC intends to ratify ASCOBANS as soon as possible,[25] but it has not yet submitted a proposal for ratification to the Council.

---

[23]   *Report of First MoP, supra* note 17, at 2; *Report of Second MoP, supra* note 8, at 2, 25; *Report of AC5, supra* note 18, at 12; *Report of AC6, supra* note 9, at 133; *Report of AC7*, at 16; *Report of Third MOP*, at 35.

[24]   Art. 1.2(d). Curiously, there is no requirement that any member state of such an organization must be a range state, but common sense suggests that this must be so.

[25]   *Report of First MoP, supra* note 17, at 2; *Report of Second MoP, supra* note 8, at 2, 25.

The main reasons for this lack of action appear to be a shortage of resources (in terms of personnel to participate in ASCOBANS' meetings) and other priorities.[26]

Unless and until the EC becomes a party to ASCOBANS, those state parties that are also members of the EC (currently all but one of the parties—and even that party, Poland, has applied for membership of the EC) are faced with a potential dilemma. Since some of the conservation measures that parties are required to take under ASCOBANS fall within the exclusive competence of the EC (notably matters relating to fisheries and fishing methods), EC member states run the risk of being unable to fulfil their commitments under ASCOBANS because the measures required fall outside their competence.[27] Up to the present time, this problem has not arisen because the commitments under ASCOBANS have been very vague, but as they become increasingly more precise (as explained below), so the risk referred to becomes more likely to materialize. If the EC does become a party to ASCOBANS, there will be a problem for third states in knowing which aspects of ASCOBANS fall within the competence of the EC and which within that of its member states. Unlike the case under some other treaties, such as the United Nations Convention on the Law of the Sea,[28] the EC is not required, on

---

[26]   *Report of AC5, supra* note 18, at 12; *Report of AC6, supra* note 9, at 13; *Report of AC7, supra* note 6, at 16; ASCOBANS, *Report of Third MOP, supra* note 20, at 35.

[27]   However, it should be noted that the under Article 2.2 of ASCOBANS the parties are to apply ASCOBANS' conservation measures "in accordance with [their] international obligations." The meaning of this phrase is obscure. It is possible that it would permit an EC member state to argue that it had not applied an ASCOBANS' conservation measure because to do so would not be in accordance with its international (EC) obligations as the measure in question was within the EC's exclusive competence and it would be contrary to EC law for a member state to adopt a measure in the field in question. Furthermore, Article 8.2 of ASCOBANS provides that its provisions "shall in no way affect the rights and obligations of a Party deriving from any other existing treaty," which could also be used by an EC member state to justify not applying an ASCOBANS' measure. Since all but one of ASCOBANS' parties are members of the EC, this would mean that ASCOBANS' measures falling within the EC's field of competence would be largely ineffectual.

[28]   United Nations Convention on the Law of the Sea, Dec. 10, 1982. 1833 U.N.T.S. 3. *See* Annex IX, art. 5.

becoming a party to ASCOBANS, to state how competence is divided between it and its member states. Some indication of this division might evolve in practice, as at meetings of the parties the EC has the exclusive right to vote (in place of its member states) in matters within its competence, and vice versa.[29] In the early years of ASCOBANS, the Secretariat found it difficult to establish good relations with the EC, while the Advisory Committee found the EC's attitude towards ASCOBANS "not helpful."[30] More recently, the situation appears to have improved.[31]

6.   INSTITUTIONAL ARRANGEMENTS

Most multilateral environmental agreements concluded since the early 1970s contain a common pattern of institutional arrangements, comprising regular meetings or conferences of the parties, one or more subsidiary bodies and a secretariat. These mechanisms facilitate in developing or adapting the normative content of the agreement in the light of changing circumstances and to oversee implementation of and compliance with the agreement by the parties.[32] ASCOBANS follows this trend, its institutional arrangements comprising regular Meetings of the Parties, an Advisory Committee and a Secretariat.

Meetings of the Parties are to be held not less than once every three years to "review the progress made and difficulties encountered in the

---

[29]   Art. 6.3 of ASCOBANS; Rule 13(1) of the Rules of Procedure for the Meeting of Parties to ASCOBANS (in *Report of Third MoP*, *supra* note 20, at 25). For a fuller discussion of questions of competence in relation to treaties in the marine area, *see* Robin R. Churchill, *The European Community and its Role in Some Issues of International Fisheries Law*, in DEVELOPMENTS IN INTERNATIONAL FISHERIES LAW 533, 536–46 (Ellen Hey ed., 1999).

[30]   *Report of Second MoP*, *supra* note 8, at 25; *Report of AC4*, *supra* note 18, at 15. *Cf. Report of AC3*, *supra* note 18, at 6. Note also the concerns expressed by WWF and other NGOs about the role of the EC in *Report of Second MoP*, *supra* note 8, at 4.

[31]   *Report of AC6*, *supra* note 9, at 13.

[32]   Further on this matter, *see* Robin R. Churchill & Geir Ulfstein, *Autonomous Institutional Arrangements in Multilateral Environmental Agreements: A Little-Noticed Phenomenon in International Law*, 94 AM. J. INT'L L. 623 (2000).

implementation and operation" of ASCOBANS.[33] So far three Meetings have been held—in 1994, 1997 and 2000. The Meeting of the Parties is the decision-making body of ASCOBANS. It decides on the measures required for the further implementation of ASCOBANS, adopts the budget and formulates terms of reference for the Advisory Committee and Secretariat.

Under ASCOBANS, the function of the Committee is to "provide expert advice and information to the Secretariat and the Parties on the conservation and management of small cetaceans and on other matters in relation to the running" of ASCOBANS.[34] The role of the Advisory Committee has been described by its chair as "driving forward the objectives" of ASCOBANS between Meetings of the Parties.[35] In practice, many of the decisions taken by the Meetings of the Parties are based on proposals from the Advisory Committee. The Committee consists of one representative from each party and normally meets annually. As of January 2001, seven meetings had been held. Meetings are usually also attended by scientific experts and representatives of both IGOs and NGOs with the aim of exchanging information and developing cooperation. In general, the Committee appears to have worked reasonably well. However, a frequent comment at meetings has been that its operations have been hampered by the lack of a sufficient breadth of expertise among its members and by insufficient time.[36]

The Secretariat is tasked with promoting and coordinating activities undertaken under ASCOBANS, providing advice and support to the parties, and servicing Meetings of the Parties and meetings of the Advisory

---

[33]   ASCOBANS art. 6.1.

[34]   Art. 5.1. *See also* the Committee's terms of reference contained in Resolution 5 adopted by the First Meeting of the Parties, *Report of First MoP*, *supra* note 17, at 37; and Resolution on Activities of the Ascobans Advisory Committee 1997–2000, adopted by the Second Meeting of the Parties, *Report of Second MoP*, *supra* note 8, at 63; Resolution No. 8. Activities of the ASCOBANS Advisory Committee 2001–2003, adopted at the Third Meeting of the Parties, *Report of Third MoP*, *supra* note 20, at 105. *See also Report of AC7*, *supra* note 6, at 2, 21, 108.

[35]   *Report of Second MoP*, *supra* note 8, at 40.

[36]   *Id.*; *Report of AC5*, *supra* note 18, at 4; *Report of Third MoP*, *supra* note 20, at 44.

Committee.[37] The Secretariat is a small body, consisting merely of the Executive Secretary and an assistant. The effectiveness of the Secretariat has been adversely affected in recent years by a high turnover of staff and the upheaval attendant upon relocating the Secretariat from the United Kingdom to Bonn in 1998, where, since the beginning of 2001, it has been integrated into the Agreements Unit of UNEP/CMS.

In addition to the above institutions, each party is to designate a Coordinating Authority, which is to serve as a contact point for the Secretariat and the Advisory Committee in their work and through which the activities of each party are to be coordinated and monitored.[38]

7.   THE CONSERVATION OBLIGATIONS OF PARTIES TO ASCOBANS

7.1.   Conservation Objectives

A preliminary, and very important, question is the appropriate conservation objectives for small cetacean species of the Baltic and North Seas. In Article 2.1, ASCOBANS simply defines the objectives as being cooperation to "achieve and maintain a favourable conservation status for small cetaceans." The phrase "favourable conservation status" is not defined in ASCOBANS. However, it is defined in the Bonn Convention. Article I(1)(c) of the latter provides that a "conservation status" is to be taken as "favourable" where (1) "population dynamics data indicate that the migratory species is maintaining itself on a long-term basis as a viable component of its ecosystem;" (2) the range of the species is neither being nor likely to be reduced; (3) there is and will be sufficient habitat to maintain the population on a long-term basis; (4) the distribution and abundance of the species "approach historic coverage and levels to the extent that potentially suitable ecosystems exist and to the extent consistent with wise wildlife management." It is reasonable to suppose this definition should apply to ASCOBANS. However, in practice its application to small cetaceans is not easy. It would require considerable scientific knowledge of their populations and migratory range, both past and present. As indicated earlier, sufficient knowledge of these

---

[37]   ASCOBANS arts. 4.1, 4.2, 4.3.

[38]   ASCOBANS arts. 2.3, 3.1.

matters does not currently exist. Secondly, conditions (1) and (4) are not clear, and raise as many questions as they answer.[39]

At their Second Meeting in 1997 ASCOBANS' parties tried to be more specific about conservation objectives. The Working Group on Scientific Matters recognized that

> in practice, it is necessary to have specific target population levels so that the status of a stock and the effectiveness of conservation measures can be evaluated. It was agreed that a suitable interim objective would be to restore populations to, or maintain them at, 80% of the carrying capacity.[40]

The Group also recognized that "while it is difficult, and perhaps impossible, to determine carrying capacity, such a theoretical target level will allow the development and application" of a longer-term approach, to be developed by the Advisory Committee, which will "take into account the uncertainty which is inevitably inherent in the data required to assess the status of stocks."[41] This approach was endorsed by the Meeting of the Parties, which adopted a resolution in which it agreed:

(1) that the aim of ASCOBANS can be interpreted as "to restore and/or maintain biological or management stocks of small cetaceans at the level they would reach when there is the lowest possible anthropogenic influence"—a suitable short-term practical sub-objective is to restore and/or maintain stocks/populations to 80 percent or more of the carrying capacity;

(2) that the general aim should be to minimize (*i.e.*, to ultimately reduce to zero) anthropogenic removals within some yet-to-be-specified time frame, and that intermediate target levels should be set;

(3) that the longer term approach, which involves *inter alia* taking into account uncertainty in the available data, should be developed by the Advisory Committee.[42]

---

[39]   For fuller discussion of the difficulties raised by the Bonn Convention's definition of "favourable conservation status," *see* P.W. BIRNIE, INTERNATIONAL REGULATION OF WHALING 514–15 (1985).

[40]   *Report of Second MoP, supra* note 20, at 43.

[41]   *Id.*

[42]   *Id.* at 59. Resolution on Incidental Take of Small Cetaceans.

The "longer-term approach" called for has not yet been developed. Instead the Advisory Committee and the Meeting of the Parties, more sensibly perhaps in view of the inherent difficulty of defining conservation objectives, have concentrated on developing concrete conservation measures.

7.2. Conservation Measures

ASCOBANS contains a Conservation and Management Plan, to be applied by each party "within the limits of its jurisdiction." The conservation measures required by this Plan include preventing pollution harmful to small cetaceans; developing modifications to fishing gear and fishing practices in order to reduce by-catches and to prevent fishing gear from getting adrift or being discarded at sea; regulating activities which seriously affect the food resources of small cetaceans; preventing other significant disturbance, especially of an acoustic nature; prohibiting the intentional taking or killing of small cetaceans; and requiring the immediate release of animals caught alive and in good health. Other elements of the Plan include cooperation to conduct research into the population and migratory patterns of small cetaceans, the location of areas of special importance to their survival and the identification of threats to their well-being; the establishment of an efficient system for reporting, retrieving and analyzing animals which have stranded or been caught as by-catches; and the provision of information to fishers and to the public to engender their support for the aims of ASCOBANS.

The conservation measures contained in the Plan are notable for being couched in vague, flexible and hortatory, rather than prescriptive, language (the parties are to "work towards" such measures). Nijkamp and Nollkaemper suggest that the reason for this is the lack of knowledge of small cetaceans of the region.[43] This may not be the only reason. In the Mediterranean and Black Seas, where there would seem to be equal scientific uncertainty, the states concerned have succeeded in agreeing on rather more precise and prescriptive measures in ACCOBAMS' Agreement. In any case, the parties to ASCOBANS have recognized the need, even before complete scientific certainty is established, to con-

---

[43]   Nijkamp & Nollkaemper, *supra* note 4, at 290–91.

cretize and prioritize the conservation measures contained in the Conservation and Management Plan. This has been done by the Meetings of the Parties and the Advisory Committee. Three types of measure have been singled out for priority action: those concerning by-catches, pollution and reduction of disturbance to habitat.

7.2.1. By-Catches

The Second Meeting of the Parties in 1997 adopted a resolution which, apart from calling for more information on levels of by-catches and more research into methods to reduce by-catches, recommended parties to ASCOBANS ensure that the total anthropogenic removal of harbor porpoises in the central and southern North Sea was reduced as soon as possible to less than 2 percent of the current abundance estimate per year, preferably by 2000. The ultimate aim would be to reduce anthropogenic removals of all small cetaceans to zero.[44] However, not long after this resolution was adopted, it became clear that the figure of 2 percent was too high. At the Sixth Meeting of the Advisory Committee, held in 1999, a joint IWC/ASCOBANS Working Group on Harbor Porpoises reported that ASCOBANS' interim objective of maintaining or restoring population size to 80 percent of carrying capacity, was not likely to be met by reducing annual by-catch to 2 percent of estimated abundance: by-catch needed to be reduced further, to not more than 1.7 percent.[45] Accordingly, the Third Meeting of the Parties in 2000 adopted a resolution recommending that by-catch of marine mammals should be reduced "as soon as possible" below 1.7 percent of the available estimate of abundance, and that in the case of harbor porpoises in the central and southern North Sea, by-catch should be reduced "without delay," although no figure for reduction was specified.[46]

In fact, the parties have not achieved even the 1997 figure of 2 percent by-catch,[47] and generally, the parties have taken little action to set

---

[44]   *Report of Second MoP, supra* note 8, at 59–60.

[45]   *Report of AC6*, at 8; *Report of Third MoP, supra* note 20, at 42, 95.

[46]   Resolution No. 3. Incidental Take of Small Cetaceans. *Report of Third MoP, supra* note 20, at 93.

[47]   *Id.* at 5, 42, 95.

in motion practical steps to reduce by-catch. Only Denmark (which is responsible for about 70 percent of all small cetacean by-catches by ASCOBANS' parties[48]) has so far elaborated a comprehensive policy for reducing by-catch, although the policy has not yet been put into effect.[49] Other parties will doubtless be helped by the recommendations for mitigation strategies (such as the use of pingers to deter small cetaceans from fishing nets, modification of fishing gear, and escape panels), which the Advisory Committee was requested by the Third Meeting of the Parties to make to parties by 2002,[50] as well as by the recovery plan for harbor porpoises in the Baltic Sea (where porpoises are in an even more parlous state than in the North Sea), which the Advisory Committee is to recommend by 2001.[51] The Meeting of the Parties has also recognized that as responsibility for fisheries management and conservation in much of ASCOBANS' area lies with the EC, it will be necessary to collaborate with the EC if ASCOBANS' objectives are to be met. Collaboration will also be essential if by-catch mitigation strategies are to be integrated into the EC's Common Fisheries Policy (a point discussed further in Section 9 below).[52]

7.2.2. Pollution

The Second Meeting of the Parties adopted a resolution in which it called on parties to carry out more research into the effects of pollution on small cetaceans, taking into account work done by the International Whaling Commission, and to strive, within existing regional marine pollution organizations, for a significant reduction of pollution emissions and sources in ASCOBANS' area.[53] Reasonable progress appears to have been made in this regard.[54]

---

[48]   ASCOBANS, *Fourth Annual Compilation of National Reports*, at 11–13 (2000).

[49]   *Report of AC6, supra* note 9, at 9–11; *Report of AC7, supra* note 6, at 6.

[50]   Resolution No. 8. Activities of the ASCOBANS Advisory Committee 2001–2003. *Report of Third MoP, supra* note 20, at 105.

[51]   *Id.* at 107.

[52]   *Id.* at 10, 12.

[53]   *Report of Second MoP, supra* note 8, at 61; *see also id.* at 44.

[54]   *Report of Third MoP, supra* note 20, at 41–42.

7.2.3. Reduction of Disturbance to Habitat

This issue has a number of aspects. As regards disturbance caused by seismic surveys of the continental shelf, the United Kingdom is the only party to have drawn up guidelines to reduce such disturbance. At its Sixth Meeting in 1999, the Advisory Committee recommended that other parties adopt these or similar guidelines,[55] a recommendation endorsed by the Third Meeting of the Parties.[56] The Third Meeting of the Parties also invited the parties to introduce codes of conduct and similar measures to reduce disturbance from military activities.[57] The United Kingdom is also the only party to have produced guidelines on reducing disturbance from whale watching: these, too, have been commended to other parties by the Advisory Committee.[58] Finally, as regards the establishment of protected areas, the Action Plan adopted by the First Meeting of the Parties called for the establishment of criteria to define protected areas for small cetaceans.[59] Such criteria have still not been elaborated, even though the Second Meeting of the Parties called on the Advisory Committee to do so by 2000.[60] Instead, the Advisory Committee and the Third Meeting of the Parties have decided that it would be more sensible to utilize the criteria for protected areas currently being elaborated by other international organizations, notably the EC, the Paris Commission, the Helsinki Commission and the Bern Convention (see further Section 9, *infra*). Once such criteria have been elaborated, a working group established by the Advisory Committee at its Sixth Meeting would consider whether further action was necessary under ASCOBANS.[61] In the meantime one ASCOBANS' party—Germany—

---

[55]   *Report of AC6, supra* note 9, at 12.

[56]   Resolution No. 4, Disturbance, *Report of the Third MoP, supra* note 20, at 97. For a summary of the guidelines, see ASCOBANS, *Triennial Report by the United Kingdom*, Document MOP3/Doc.15d(P), at 27 (2000).

[57]   *Id.*

[58]   *Report of AC6, supra* note 9, at 12. For a summary of these guidelines, *see* ASCOBANS, *supra* note 56, at 27.

[59]   *Report of First MoP, supra* note 17, at 28.

[60]   *Report of Second MoP, supra* note 8, at 63.

[61]   *Report of AC6, supra* note 9, at 25; *Report of AC7, supra* note 6, at 34; *Report of Third MoP, supra* note 20, at 5, 44.

has established a protected area for small cetaceans off a section of its North Sea coast.[62]

While the measures just described have put some flesh on the bare bones of ASCOBANS' Conservation and Management Plan, they are nevertheless generally lacking in precision and are merely aspirational in nature. Furthermore, being described as recommendations, they are hortatory rather than binding. Nearly seven years after the entry into force of ASCOBANS, it is difficult to resist the impression that the conservation measures so far adopted by ASCOBANS represent moves towards action, rather than effective action itself.

8.   MONITORING PARTIES' COMPLIANCE WITH THEIR CONSERVATION OBLIGATIONS

It has become common in international environmental agreements to provide mechanisms to monitor parties' implementation of and compliance with their obligations. In a few cases additional machinery has been established in order to address instances of non-compliance.[63] Usually parties report on their activities, and ASCOBANS uses this system of monitoring. Under Article 2.5, each party is to submit a "brief report" each year to the Secretariat on progress made and difficulties experienced in implementing ASCOBANS. In the early years of ASCOBANS, this reporting system did not function very well, with many reports submitted late or incomplete,[64] in spite of guidelines that were adopted to standardize reporting,[65] but by 1999 the system was working reasonably well. The Secretariat is instructed to send a summary of the reports to the Coordinating Authority of each party: to date four such summaries of reports have been published. The Secretariat will also consider the reports when preparing its own report on the functioning of ASCOBANS, which it presents to each Meeting of the

---

62   *Report of AC6, supra* note 9, at 15–16; *Report of AC7, supra* note 6, at 15.

63   *See* further on this question, Churchill & Ulfstein, *supra* note 32, at 643–47.

64   *Report of Second MoP, supra* note 8, at 24; *Report of AC4, supra* note 18, at 17.

65   *Report of First MoP, supra* note 17, at 24; *Report of Second MoP, supra* note 8, at 63.

Parties. Unlike some other environmental treaties, ASCOBANS does not contain a provision for review of the parties' annual reports by the Meeting of the Parties or another body. In practice, the Advisory Committee and the Meetings of the Parties do refer in passing to the action taken (or more commonly not taken) by the parties as revealed by these reports. However, as there is no formal mechanism under ASCOBANS for reviewing parties' reports, it follows that there is also no provision for action by the parties where a report reveals that a party has not taken adequate action to implement ASCOBANS' conservation measures. Currently, there would be little point in ASCOBANS having such a noncompliance mechanism. First, the measures are recommendatory in nature, rather than being strictly legally binding, so that the question of breach of obligation does not really arise. Second, a number of the measures are sufficiently imprecise that it would be difficult to assess party compliance.

9.   RELATIONS WITH OTHER INTERNATIONAL ORGANIZATIONS AND INSTRUMENTS

As has already been hinted, the activities of other international organizations and instruments are relevant to the conservation of small cetaceans in ASCOBANS' area. It is worth identifying such organizations and instruments and explaining their relevance to ASCOBANS, although for reasons of space such discussion must be very brief. The main organizations and instruments concerned are as follows.

- *The International Whaling Commission (IWC).* As has been explained earlier in this book, the IWC is the premier organization for the regulation of whaling at the global level. It is, however, controversial whether the IWC has the competence to adopt regulatory measures for small cetaceans, and its members are divided on this question.[66] In practice, the IWC has adopted

---

[66]   Further on this issue *see* Birnie, *supra* note 39, at 425, 470, 473–74, 605–606 and 630–31; P.W. Birnie, *Small Cetaceans and the International Whaling Commission*, 10 GEO. INT'L ENVTL. L. REV. 1(1997); W.T. BURKE, THE NEW INTERNATIONAL LAW OF FISHERIES 292–94 (1994); William C. Burns, *The International Whaling Commission and the Regulation of the Consumptive and*

some measures for a few species of small cetaceans, but none for the species to which ASCOBANS applies. Nevertheless, the IWC has conducted research on small cetaceans for many years. Two of its research programs are of particular relevance to ASCOBANS. The first is a working group developing a framework for modeling harbor porpoise stocks in the North Sea and adjacent waters; the second is an investigation of pollution cause-effect relationships in cetaceans, called Pollution 2000+, which is to run from 2000 to 2005 One area of particular focus will be harbor porpoises in the Baltic and North Seas.

- *International Council for the Exploration of the Sea (ICES)*. ICES is also concerned with research. Founded almost a century ago, it is the oldest and perhaps foremost international organization concerned with marine scientific research. Its remit is to promote and coordinate research into fisheries and marine pollution in the North-East Atlantic, the North Sea and Baltic. It has a working group investigating the habitats of marine mammals, which in 1998 published a report on pollution impacts on marine mammals.

- *The European Community (EC)*. The EC is in the unusual position of being a fellow international organization of ASCOBANS, a potential party to ASCOBANS, and able to legislate directly (through regulations) and indirectly (by means of directives) on behalf of all but one of the present parties to ASCOBANS. The EC's work is relevant to the three main areas of concern to ASCOBANS: by-catch, pollution and disturbance to habitat. As regards by-catch, the EC has the exclusive competence to manage fisheries in the waters of its member states, the latter's role in this regard being limited principally to one of implementation and enforcement. The EC has adopted a number of measures that may help to reduce by-catch of small cetaceans. It has prohibited the use of drift nets in excess of 2.5 km since 1993,[67] and

---

Non-Consumptive Uses of Small Cetaceans: The Critical Agenda for the 1990s, 13 WIS. INT'L L.J. 105, at 111–25 (1994); and Rose & Crane, *supra* note 3, at 175–78.

[67]   Regulation 345/92, *Official Journal of the European Communities (OJEC)* 1992 L42/15. Repealed and replaced by arts. 11 and 12 of Regulation 894/97, *OJEC* 1997 L132/1. art. 12 has in turn been repealed and replaced by Art. 24(2) of Regulation 850/98, *OJEC* 1998 L125/1.

from 2002 this prohibition will extend to all drift nets of any length.[68] The EC has also prohibited the encircling of any school or group of marine mammals by using purse-seine nets and prohibited the use of such nets when fishing for tuna in certain areas.[69] None of these measures apply to the Baltic, however.[70] Secondly, the EC has adopted a considerable amount of legislation (in the form of directives) to reduce marine pollution from land-based sources. Thirdly, as regards habitat disturbance, Articles 3 and 4 of the Habitats Directive[71] require member states to designate sites as special areas of conservation (SACs) for two species of small cetacean, the bottlenose dolphin and harbor porpoise, to the extent that there are clearly identifiable areas representing the physical and biological factors essential to their life and reproduction. The Commission is currently evaluating proposals for SACs put forward by member states. In addition, Article 12 of the Directive requires member states to prohibit the deliberate killing and disturbance of all species of small cetaceans, as well as damage to their breeding sites. Member states are also required to monitor the incidental killing of small cetaceans and take measures to ensure that such killing does not have a significant negative impact on the relevant species.

- *The Paris Commission.* The Paris Commission was established by the 1992 Convention for the Protection of the Marine Environment of the North-East Atlantic[72] and is the single successor to

---

[68]   Reg 1239/98, *OJEC* 1998 171/1.

[69]   Regulation 850/98, *supra* note 67, arts. 24(1) and 33, repealing and replacing earlier legislation.

[70]   The EC has adopted a special set of fishery conservation measures for the Baltic: regulation 88/98, *OJEC* 1998 L9/1. The only provision of this Regulation that is relevant to reducing by-catch of small cetaceans is art. 9(2), which prohibits anchored floating nets and drift nets in excess of 35 meters when fishing for salmon and sea trout.

[71]   Directive 92/43 on the Conservation of Natural Habitats and of Wild Fauna and Flora, *OJEC* 1992 L206/7. Further on the Directive and cetaceans, *see* P.G.G. Davies, *Legality of Norwegian Commercial Whaling under the Whaling Convention and its Compatibility with European Law*, 43 INT'L & COMP. L.Q. 270, 279–83 (1994).

[72]   32 I.L.M. 1072 (1993). Further on the Commission and Convention, *see* E.

the Oslo Commission (established by the 1972 Convention for the Prevention of Marine Pollution by Dumping from Ships and Aircraft[73]) and the Paris Commission (established by the 1974 Convention for the Prevention of Marine Pollution from Land-based Sources[74]). The 1972 and 1974 Conventions contain provisions limiting pollution by the dumping of waste at sea and from land-based sources, respectively, and these provisions have been supplemented by a considerable number of measures adopted by the Oslo and Paris Commissions. These measures remain in force and are supplemented by further measures that have been and are being adopted by the new Paris Commission, whose area of application covers all ASCOBANS' waters apart from the Baltic. Many of the provisions of the Conventions and Commission measures concern pollutants that are harmful to marine mammals. In 1998, the Commission's role was expanded when it adopted an annex to the 1992 Convention for the protection and conservation of ecosystems and biological diversity.[75] At the same time, the Commission adopted a strategy to realize the aims of the new annex.[76] The Commission is currently engaged in developing criteria for the designation of protected areas (including areas for the protection of small cetaceans).

• *The Helsinki Commission.* The Helsinki Commission, originally established by the 1974 Convention on the Protection of the Marine Environment of the Baltic Sea Area[77] (now replaced by

---

Hey, T. IJlstra & A. Nollkaemper, *The 1992 Paris Convention for the Protection of the Marine Environment of the North-East Atlantic: A Critical Analysis*, 8 INT'L J. MARINE & COASTAL L. 1 (1993); Louise de La Fayette, *The OSPAR Convention Comes into Force: Continuity and Progress*, 14 INT'L J. MARINE & COASTAL L. 247 (1999).

[73]   11 I.L.M. 262 (1972).

[74]   13 I.L.M. 352 (1974).

[75]   For the text *see* <http://www.ospar.org/eng/html/convention/ospar_conv10. html> (Apr. 6, 2001). On this development, *see* de La Fayette, *supra* note 72, at 265–70.

[76]   Text at <http://www.ospar.org/eng/html/sap/speciestrat.htm> (Apr. 6, 2001).

[77]   13 I.L.M. 594 (1974).

the 1992 Convention of the same name[78]) performs a role for the Baltic similar to that of the Paris Commission. Not only has the Helsinki Commission adopted a number of measures to reduce pollution, supplementing provisions of the 1974 and 1992 Conventions, but it has also been concerned with the protection of marine habitats. The 1992 Convention calls on its parties to take appropriate measures to conserve natural habitats and biodiversity. It also calls on the Commission to elaborate guidelines and criteria for this purpose, which the Commission has recently begun doing. In addition, in 1996 the Commission adopted Recommendation 17/2 concerning the protection of Baltic harbor porpoise populations, which recommends that parties to the Convention accord the highest priority to avoiding by-catches of porpoises and consider the establishment of protected areas for porpoises.[79]

- *The North Sea Conferences*. Since 1984, the North Sea states have met every few years in conferences to discuss the problems of pollution in the North Sea, each such conference ending in a declaration. These declarations, which are soft-law instruments, contain a mixture of elements: undertakings to ratify and effectively implement existing treaties; proposals to press for certain action to be taken under existing treaties and organisations (including the EC and the Paris Commission); and undertakings to take action outside existing agreements.[80] All North Sea conferences since the third in 1990 have addressed wildlife protection issues. As indicated earlier, the third Conference provided

---

[78]   22 L. SEA BULL. 54 (1993). Further on the Conventions and Helsinki Commission, *see* P. Ehlers, *The Helsinki Convention 1992: Improving the Baltic Sea Environment*, 8 INT'L J. MARINE & COASTAL L. 1 (1993); M. FITZMAURICE, INTERNATIONAL LEGAL PROBLEMS OF THE ENVIRONMENT AND PROTECTION OF THE BALTIC (1992); and O. Greene, *Implementation Review and the Baltic Sea Regime, in* THE IMPLEMENTATION AND EFFECTIVENESS OF INTERNATIONAL ENVIRONMENTAL AGREEMENTS: THEORY AND PRACTICE 177 (D.G. Victor, K. Ranstiala & E.B. Stolnokoff eds., 1998).

[79]   The text of the recommendation can be found at <http://www.helcom.fi> (Apr. 5, 2001).

[80]   Further on the North Sea conferences, *see* P. Ehlers, *The History of the International North Sea Conferences, in* Freestone & IJlstra, *supra* note 3, at 3.

the impetus for the negotiation of ASCOBANS. The declaration adopted at that Conference also called for an improvement in the protection of marine wildlife and the adoption of a common and coordinated approach to developing species and habitat protection.[81] This call was repeated, emphasized and developed in the declaration adopted at the fourth Conference held in 1995.[82] It was in response to this call that the members of the Paris Commission adopted the annex to the 1992 Convention (referred to above), which deals with the protection of species and habitats. The declaration went on to recommend that the fisheries policies of the EC and Norway should be formulated to minimize by-catches and other negative impacts on marine mammals.[83] This was followed by an Intermediate Ministerial Meeting on the Integration of Fisheries and Environmental Issues held in 1997, which adopted a Statement of Conclusions[84] in which the ministers resolved to integrate further fisheries and environmental protection and *inter alia* agreed that "fishing practices should be adjusted to minimise the deterioration of sensitive habitats and unacceptable incidental mortality generated by such practices," and invited the competent authorities to consider the application of measures to achieve this objective.[85]

- *The Bern Convention.* The Bern Convention on the Conservation of European Wildlife and Natural Habitats of 1979[86] requires its parties to ensure the special protection of all species listed in Appendix II, which includes all species of small cetaceans found in ASCOBANS' area. In particular, the parties are to prohibit the deliberate killing of such species, the deliberate damage to or destruction of their breeding or resting sites, and their deliber-

---

[81]   Para. 39 and Annex 5 of the Declaration, *reproduced in* Freestone & IJlstra, *supra* note 13, at 3.

[82]   Paras 1–9 of the Declaration, which can be found at <http://odin.dep.no/nsc/esbjerg.html#MINISTERIAL> (Sept. 9, 1998).

[83]   *Id.* at para. 16.

[84]   In <http://odin.dep.no/nsc/soc. html> (Sept. 9, 1998).

[85]   Para. 9.

[86]   1284 U.N.T.S. 209.

ate disturbance, particularly during the breeding season; and in general are to conserve their habitats. Hitherto, the Convention has had little practical application as far as the conservation of small cetaceans is concerned, although recently the Convention institutions have proposed a network of protected areas for the species listed in Appendices I and II.[87]

Other organizations that have some relevance to the work of ASCOBANS include the International Baltic Sea Fisheries Commission,[88] the North Atlantic Marine Mammal Commission,[89] and the Wadden Sea Secretariat.[90]

The activity of the organizations and instruments briefly surveyed above obviously has considerable implications for ASCOBANS. What should the consequences of this activity be for the role of ASCOBANS? Clearly there is no point in ASCOBANS duplicating this work, as it has itself acknowledged.[91] Obviously, it should draw on the work of these organizations in carrying out its own activities, and in general collaborate with the organizations concerned by informing them of the work it does, inviting them to attend ASCOBANS' meetings, and attending their meetings—all of which in practice ASCOBANS does. Furthermore, given that one or more members of ASCOBANS are parties to each of the organizations and instruments concerned, ASCOBANS' members

---

[87]   Further on the Bern Convention and the conservation of small cetaceans, *see* Birnie, *supra* note 10, at 268–69; Churchill, *supra* note 1, at 230–31; and R.R. Churchill and J. Gibson, *The Implementation of the North Sea Declarations by the United Kingdom: An Assessment*, Third North Sea Conference, Greenpeace Paper 17, at 149–88 (1990).

[88]   Established by the Convention on Fishing and Conservation of the Living Resources of the Baltic Sea and Belts, Gdansk, Sept. 13, 1973, 1090 U.N.T.S. 54.

[89]   Established by the Agreement on Cooperation on Research, Conservation and Management of Marine Mammals in the North Atlantic, Apr. 9, 1992. 26 L. SEA BULL. 66 (1994).

[90]   Established by the Administrative Agreement on a Common Secretariat for the Cooperation on the Protection of the Wadden Sea, 1987, *reproduced in* Freestone & IJlstra, *supra* note 13, at 261.

[91]   ASCOBANS art. 5; *Report of Second MoP, supra* note 8, at 65; *Report of Third MoP, supra* note 20, at 103.

should take initiatives in or maintain pressure on other organizations to take into account the conservation needs of small cetaceans in their work. Again, this in practice happens. Ultimately, given the range of activities of these other organizations, the question must be asked whether there is a meaningful autonomous role for ASCOBANS. As regards pollution, the IWC and ICES are conducting research on the consequences of pollution for small cetaceans. Thus, it would seem that there is no independent role for ASCOBANS in this context, nor does it have the capacity to carry out such research. Concrete measures to reduce pollution are being taken primarily by the EC, the Paris Commission and the Helsinki Commission, so there is little work here for ASCOBANS either, although it could in theory recommend measures for pollutants not covered by such organizations or highlight pollutants particularly injurious to cetaceans. As regards disturbance to habitat, the designation of protected areas would seem best left to other organizations. However, there does seem to be an independent role for ASCOBANS in taking measures to reduce disturbance from seismic testing, military activities and whale watching, which it has already done to some degree. Finally, as regards by-catches (the most important cause of small cetacean mortality), the actual adoption of mitigation measures is most likely to occur through the EC. However, there is nothing to prevent ASCOBANS from adopting its own measures, which would be necessary for Poland (the one current member of ASCOBANS not at present a member of the EC), although for EC members there could be problems if this were done (as explained earlier). Moreover, it would certainly seem that ASCOBANS could and should continue to set a target for reducing by-catch, as this is unlikely to be done by any other organization. In addition, ASCOBANS could more effectively require parties to report on the levels of by-catch and strandings, although in the past ASCOBANS' members have complained that this is an onerous obligation that, in any case, is required under the EC's Habitats Directive. It could also promote and coordinate research into measures to reduce by-catches. More generally, ASCOBANS is responsible for abundance surveys of small cetaceans in the ASCOBANS' area and indeed is currently planning the next such survey as a follow-up to SCANS (referred to earlier in Section 2).[92] Overall, therefore,

---

[92]   Resolution No. 5, Monitoring, Status and Population Studies, *Report of Third MoP, supra* note 20, at 99.

there would seem to be an autonomous role for ASCOBANS, albeit a fairly limited one. At least as important is its role as a catalyst for action in other organizations.

Insofar as ASCOBANS has an autonomous role, the question arises as to whether the ASCOBANS' institutions could, if they so wished, adopt legally binding measures. As already seen, the obligations of the Conservation and Management Plan are couched in non-mandatory language ("work towards," "shall endeavour," "should") and the subsequent measures adopted by the Meetings of the Parties and Advisory Committee are all in the form of recommendations. Article 6.1 of ASCOBANS sets out the powers of the Meeting of the Parties. The only power which is relevant in the present context is that given in paragraph (d), which provides that the Meeting of the Parties has the competence to "decide upon . . . any other item relevant to this agreement." This wording would seem to allow the Meeting of the Parties to adopt binding measures if it so wished. However, this power is subject to a procedural limitation. Decisions must be proposed by a party or by the Secretariat and circulated to members of ASCOBANS at least 90 days before the meeting. A way of circumventing this procedural requirement would be to argue that ASCOBANS has an implied power to take binding decisions. It has been argued that meetings of the parties to international environmental agreements may have implied powers to take certain kinds of decisions.[93] In the case of ASCOBANS it could be argued that the need for the agreement to function effectively and protect small cetaceans implies a power in the Meeting of the Parties to adopt legally binding measures. Turning to the powers of the Advisory Committee under Article 5 of ASCOBANS, it has no power to take binding decisions, as is also the case with advisory bodies in other international environmental agreements.

10. AN EVALUATION OF ASCOBANS

ASCOBANS is more than just an agreement. Because it provides for institutions with a continuing role in developing the normative content of the agreement and overseeing its implementation, it may be described, in international relations terminology, as a regime. As with any inter-

---

[93]   Churchill & Ulfstein, *supra* note 32, at 639–40.

national environmental regime, ASCOBANS can be evaluated on the basis of various criteria, such as the level of participation in the regime, the functioning of its institutions, its practical impact so far on the environmental problem at issue, its potential for addressing that problem in the future, and its impact on the behavior of its parties.[94]

As regards the first of these criteria, the level of participation in the regime by the relevant range states is not very high. None of the coastal states of the eastern Baltic apart from Finland is a party. This may not matter much as there are apparently few small cetaceans in the eastern Baltic. On the other hand, non-participation by the eastern Baltic states may mean that small cetaceans will not (re)establish themselves in these waters. Non-participation by France, Norway and the EC is more serious, although Norway's non-participation is mitigated by the fact that it participates actively in the work of ASCOBANS' bodies as an observer and collaborates with the parties on research.

As regards ASCOBANS' institutions, they appear to be functioning reasonably well, although there is some room for improvement. The Advisory Committee needs more resources in terms of expertise and time if it is to do its job effectively; parties need to include more details in their annual reports to ASCOBANS and submit them more promptly; and some effective mechanism for monitoring these reports needs to be developed.

Turning to the third criterion, one way to judge the practical impact of ASCOBANS so far would be to assess whether the population status of small cetaceans in the Baltic and North Seas has changed since ASCOBANS came into force. However, the data to do this does not exist. Even if it did, the time frame would be too short to be meaningful, and in any case it would

---

[94]   There is a considerable theoretical literature, particularly by political scientists, on how the effectiveness of international environmental regimes is to be assessed. *See, for example*, REPORT OF THE CONCERTED ACTION ON THE EFFECTIVENESS OF INTERNATIONAL ENVIRONMENTAL AGREEMENTS (M.L. Honkanen *et al.*, eds., 1999); PROCEEDINGS FROM THE 1999 OSLO CONCERTED ACTION WORKSHOP ON THE EFFECTIVENESS OF GLOBAL AND REGIONAL ENVIRONMENTAL AGREEMENTS (J. Wettestad *et al.*, eds., 2000); and D.F. Sprinz & C. Holm, *The Effect of Global Environmental Regimes: A Measurement Concept*, 20 INT'L POL. SCI. REV. 359 (1999). This section draws to some extent on this literature.

be difficult to assess to what extent ASCOBANS was the cause of any change in population status that had occurred. Instead, all that can be attempted is to assess the effectiveness of the measures taken so far to reduce threats to cetaceans. As regards by-catch, targets have been set (which initially proved too high) but have not been met, and there is little evidence that anything effective has been done to reduce by-catch. Investigation and monitoring of the levels of by-catch are also inadequate. Research on pollution is on-going, and some measures to reduce pollution have been taken by other international organizations, but this is a long-term program of action and it will also be a long time before it makes a significant difference to the quality of the marine environment. As regards disturbance to habitat, protected areas are yet to be designated (apart from one off the German coast) and other measures to reduce disturbance (such as guidelines for seismic surveys) are essentially voluntary and are of limited application. Overall, it must be said that in ASCOBANS (as in some of the other organizations discussed in Section 9) there has been much said, but considerably less done in the way of effective action to address the conservation needs of small cetaceans.

As far as ASCOBANS' future is concerned, there are fundamental questions about its role. To a considerable degree ASCOBANS is a body coordinating and stimulating action in other fora, rather than a body with an autonomous program of action of its own. To the relatively limited degree that it has an autonomous role, there is doubt about its legal competence and political will to adopt legally binding measures, which surely one day ought to replace the present recommendations if ASCOBANS is to be truly effective.

Finally, as regards the effect of ASCOBANS on the behavior of its members, this appears to be fairly limited. Denmark has elaborated (even if it has not yet implemented) a policy on by-catch, and Sweden is also in the course of elaborating a similar policy, but otherwise ASCOBANS appears to have had little effect on its parties' behavior as far as by-catch is concerned. In the case of habitat disturbance, ASCOBANS has had some effect on the United Kingdom (which has developed guidelines on seismic surveys and whale-watching) and on Germany, which has designated a protected area, but otherwise there appears to have been little effect on the other parties. Pollution issues, as already seen, are largely being dealt with in other fora so that one would not expect ASCOBANS to have had any real influence on its parties' behavior in this area. Overall, therefore, it must be concluded that ASCOBANS' influence on the behavior of its parties so far has been modest.

At each Meeting of the Parties so far, the chair of the Advisory Committee has produced a personal checklist on the performance of ASCOBANS' institutions and parties during the preceding triennium in relation to each area of ASCOBANS' activities. He uses a scoring system where "double plus" means the activity was addressed sufficiently, "plus" partly sufficiently, "minus" partly but not sufficiently, and "double minus" means the activity was not addressed at all. He never gives a mark for ASCOBANS' activities as a whole. Were one to give a mark to ASCOBANS' performance generally over the seven years since it came into force, "minus" might not be inappropriate.

In an ideal world, the existence of ASCOBANS as a separate regime would not be necessary. The issues that it seeks to address—the variety of threats to the well being of small cetaceans—are part of the general environmental problems of the Baltic and North Seas. Unfortunately, the response of states in the region, especially those in the North Sea, has been piecemeal. This has resulted in a patchwork of international agreements and organizations to deal with different aspects of the problem rather than the establishment of a single overarching body to tackle the environmental problems of the North Sea, which some writers[95] have advocated.[96] Were such a body ever to come into existence (which seems unlikely), ASCOBANS should be incorporated within it.

---

[95]   See, e.g., P.W. Birnie, *The North Sea: A Challenge of Disorganised Opportunities?, in* GREENWICH FORUM V. THE NORTH SEA: A NEW INTERNATIONAL REGIME? 3 (Donald Cameron Watt ed., 1980); Jean-Paul Ducrotoy & Michael Elliott, *Interrelations between Science and Policy-Making: the North Sea Example,* 34 MARINE POLLUTION BULL. 686 (1997). G. Peet, *Sea Use Management for the North Sea, in* UN CONVENTION ON THE LAW OF THE SEA: IMPACT AND IMPLEMENTATION 430 (E.D. Brown & R.R. Churchill eds., 1987); THE NORTH SEA: CHALLENGE AND OPPORTUNITY 241–45 (M.M. Sibthorp ed., 1975).

[96]   In the North Sea, the North Sea Conferences have fulfilled a limited role in this context. The EC could establish a framework for ecosystem management if all coastal states in the Baltic and North Seas were to become members.

11. CONCLUSION

The populations of the small cetacean species found in the Baltic and North Seas have declined significantly in recent decades. The main causes of this decline are the incidental catching of a large number of small cetaceans in fishing nets, pollution and habitat disturbance. ASCOBANS was established as an agreement under the Bonn Convention on Migratory Species to address this population decline. Although it has been in force for seven years, ASCOBANS has so far failed to take effective action to tackle the causes of small cetacean population decline. There appear to be several reasons for this failure. First, there are weaknesses in ASCOBANS' agreement itself. The conservation obligations of the parties are formulated in language that is vague and hortatory rather than precise and prescriptive. Second, there is still much uncertainty about the population status and some of the causes of population decline of the small cetaceans of the Baltic and North Seas. This scientific uncertainty is particularly relevant when seeking to establish conservation objectives (and in particular to set a target for by-catch reduction) and when seeking to adopt measures to tackle pollution and disturbance to habitat. Third, there seems to be a lack of political will by the states parties to ASCOBANS, both individually and when acting collectively through the ASCOBANS Meeting of the Parties, to adopt precise and legally binding conservation measures, possibly perhaps because such measures would inevitably have an adverse impact on the beleaguered fishing industries of northern Europe. Finally, ASCOBANS does not exist in isolation. There are quite a number of other inter-governmental bodies whose activities bear on the conservation of small cetaceans. Some of these bodies, notably the EC in respect to by-catch mitigation measures and the Helsinki and Paris Commissions in respect to pollution prevention, are better positioned than ASCOBANS to take effective action to counter the population decline of the small cetaceans of the Baltic and North Seas. For the sake of these marine mammals, we must hope that these bodies and ASCOBANS act before it is too late.

Part IV

Anthropogenic Threats to Cetaceans

Chapter 11
Evaluating the Threat from Pollution to Whales

Mark P. Simmonds[1]

1. INTRODUCTION

One important consideration in the development of any effective protection regime for marine mammals is the threat posed by chemical pollution, and particularly the persistent, bioaccumulative and toxic pollutants that are now pervasive in the marine environment. It has long been known that many marine mammals accumulate high concentrations of such compounds. This is well established for polychlorinated biphenyls (PCBs) and certain pesticides, including dichlorodiphenyl-trichlororthane (DDT),[2] and it is becoming increasingly clear that a range of less well-known organic contaminants such as organic tin compounds also accumulate in the tissues of marine mammals.[3] However, it is critical to evaluate the consequences of this accumulation in marine mammals as opposed to merely verifying the presence of the toxic compounds.

It is also well-established that levels of organochlorines and metals are usually lower in baleen whales than in toothed cetaceans. This can be explained by differences in feeding habits and feeding areas. For example, the lower concentrations reported in fin and sei whales are

[1]   Natural Resources Institute, The University of Greenwich, Chatham Maritime, Kent, ME4 4TB, UK, msimmond@wdcs.org. With grateful thanks to Dr. Koichi Haraguchi, Sarah Dolman, Berenice Goddard and Dr. Nicola Kemp for their kind help in the preparation of this Chapter.

[2]   For a recent review *see* P.J.H. Reijnders *Organohalogen and heavy metal contamination in cetaceans: observed effects, potential impact and future prospects, in* THE CONSERVATION OF WHALES AND DOLPHINS 205–217 (M.P. Simmonds & J.D. Hutchinson eds., 1996).

[3]   *See* S. Tanabe, *Butyltin Contamination in Marine Mammals—A Review*, 39 MARINE POLLUTION BULL. 62–72 (1999).

principally attributable to their feeding on planktonic crustaceans which are relatively low on the food web and, hence, less contaminated.[4]

Organochlorine contaminants, particularly PCBs and DDT, have long been implicated in reproductive and immunological disorders in aquatic and other mammals. A variety of abnormalities in a range of aquatic mammal populations have been associated with high tissue burdens of PCBs. The potential and observed effects of such pollutants have been reviewed by various authors[5] and range from impacts at the level of the health of the individual to impacts on populations (see Figure 1 in the Appendix). In fact, since 1968, at least 16 species of aquatic mammals have experienced pollution-associated major stranding episodes, reproductive impairment, endocrine and immune system disturbance, other forms of population instability, or have been affected with serious infectious diseases.[6]

The study of the toxicology of chemical mixtures in environmental samples, including marine mammal tissues, is a developing field.[7] This is also a complex topic because of the difficulties involved in correlating the effects of single compounds (*e.g.*, DDT or methylmercury), or families of compounds (*e.g.*, the PCBs), with health risks for wildlife or humans, including *inter alia:*

- In the wider environment, pollutants exist in very complex mixtures (which is reflected in the tissues of organisms);

---

[4]   Reijnders *supra* note 2, at 209. Toothed cetaceans, feeding higher in the trophic web, show higher levels of contamination and, generally, small odontocetes such as bottlenose dolphins and harbor porpoises show the highest levels of contamination—giving rise to considerable concern about impacts on their health.

[5]   For example, M.P. Simmonds, K. Hanly & S.J. Dolman, *Cetacean contaminant burdens: regional examples*, Paper submitted to the Scientific Committee of the International Whaling Commission SC/51/E13. (1999); T. Colborn & M. Smolen, *An epidemiological analysis of persistent organochlorine contaminants in cetaceans*, 146 REV. ENVTL. CONTAMINATION & TOXICOLOGY 91–172 (1996).

[6]   Colborn & Smolen, *supra* note 5, at 93.

[7]   *See* H. Hansen, C.T. de Roas & H. Pohl *et al., Public health challenges posed by chemical mixtures*, 106 ENVTL. HEALTH PERSPECTIVES 1271 (1998).

- Analytical data typically identifies some, but not all, of the chemicals present; and
- The chemicals of particular concern (*i.e.*, the organochlorines) originate as complex mixtures of congeners. However, because of environmental degradation and other factors, their concentrations in environmental and biological samples are very different from the original mixtures.

Among the pollutants of most concern, due to their tendency to bioaccumulate in both human fat and the ample fatty tissues of marine mammals are the dioxins, a term given to the 75 possible congeners (structurally related compounds) of the polychlorinated dibenzo-*p*-dioxin group (PCDD).[8] Other important bioaccumulating contaminants include the 135 possible congeners of the polychlorinated dibenzofurans (PCDFs). PCDDs and PCDFs (PCDDs/Fs) which are found as contaminants in industrial chemicals, and are also formed in combustion processes and may, therefore, be emitted by incinerators.[9]

2.   TOXIC EQUIVALENCY

Risk assessment for marine and other wildlife can be informed by the approach taken for human health with respect to dioxins and related compounds. Humans are mainly exposed to dioxins and dioxin-like compounds through diet, with food from animal origins being the major source.[10] Risk evaluation for humans is made particularly difficult by the complex nature of PCDD, PCDF and PCB mixtures and the varying potency of the compounds concerned. In response to this problem, the concept of "toxic equivalency" has been developed to facilitate risk assessment and regulatory control.[11] This methodology uses one particular dioxin, 2,3,7,8-tetrachlorodibenzo-*p*-dioxin (TCDD), which is

---

[8]   J.R. Startin, *Polychlorinated dibenzo-p-dioxins, polychlorinated dibezofurans, and the food chain*, FOOD CONTAMINANTS—SOURCES AND SURVEILLANCE 21, 21 (C.S. Creaser & R. Purchase eds., 1991).

[9]   *Id.*

[10]   R.F.X. van Leeuwen, M. Feeley & D. Schrenk *et al., Dioxins: WHO'S tolerable daily intake (TDI) revisited*, 40 CHEMOSPHERE 1095, 1096 (2000).

[11]   Startin, *supra* note 8, at 27.

thought to be particularly toxic, as the standard. Toxic equivalency factors (TEFs) are calculated to express the toxicity of other dioxins, furans and certain dioxin-like PCBs, relative to TCDD.[12] This allows a risk evaluation to be made for an entire category of important toxic chemicals that are similar in structure and biological effects. For example, the most toxic PCDD/Fs and most toxic PCBs bind to the Ah-receptor and thereby cause similar biological effects, although the compounds differ considerably in potency.[13]

The total concentration of PCDD/Fs and dioxin-like PCBs in a sample is now most often expressed as the TCDD (or Toxic) Equivalency (TEQ). In the calculation of TEQ values, each congener is given a value based on its toxicity compared with TCDD. The concentration of each compound in a sample is multiplied by this value, and the products (the TEFs) for all the compounds are summed to give the total TEQ concentration. In many environmental samples the PCBs, rather than the dioxins or furans, are the major contributors to total TEQ values,[14] reflecting their toxicological importance.

However, it is important to note that this approach quantifies the likely contribution to toxicity from all the compounds with dioxin-like structures and activities and, while dioxins have become the major focus of international concerns (see below), risk is better determined by considering all the similar compounds together. PCDD/Fs and dioxin-like PCBs are referred to here as the "Dioxin Group." TEFs are presently ascribed to seven dioxins, ten furans, four non-ortho PCBs and eight mono-ortho PCBs.[15]

Several different TEF schemes have been developed for Dioxin Group compounds,[16] and until recently it has not been clear to what

---

[12]   M. Van den Berg, L. Birnbaum & B.T.C. Bosveld, *et al.*, *Toxic Equivalency Factors (TEFs) for PCBs, PCDDs, PCDFs for Humans and Wildlife*, 106 ENVTL. HEALTH PERSPECTIVES 775, 775 (1998).

[13]   B. Brunström & K. Halldin, K. *Ecotoxicological risk assessment of environmental pollutants in the Arctic*, 112(3) TOXICOLOGY LETTERS 111, 112 (2000).

[14]   *Id.* at 113.

[15]   See table 1, van Leeuwen *supra* note 10, at 1099.

[16]   Startin, *supra* note 8, at 27.

extent human and wildlife TEFs are compatible. However, Van den Berg *et al.*[17] have reviewed existing TEFs and concluded that the same TEFs can be applied to humans and to other (wild) mammals, providing that they are calculated in the same way, but not to fish and birds (for which they developed separate sets). We therefore now have a tool which may be used to inform risk assessment for marine mammals, including cetaceans.

Clearly this should be done with caution, noting that uncertainties exist in the use of the TEF concept for human risk assessment. It should also be noted that the use of values for the highly toxic dioxin TCDD alone, as a measure of PCDDs, PCDFs and PCBs, would severely underestimate risk.[18] A reassessment of TEFs for mammals has recently been completed by a World Health Organization (WHO) expert group. Since December 1990, when WHO established a tolerable daily intake (TDI) of 10 pg/kg bw (body weight)[19] for TCDD, new epidemiological and toxicological data have emerged relating to neurodevelopmental and endocrine effects of dioxins. Subsequently, the European Center for Environment and Health of WHO and the International Program on Chemical Safety organized a formal reassessment of TDI. The experts used a tiered approach in which results of animal toxicity studies, especially those involving (sub)chronic exposure, were given more weight than results of *in vitro* or biochemical studies. To safely determine a TDI, expressed as a TEQ, a "composite uncertainty factor of 10" was used and a new TDI range of 1–4 pg TEQs/kg bw was established. Contributors to the reassessment exercise also recognised that subtle effects might "already be occurring in the general [human] population

---

17   Van den Berg *supra* note 12, at 776.

18   Van Leeuwen *supra* note 10, at 1098.

19   In this text, a number of very small concentrations are expressed in units that may be compared one with another. The TDI considered here is a dose expressed in terms of body weight—*i.e.*, 10 pg/kg bw means that the limit is ten "picogrammes" per kilogram of weight of the individual. So, a 40 kilogram individual would have a 400 picogram limit.

As some other units are also used, here is an explanation of all units: mg is milligram, *i.e.*, 1×10^{-3}g or 0.001g; μg is microgram, *i.e.*, 1×10^{-6}g; ng is nanogram, *i.e.*, 1×10^{-9}g; pg is picogram *i.e.*, 1×10^{-12}g; and parts per million is numbers of milligrams of a substance in a kilogram of sample.

in the developed countries at current background levels of exposure to dioxins and dioxin-like compounds."[20]

3.   EXISTING RISK ASSESSMENT EVALUATIONS FOR CETACEANS

Before considering the application of TEQs to marine mammals further, it will be helpful to review other risk assessments that have been made to date. Various authors have tried to ascribe toxicological significance to contaminant burden data from cetaceans. For example, it has been suggested that total PCB concentrations of more than 50 parts per million (ppm—wet weight) may cause a health risk to cetaceans,[21] but such a threshold has not been reviewed in the light of more recent toxicological studies.[22]

O'Shea and Brownell[23] also reviewed organochlorine and metal contamination in baleen whales and commented on the conservation implications of these pollutants. Although most blubber residue concentrations were less than 5 ppm (with relatively small sample sizes), they also report a number of higher values:

- 590 ppm total DDT (wet weight) and 28 ppm (wet weight) PCBs in the blubber of an adult female minke whale, *Balaenoptera acutorostrata*, that stranded in California in 1979;

---

[20]   Van Leeuwen *supra* note 10, at 1100.

[21]   R. Wageman & D.C.G. Muir, *Concentrations of heavy metals and organochlorines in marine mammals of northern waters: overview and evaluation*, Canadian Technical Report of Fisheries and Aquatic Sciences. No 1279, Western Region, Department of Fisheries and Oceans, Canada (1984).

[22]   Within cetacean taxa, 50 ppm is a relatively high level of contamination. Many small cetaceans found in polluted coastal environments would have body burdens exceeding this but it is a higher concentration than typically reported from baleen whales—some examples of levels reported from baleen whales are provided *infra*.

[23]   T.J. O'Shea & R.L. Brownell, Jr., *Organochorine and metal contaminants in baleen whales: a review and evaluation of conservation implications*, 154 SCI. TOTAL ENV'T 179–200 (1994).

- 61.9 ppm (lipid weight) total DDT and 47.3 ppm (lipid weight) PCBs in the blubber of an immature North Atlantic fin whale, *Balaenoptera physalus*;

- PCBs at 10 ppm (wet weight) in gray whale, *Eschrichtius robustus*, blubber from Washington, 10 ppm PCBs in the blubber of a fin whale and 27 ppm in a minke whale (both wet weights and both from the St. Lawrence estuary in 1979); and

- total DDT at 23 and 7.6 ppm (wet weights) in two humpback whales, *Megaptera novaeangliae*, sampled off Nova Scotia and New Jersey (sampled in the early 1970s).

O'Shea and Brownell stated that there is "no firm basis to conclude that the contaminants reviewed . . . have affected baleen whale populations"[24] and "it would be speculative and alarmist to focus on contaminants investigated so far as important factors influencing the status and conservation needs of baleen whales."[25] However, such a conclusion now stands to be tested against approaches being applied to other mammals, including humans and new toxicological data.

Weisbrod *et al.* provide an insightful look into bioaccumulation and organochlorine exposure in the endangered Northwest Atlantic right whale, *Eubalaena glacialis*, and take much the same view as O'Shea and Brownell.[26] They consider a population of whales, which despite the cessation of commercial whaling, has shown no recovery and is now regarded as critically endangered. Several factors have contributed to the continued vulnerability of this population, including ship collisions, entanglement in fishing gear, unsustainable population size, and exposure to pollutants.[27] Weisbrod *et al.* report total PCB values of 5.7 +/- 8.9 ppm lipid weight and total pesticides of 11.4 +/- 15.4 ppm lipid weight.[28] They compare the total PCB (TPCB) concentrations in right

---

[24]   *Id.* at 179.

[25]   *Id.* at 195.

[26]   A.V. Weisbrod *et al.*, *Organochlorine exposure and bioaccumulation in the endangered Northeast Atlantic right Whale* (Eubalaena glacialis) *population*, 19 ENVTL. TOXICOLOGY CHEM. 654, 665 (2000).

[27]   *Id.* at 654.

[28]   *Id.*

whale biopsy samples with those determined in captive seals, free-ranging seals, and beluga whales, *Delphinapterus leucas*. In these animals, blubber TPCB concentrations of >20 ppm fresh weight correlate with immune problems and >60 TPCB ppm fresh weight with endocrine and other alterations. They conclude "because concentrations were lower than those found in marine mammals affected by PCBs and DDTs, we do not have evidence that the endangered whales bioaccumulate hazardous concentrations of organochlorines."[29]

If one takes the lower limit (20 ppm) as potentially indicative of a level at which physiological problems might start, is the observed mean and range of concentrations in the right whales (5.7 +/-8.9ppm), sufficiently below this limit to safely assume that the population is not being adversely impacted by pollution? Furthermore, should we assume that a "safety limit" based on PCBs alone is entirely adequate? (Note that in this case, lipid and fresh weights are being compared and this will make a small difference.) These questions are considered further below in the light of other approaches used in human risk assessment.

4.   NEW RISK ASSESSMENTS

4.1   Human Examples

Human health assessments have tended to focus on food contamination rather than tissue burdens. Food surveys in industrialized countries have shown an average daily human intake of PCDDs and PCDFs in the order of 1–3 pg I-TEQ kg/bw for a 60 kg adult.[30] If dioxin-like PCBs are also included, the daily total TEQ intake can be higher by a factor of 2–3. It is also well established that special consumption habits, particularly diets low in animal fat or high in highly contaminated food stuffs, may lead to lower or higher TEQ intake values, respectively.[31]

---

[29]   *Id.* at 665.

[30]   Van Leeuwen *supra* note 10, at 1096. I-TEQ here refers to the "International TEQ standard"—an earlier version of a TEQ scheme which does not take PCBs into account.

[31]   The ubiquity and toxicity of these environmental pollutants has raised considerable concern about their chronic impacts on human health, and especially their potential as carcinogens.

Dioxins and PCBs were recently at the center of a major international health scare in Europe. In May 1999, the Belgian authorities informed the European Commission that a batch of animal feed was discovered to have been contaminated by a substance containing high concentrations of dioxin and PCBs.[32] This massive contamination of feed then led to further contamination of certain products of animal origin intended for human consumption, affecting some 25 percent of domestic farms in Belgium. The Minister of Public Health advised that such products be removed from public sale. Further, the European Commission prohibited the sale of suspect Belgian beef and pork and products derived from them. PCBs were also identified in the contaminated foods. No permanent levels for dioxin contamination have been set for individual food products, but this outbreak led a European Commission working group to establish temporary action levels of maximum PCBs of 0.2 ppm for poultry and poultry products and of 0.1 ppm for milk and milk products. Because of the contamination, huge numbers of pigs, chickens and other animals had to be destroyed. In fact, by October 1999, some 6×10^7 kg of dead animals had been disposed of.[33]

The contaminant levels that reached human consumers are unclear, although levels in two chickens were reported as 958 and 775 ppt (parts per trillion) dioxins and 400 ppm of PCBs.[34] Experts from WHO's European Center for Environment and Health and the University of Urecht stated that while cancer risk may be low, neural and cognitive development, the immune system, and thyroid and steroid hormones may be affected.[35] They emphasised the risk to the young and unborn and recommended that people at risk should be monitored for the next ten years.

---

[32]   Commission Decision (1999/449/EC) of July 9, 1999 on protective measures with regard to contamination by dioxins of certain products of animal origin intended for human or animal consumption (OJ L175, 10.7.1999 at 70).

[33]   PCB and dioxin contamination in the feed and food chain in Belgium. Statement from the Federal Ministry of Agriculture and Federal Ministry of Health, <http://dioxin.fgov.be/pe/ene00.htm> (1999).

[34]   D. Mackenzie, *Belgian meat dioxin incident may have serious consequences for EU children*, NEW SCI. June 12, 1999, at 10.

[35]   *Id.*

Concerns about what might be termed "background exposure levels" to such compounds are more typical. For example, Japanese researchers have recently commented on the assessment of human health risk from dioxins in Japan, with particular reference to fish as a source of contaminants.[36] They noted that "since the Japanese consume an average of approximately 90g of fish and shellfish a day . . . and in addition, there exist as many as 2000 municipal solid waste incinerators (MSWIs) in a small overpopulated country, the human health risk caused by dioxins has become a growing public concern. Furthermore, a large quantity of dioxins included in some herbicides was emitted into the environment in Japan during the 1960s and 1970s and dioxins in the aquatic environment are still largely influenced by the use of some herbicides in the past. . . ."[37]

The same researchers evaluated the human health risk from dioxins for four groups:

(1) the general population;
(2) the local residents living near a municipal solid waste incinerator;
(3) a group that ate relatively large quantities of fish; and
(4) the infants and fetuses of all three other groups.

The researchers used a number of parameters in their probabilistic risk assessment and concluded that the margin of exposure values for dioxin-induced neurobehavioral effects on infants and fetuses suggested a "considerable risk" for the infants and consumers of groups 1 and 3. They also commented that "although fish and shellfish are very important in the Japanese diet as high quality protein foods, approximately one-half of the total daily intake of dioxins by the general population corresponds to the intake via fish ingestion. However, the urgent countermeasures [recently enacted by the Ministry of Health and Welfare to reduce dioxins emitted from MSWIs] are not expected to lead to sufficient reduction in dioxin levels in fish in the near future. This contradiction for fish ingestion in terms of dioxin intake must be examined by

---

[36]   *See* K. Yoshida, S. Ikeda & J. Nakanishi, *Assessment of human health risk of dioxins in Japan*, 40 CHEMOSPHERE 177–85 (2000).

[37]   *Id.* at 177.

determining whether the anticipated benefits outweigh the potential risk to Japanese [people]."

In what might be judged as a quirk of fate, Haraguchi *et al.*[38] recently identified cetacean products (*i.e.*, "whale meat") as another major marine source of TEQs to Japanese consumers and this is discussed further below as it provides a new source of information about TEQs in cetaceans. Will the high levels of contaminants found in cetaceans, and which pose a threat to their health, now prove a form of protection from the cetaceans' human consumers?

Studies have also been made of several indigenous communities found to have a high dependency on contaminated aquatic products. For example, the fishing people from the lower North Shore of the St. Lawrence River, Canada have been considered.[39] Blood plasma levels of dioxin-like compounds in this community were approximately eight times higher than levels seen in urban residents.

4.2. Marine and Other Mammals

It has been stated that "for the effective protection of marine mammals, it is necessary to know the potential hazard of persistent, bioaccumulative and toxic pollutants to which they are exposed."[40] However, while a shared feature of TCDD toxicity in all species investigated is atrophy of the thymus gland, numerous toxicity studies have shown that TCDD and the structurally related PCDDs, PCDFs and planar PCBs can cause a wide range of toxic effects. There is also a remarkable interspecies variation in the target organs and toxicity.[41]

---

[38] K. Haraguchi *et al., Levels and human health significance of dioxins and coplanar PCBs in cetacean products sold in Japan*, Dioxin 2000: 20th International Symposium on Halogenated Environmental Organic Pollutants and POPs: Monterey, California (paper submitted).

[39] J.J. Ryan & C. Laliberté, *Dioxin-like Compounds in Fishing People from the Lower North Shore of the St. Lawrence River, Québec, Canada*, 52 ARCHIVES ENVTL. HEALTH 309, 309 (1997).

[40] K. Kannan *et al., Toxicity reference values for the toxic effects of polychlorinated biphenyls to aquatic mammals*, 6 HUMAN & ECOLOGICAL RISK ASSESSMENT 181, 181–82 (2000).

[41] J.G. Vos, *Immunotoxicity of dioxins in seal and flounder*, DIOXINS & DIOXIN-

In a follow-up study to the 1988 seal mortality in Europe, two groups of harbor seals, *Phoca vitulina*, were fed herring originating from either the highly contaminated Baltic Sea or the relatively unpolluted Atlantic Ocean, in order to investigate possible pollution-induced immunosuppression that might have been associated with the die-off.[42] A variety of parameters linked to immune function were monitored over a two-year period. Natural killer (N-K)-cell activity and T-cell mitogen-induced proliferative responses in peripheral blood were found to be significantly lower in the seals fed the more contaminated fish. These results indicated that the contaminants in the Baltic herring were immunotoxic. Further evidence for this conclusion was provided by data showing impaired mixed lymphocyte reactions and antigen-specific lymphocyte proliferative responses in the Baltic group. Thus, in the harbor seals, resistance to viral infection is impaired by exposure to pollutants at levels occurring in the wild (as N-K and T-cells play a major role in defense against viral infections). An additional indication that PCB-like compounds caused the toxic effects in Baltic seals was an observed decline in vitamin A levels.

The estimated daily intakes of TEQs by the Baltic seal group were ten times higher than those of the Atlantic group, leading to a blubber concentration in the Baltic seals of 286 +/-17 ng TEQ/kg lipid weight (mainly derived from mono-ortho PCBs). This can be compared with 90 +/-6ng TEQ/kg lipid weight in the Atlantic seals. From published data it appears that many free-ranging harbor seal populations inhabiting areas of Europe and North America have PCB or TEQ blubber concentrations at, or above, those observed in the Baltic group in the study.[43]

Brunstrom and Halldin[44] have recently considered the ecological risk assessment of environmental pollutants in Arctic wildlife (although they did not include cetaceans). They compare TEF/TEQ and other data from wildlife using the concept of lowest-observed-adverse-effect-level

---

Like Chemicals, Abstract Volume, University of Zurich, Zurich, Switzerland, at 50 Mar 3, 2000.

[42]   Reviewed in *id.*, at 53–55.

[43]   Vos, *supra* note 41, at 55.

[44]   Brunström & Halldin, *supra* note 13, at 113–16.

(LOAEL). They report that four different LOAELs have been calculated for mink, *Mustela vison:* 0.13 mg PCB/kg bw/day; 2.4 ng TEQs/kg bw/day; 3.9 ng TEQs/kg bw/day and 10ng TCDD/kg bw/day. They also review the NOAELs (no-observed-adverse-effect-levels) reported for various animals. For example a NOAEL of 0.13 ng TCDD/kg calculated for Rhesus monkeys, *Macca mulatta*. They conclude that the NOAEL values of 0.13–1.0 ng TEQs/kg day, as suggested from the various studies in mammals, would mean that a polar bear, *Ursus maritimus*, weighing 200 kilograms would be exposed to TEQs at the NOAEL, if it consumed only 0.6–5 kg of ringed seal, *Phoca hispida*, blubber (containing 40 pg TEQ/g) per day. From these calculations, it can be deduced that polar bears are among those species that are already likely to be at exposure levels to dioxin-like compounds that would cause adverse affects. This is further supported by other data showing that PCB levels in polar bear fat range from 5 to 80 ppm with a significant negative association between retinol (a precursor for vitamin A) and PCBs in the blood plasma.[45] This situation would seem to be analogous to that seen in the studies on seals mentioned above.

There are few studies that consider TEQs and dioxin-like compounds in cetaceans, although Carvan *et al.*[46] did so when they exposed dolphin kidney cells *in vitro* to TCDD (which inhibited cell proliferation). More recently, Binh Minh *et al.*[47] reported on contamination by persistent organochlorines in small cetaceans from Hong Kong coastal waters and found extremely high concentrations of DDT and PCBs in some specimens. The relatively high concentrations of PCBs prompted them to examine the residue levels of highly toxic coplanar PCBs and they used the toxic equivalency approach for the assessment of the mono- and non-ortho coplanars. Using this method, they were able to identify which

---

[45]   J.U. Skaare, *et al.*, *Organochlorines in top predators at Svalbard—occurrence, levels and effects*, 112(3) TOXICOLOGY LETTERS 103, 107 (2000).

[46]   *See* M.J. Carvan III *et al.*, *Effects of benzo(a)pyrene and tetrachlorodibenzo(p)dioxin on fetal dolphin kidney cells: inhibition of proliferation and initiation of DNA damage*, 30(1) CHEMOSPHERE 187–98 (1995).

[47]   *See* T. Binh Minh *et al.*, *Contamination by persistent organochlorines in small cetaceans from Hong Kong coastal waters*, 39 MARINE POLLUTION BULL. 383–92 (1999).

of the PCBs provided the greatest contribution to the TEQ, and hence to deduce which were the most potentially hazardous pollutants for the cetaceans of Hong Kong. In this case, they identified the mono-ortho congener IUPAC-118 (2,3',4,4',5-pentachlorobiphenyl) and non-ortho congener IUPAC-126 (3,3',4,4',5-pentachlorobiphenyl) as the greatest concerns with respect to long term toxic effects.

The TEQ data presented by Binh Minh *et al.* do not include PCDD/F values and, therefore, cannot be compared with the other data considered here. However, the TEQs calculated for individual congeners in hump-backed dolphins, *Sousa chinensis*, are very high, *i.e.*, 290 pg/g(PCB 169) and 200 pg/g (PCB 118). For finless porpoises, *Neophocaenoides*, the maximum values are 92 pg/g(PCB 169) and 380 pg/g (PCB 118). All values are pg/g (wet weight) and they demonstrate that summed TEQs in the Hong Kong animals would be relatively high.

Kannan *et al.*[48] further develop the NOAEL and LOAEL approach to determine a number of threshold concentrations derived from the published results of "semi-field," or field, toxicity studies and observations.[49] Their results are summarized in Table 1 (see Appendix).

The minimum dose where adverse effects are likely to occur is probably between the NOAEL and the LOAEL and a series of thresholds for toxic consequences can be proposed by calculating the geometric mean between the two values.[50] Examples of these thresholds are presented in Table 2 (see Appendix).

Based on the dietary thresholds for mink and otters, *Lutra lutra*, (see Table 2), the author suggests 1.6pg TEQ/g wet weight as an estimate for dietary risk assessment purposes.[51]

There is some uncertainty inherent in their proposed thresholds.[52] For

---

[48] Kannan, *supra* note 40, at 189–95.

[49] "Semi-field" refers to studies where food items collected in the wild were used to feed animals held in captivity.

[50] Kannan, *supra* note 40.

[51] This is also within the range of 0.79–2.4 pg/g wet weight for dietary TEQs, proposed for mustelids and pinnipeds by Environment Canada.

[52] Kannan, *supra* note 40, at 195.

example, the toxicity end points used in most mink studies to generate thresholds were reproductive effects, whereas those in seals were immune system effects. Other, even more sensitive, end-points may exist. Recently, for example, it has been shown that bacculum size was negatively correlated with hepatic PCB concentration above 0.02µg/g, wet weight in mink—some ten times lower than the level that affects survival. Moreover, the immune toxicity observed in seals was hypothesised to be mediated by an AhR mechanism, but other immunotoxicants, such as organotin compounds that act through a non-AhR-mediated mechanism, can also contribute to the effects observed. Furthermore, TEFs do not yet take into account the toxicity of all the compounds present in biological samples—only those with dioxin-like activities.

As noted earlier, there are few sources of equivalent TEQ data from cetaceans with which to make comparisons. However, Muir *et al.*[53] reported on TEQs for beluga whales from the highly polluted St. Lawrence estuary. Unsurprisingly, TEQs were very high, averaging 330 ng/kg in females and 1,400 ng/kg in males (wet weights).[54] Muir *et al.*[55] commented on the variations seen between the concentrations of various PCB congeners in the belugas and those reported from other cetacean species. For example, the total concentrations of the non-ortho PCBs CB77, 126 and 169 in the St. Lawrence animals, were 10–20 times higher than those seen in Arctic beluga, but similar to those seen in killer whales, *Orcinus orca*, harbor porpoises, *Phocoena phocoena*, and Dall's porpoises, *Phocoenoides dalli*, on the Canadian west coast. Porpoises from the southern Baltic Sea had lower total non-orthos and a different pattern, with CB77 predominating. Variations seen in non-ortho PCBs may reflect geographic differences in sources of PCBs, age

---

[53]   D.C.G. Muir *et al.*, *Persistent Organochlorines in Beluga Whales* (Delphinapterus leucas) *from the St. Lawrence River Estuary-I. Concentrations and Patterns of Specific PCBs, Chlorinated Pesticides and Polychlorinated Dibenzo-p-Dioxins and Dibenzofurans*, 93(2) ENVT'L POLLUTION 219, 219, 226 (1996).

[54]   Again congeners IUPAC-118 (2,3',4,4',5-pentachlorobiphenyl) and non-ortho congener IUPAC-126 (3,3',4,4',5-pentachlorobiphenyl) were the most important. CB126 accounted for 47 percent and 54 percent of TEQs in males and females, respectively, and CB118 was the next highest contributor.

[55]   Muir, *supra* note 53, at 226–28.

and sex differences in the relatively small number of animals studied, and also differences in metabolic capability among cetaceans.

As part of an ongoing study into contamination in whale meat sold in Japan, TEQs have recently been reported for a range of cetacean species.[56] Preliminary data are summarised in Table 3 (see Appendix). TEQ blubber values for dolphins and toothed whales are particularly high. Haraguchi *et al.* report that the relative contributions of the PCDDs, PCDFs and Co-PCBs to the total TEQs in blubber/bacon samples were 0.5 percent, 18 percent and 82 percent, respectively.[57] The TEFs for co-PCBs exceeded those for PCDD/Fs in most samples, although the PCDFs in bacon/blubber from toothed whales and dolphins made a relatively high contribution to the total toxic potency (TEQ).

It has previously been concluded that odontocetes have relatively low levels of metabolic activity towards compounds metabolised by certain isozymes (*i.e.*, the "CYP2B type")[58] and noted that the relative abundance of certain PCDD/F and non-ortho-substituted PCB congeners in St. Lawrence belugas is probably best explained by the "unusual metabolic capabilities" of these odontocetes, rather than local sources.[59] In support of this finding is the fact that the profiles of these contaminants are similar in the Arctic and St. Lawrence belugas, despite the potentially large differences in exposure. New data[60] suggest that this is also true for small cetaceans taken in the North Pacific. These data also show that many cetacean products are a substantial source of Dioxin Group contamination for human consumers. From a human health perspective, this argument needs to be considered in the context of the high dioxin intakes reported from other sources in Japan. It should also be noted that consumers in Japan cannot safely identify which cetacean product they are buying because of widespread mislabeling in the market place (*i.e.*, they cannot necessarily avoid the more contaminated meat/blubber if

---

[56]   Haraguchi *et al., supra* note 38.

[57]   The most important PCB congeners were again CB126 and CB118.

[58]   S. Watanabe, *et al., Specific profile of liver microsomal cytochrome P450 in dolphin and whales*, 27 MARINE ENVTL. RES. 51, 61 (1989).

[59]   Muir, *supra* note 53, at 230

[60]   *See*, Binh Minh, *supra* note 47; Haraguchi, *supra* note 38.

desired).[61] Nor are the Dioxin Group chemicals the only important contaminants to occur in whale products sold in Japan.[62]

Considering these data and using the seal blubber threshold proposed of 160 pgTEQ/g lipid wt,[63] it can be shown that some of the cetaceans sampled in Japan and elsewhere may be at risk based on a TEQ evaluation. However, this threshold also requires careful evaluation.

5.   OTHER CONSIDERATIONS

Uncertainty remains over the potential impacts of other compounds that are not regularly analyzed but which may be of significance for mammalian health. Such compounds do not yet have any associated advisory human health limits and toxicity data is still scant. For example, both Watanabe *et al.* and Binh Minh[64] have recently reported on the presence of tris (4-chlorophenyl)methane and tris (4-chlorophenyl)methanol in marine mammals from Asian waters. Similarly, as noted earlier, butyltins (BTs) in marine mammals have recently been reviewed.[65] Cetaceans retained higher butyltin concentrations than pinnipeds and this too is probably attributable to a low breakdown capacity in the cetaceans. A

---

[61]   This is fully explained in F. Cipriano & S.R. Palumbi, *Rapid genotyping techniques for identification of species and stock identity in fresh, frozen, cooked and canned whale products.* Paper submitted to the Scientific Committee of the International Whaling Commission. (SC/51/09).

[62]   For example, cooked cetacean livers, as sold over the counter in Japan, have been found to have a mean total mercury concentration of 275 ppm (n = 9), and the maximum level recorded of 645 ppm, is some 1,600 times above the permitted level for human consumption, *see* T. Endo, M. Sakata, K. Haraguchi *et al. Contamination of heavy metals and organochlorines in whale meat (part 2): mercury contamination.* Proceedings of the 79th Meeting of the Food Hygienic Society of Japan, Tokyo, 130 (2000).

[63]   *I.e.,* the threshold at which these authorities suggest marine mammals are at risk from dioxin-like contaminants, *see* Kannan, *supra* note 40, at 194.

[64]   M. Watanabe *et al., Contamination of tris(4-Chlorophenyl) methane and tris(4-Chlorophenyl)methanol in marine mammals from Russia and Japan: body distribution, bioaccumulation and contamination status.* 39 MARINE POLLUTION BULL. 393–98 (1999).

[65]   Tanabe, *supra* note 3.

relatively high percentage of these compounds was also found in the liver of cetaceans. Tanabe[66] suggested that "the present contamination by BTs may pose a considerable toxic threat to some coastal species of cetaceans." Le et al.[67] also recently provided data on organotin levels in cetaceans from Japanese coastal waters.

6.   CONCLUSIONS

The presence of substantial quantities of Dioxin Group compounds in marine mammals is a considerable health concern for both marine mammals and their predators/consumers (particularly those already exposed to high levels of such compounds in the rest of their diet, a category that would include polar bears and the human consumers of cetacean products).

The toxic equivalency approach has the potential to be a valuable tool in marine mammal-health risk evaluation. In its favor are the facts that (1) TEQs can be used in addition to other approaches such as those using single groups of chemicals (e.g., PCBs); and (2) it allows the toxicity of the important Dioxin Group to be assessed in samples as one unit.

However, the methodology is an emergent one that still has limitations. There remain a number of problems with the calculation and application of TEQs. In particular, this approach can only be used when appropriate data have been generated[68] and such analyses are very expensive. Moreover, TEQs cited in the literature may not be strictly comparable. For example, the new WHO TEFs will result in an approximate 10 percent increase compared to both the I-TEFs and the original WHO TEFs.[69] Furthermore, there are very few data from whales to compare with thresholds based on TEQs.

––––––––––––

[66]   Tanabe, *supra* note 3 at 62.

[67]   *See*, L.T.H. Le et al., *High Percentage of Butyltin Residues in Total Tin in the Livers of Cetaceans from Japanese Coastal Waters*, 33 ENVTL. SCI. & TECH. 1781–86 (1999).

[68]   Colborn & Smolen, *supra* note 6, at 146.

[69]   Van Leeuwen, *supra* note 10, at 1099 and as used by Binh Minh et al., *see supra* note 44.

However, the major contribution made by PCB compounds to TEQ values in three separate studies of cetaceans, as reviewed above, strongly supports their choice by the IWC[70] as an important and practical indicator group for contaminant impacts. It is also significant that the value derived by Kannan *et al*.[71] as a blubber threshold (*i.e.*, 17µg PCBs/g lipid wt) is close to that used by Weisbrod *et al*.[72] in their evaluation of risk to the North Atlantic right whale (i.e. TPCB >20 µg/g fresh weight).

But are such health PCB and TEQ thresholds adequately precautionary? For example, the >0.02 µgPCB/g affect limit found in mink and described earlier has unknown biological consequences but is still a toxic response. It is also far lower than the marine mammal hepatic threshold of 0.44 µgPCB/g,[73] which would correlate with the 17 µgPCB/g blubber threshold. If cetaceans are as sensitive to Dioxin Group chemicals as mink then the proposed blubber threshold could be an order of magnitude (or more) too high.

In fact, it is doubtful that we are yet in a position to propose safe thresholds of contamination for cetaceans. Clapham *et al*.,[74] commenting specifically on the status of the most endangered cetacean populations, noted the conclusion of O'Shea and Brownell that baleen whales were not at risk but added, "However, the manner in which pollutants negatively impact animals is complex and difficult to study, particularly in taxa (such as large whales) for which many of the key variables and pathways are unknown."

---

[70] At its meeting in 1999, the Scientific Committee of the International Whaling Commission (IWC) recommended to the Commission the establishment of an ambitious program of work to evaluate the impacts of pollutants on cetaceans. This program (POLLUTION 2000+) is described in 2 J. CETACEAN RES. & MGMT. at 209–13 (2000 Suppl.). The program was endorsed by the Commisison and component studies were initiated. PCBs are the contaminants that the program seeks to link to "biomarkers."

[71] Kannan, *supra* note 40, at 193.

[72] Weisbrod, *supra* note 18, at 660.

[73] As proposed in Kannan, *supra* note 40, at 192.

[74] P.J. Clapham *et al., Baleen Whales: Conservation issues and the status of the most endangered populations*, 29 MAMMAL REV. 35–60 (1999).

The identification and development of biomarkers by the IWC pollution programme should greatly aid the evaluation of thresholds for contaminant burden data.[75] Meanwhile, application of a precautionary approach (which might refer to a precautionary blubber concentration threshold of 1–2 µgPCB/g) would probably dictate that many cetaceans, including many baleen whales, should be classified as at risk from chemical contamination until it can be demonstrated otherwise.

APPENDIX

Figure 1: Biological Response to Environmental Pollutant Exposure Modified from Hansen *et al.*, 1998[76]

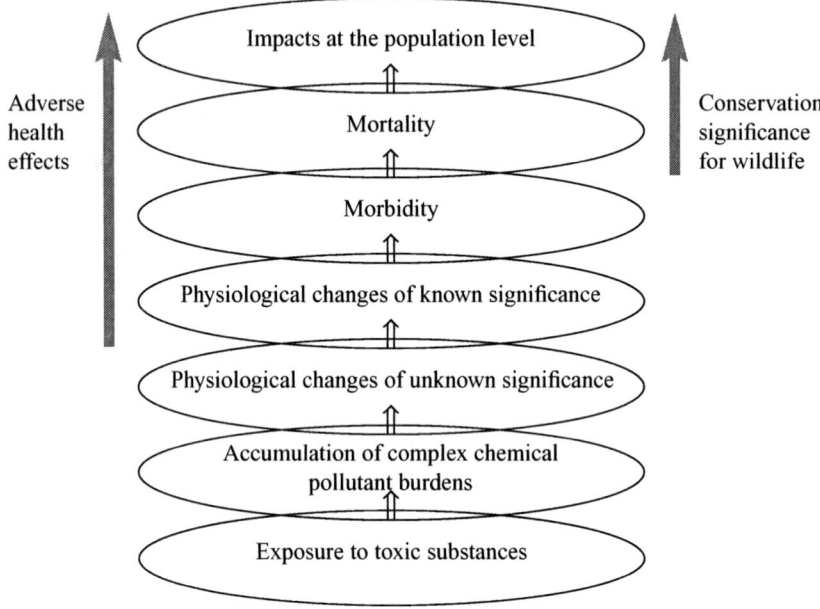

Adverse health effects

Conservation significance for wildlife

Impacts at the population level

Mortality

Morbidity

Physiological changes of known significance

Physiological changes of unknown significance

Accumulation of complex chemical pollutant burdens

Exposure to toxic substances

---

[75]   *See supra* note 67.

[76]   *Supra* note 7.

Table 1 Summary of NOAEL and LOAEL for Total PCB and TEQs Investigations with Seals and Dolphins (from Kannan *et al.*, 2000)[77]

	PCBs	TEQs
Seal daily dose NOAEL	5.2µg/kg bw/d	0.58ng/kg/bw/d
Seal daily does LOAEL	28.9µg/kg bw/d	5.8ng/kg/bw/d
Seal dietary NOAEL	100ng/g wet wt	NA
Seal dietary LOAEL	200ng/g wet wt	NA
Seal blood NOAEL	5.2µg/g bw/d lipid wt	NA
		(lipid 0.5–0.32%)
Seal blood LOAEL	25µg/g lipid wt	NA
Seal blubber NOAEL	NA	90pg/g, lipid wt
Seal blubber LOAEL	NA	286pg/g, lipid wt
Dolphin blood LOAEL (in vitro)	26ng/g wet wt	NA

Table 2 Toxic Thresholds for Tissues and Diet (from Kannan *et al.*, 2000)[78]

1. Toxic thresholds for PCBs "Marine mammal blubber"	17µg PCBs/g lipid wt
Marine mammal dietary threshold	10–150 ng/g, wet wt (mean 89 ng/p wet wt)
2. Toxic thresholds for TEQs Seal blubber	160pg TEQ/g lipid wt
Dietary threshold—mink	1.9 pg TEQ/g wet wt
Dietary threshold—otter	1.4 pg TEQ/g wet wt

---

[77]   *Supra* note 40.

[78]   *Supra* note 40.

Table 3 TEQ Levels of PCDDs/PCDFs and Coplanar PCBs in Cetacean Products Sold in Japan (from Haraguchi *et al.*, 2000)[79]
Data presented as ranges and averages (in parentheses) in pg-TEQ/g wet weight.

	Toothed whale & dolphin products products		North Pacific Minke whale		Southern Hemisphere Minke whale products	
	meat	bacon/ blubber	meat	bacon/ blubber	meat	bacon/ blubber
	n = 6	n = 13	n = 3	n = 9	n = 2	n = 5
Total TEQ	0.85–5.04 (2.97)	27.1–691 (232)	0.84–13.8 (5.2)	15.6–127.2 (57.4)	0.5–2.5 (1.5)	2.1–8.4 (4.9)

[79]   *Supra* note 38.

Chapter 12
Climate Change and the International Whaling Commission in the 21st Century

William C.G. Burns[1]

While we debate the limits that should be placed on whaling in order to protect the status of the stocks, a silent menace threatens to destroy the populations we strive to protect.

—D. James Baker[2]

1. INTRODUCTION

The 50th Meeting of the Parties to the International Convention for the Regulation of Whaling (ICRW)[3] held in Oman in May 1998 may ultimately be recognized as a watershed in the history of the International Whaling Commission's (IWC) efforts to manage and conserve cetacean species. While the primary focus of most meetings of the IWC during its first half century was on regulating the harvesting of regulated species, IWC50 was dominated by questions of how to confront perhaps the gravest long-term threat to cetaceans: environmental change. As identified by the IWC's Scientific Committee, the term "environmental change" encompasses the following: climate change, chemical pollution, physical and biological habitat degradation, effects of fisheries,

[1]   William C.G. Burns, Co-Chair, American Society of International Law—Wildlife Interest Group, El Cerrito, California, USA, jiwlp@international-wildlifelaw.org. I would like to express my thanks to Dr. Ray Gambell and Julie Creek of the International Whaling Commission for assistance in obtaining documents. An earlier version of this chapter was published in 13(2) GEO. INT'L ENVTL. L. REV. (2001).

[2]   U.S. Commissioner to the International Whaling Commission.

[3]   *See* International Convention for the Regulation of Whaling, Dec. 2, 1946, 161 U.N.T.S. 72 [hereafter Whaling Convention].

ozone depletion and UV-B radiation, Arctic issues, disease and mortality events, and the impact of noise.[4] The purpose of this chapter is to assess the implications of what could prove to be one of the direst of these threats for the viability of cetacean species—climate change—and the role of the IWC in seeking to ameliorate climate change impacts. Section 2 of this chapter will discuss the potential ramifications of climate change for cetacean species. Section 3 will outline the history of the IWC's treatment of climate change issues. Finally, Section 4 will assess the viability of the IWC's strategies to protect cetaceans from climate change in the next century and the role of other institutions, including the United Nations Framework Convention for Climate Change and the Convention for the Conservation of Antarctic Marine Living Resources.

2.   CLIMATE CHANGE AND CETACEANS

2.1.   Climate Change Scenarios

This section will briefly explain the "greenhouse effect" and projected warming trends over the next century, and then seek to assess the possible implications for cetaceans.

The surface of the Earth is heated by solar radiation emanating from the sun at short wavelengths between 0.15 and 5 μm. Each square meter of the Earth receives an average of 342 watts of solar radiation throughout the year.[5] Approximately one-third of the incoming solar radiation is reflected back to space in the form of thermal infrared, or longer-wave radiation, at wavelengths of 3–50 μm.[6] Of the remainder, a portion is

---

[4]   50th Meeting of the International Whaling Commission, *Resolution on Environmental Change and Cetaceans*, IWC Resolution 1998–6 (1998) [hereafter IWC Resolution]. Resolutions from the 50th Meeting are available on the American Society of International Law—Wildlife Interest Group website, available at <http://www.eelink.net/~asilwildlife>.

[5]   INTERGOVERNMENTAL PANEL ON CLIMATE CHANGE, CLIMATE CHANGE 2001: THE SCIENTIFIC BASIS, Contribution of Working Group I to the Third Assessment Report of the Intergovernmental Panel on Climate Change (2001), at 89 (hereafter *Climate Change 2001*).

[6]   INTERGOVERNMENTAL PANEL ON CLIMATE CHANGE, RADIATIVE FORCING OF CLIMATE CHANGE 7 (1994) [hereafter *Radiative Forcing*].

partly absorbed by the atmosphere, but most (168 watts per square meter) is absorbed by land, ocean and ice surfaces.[7]

Some of the outgoing infrared radiation is absorbed by naturally occurring atmospheric gases—principally water vapor (H_2O)—as well as carbon dioxide (CO_2), ozone (O_3), methane (CH_4), nitrous oxide (N_2O) and clouds.[8] This absorption is termed the "natural greenhouse effect" because these gases, which are termed "greenhouse gases," operate much like a greenhouse: they are "transparent" to incoming short-wave radiation, but "opaque" to outgoing infrared radiation, trapping a substantial portion of such radiation and re-radiating much of this energy to the Earth's surface.[9] This process is critical to the sustenance of life on Earth, elevating surface temperatures by about 33 degrees Celsius.[10]

In the past, the net incoming solar radiation at the top of the atmosphere was balanced by net outgoing infrared radiation, contributing to climatic stability.[11] However, with the advent of fossil fuel burning plants to support industry, automobiles and the energy demands of modern consumers, "humans began to interfere seriously in the composition of the atmosphere."[12]

The burning of fossil fuels, mainly coal, oil and gas, has soared since the beginning of the Industrial Revolution, producing approximately 6.5

---

[7]   *Climate Change 2001, supra* note 5, at 89.

[8]   MELVIN A. BENARDE, GLOBAL WARMING . . . GLOBAL WARNING 45 (1992).

[9]   Envtl. & Geographical Sci. Dep't, U. Capetown, *Climate Change—Some Basics* (1999), available at <http://www.egs.uct.ac.za/csag/faq/climate-change/faq-doc-5.html> (last visited Nov. 10, 2000).

[10]   *Id.* The "greenhouse effect" phenomenon was first described by the French scientist Fourier in 1827. *See* Spencer Weart, *From The Nuclear Frying Pan Into The Global Fire*, BULL. ATOMIC SCI. 18, 19 (June 1992).

[11]   *Scientists Remain Unanimously Concerned Over Climate Change*, 23 ECO-LOG WK., July 14, 1995, *available at* 1995 WL 2406417. "For the past 8,000 years, the world's climate has been very stable, varying only within a range of + or—1 degrees C." *Id.*

[12]   Fred Pearce, *World Lays Odds On Global Catastrophe*, NEW SCI., Apr. 8, 1995, at 4.

gigatons of carbon annually in recent years, nearly all of which enters the atmosphere as CO_2.[13] As a consequence, concentrations of carbon dioxide in the atmosphere have increased approximately 25 percent since 1850:[14] from 270–280 parts per million (ppm) by volume in pre-industrial times to over 370 ppm today, with most of the increase occurring in the past 50 years.[15] Anthropogenic activities have also resulted in substantially increased atmospheric concentrations of other greenhouse gases, including methane and nitrous oxides,[16] as well as new sources, such as chlorofluorocarbons and halons.[17]

---

13 WORLDWATCH INSTITUTE, VITAL SIGNS 2002 at 52 (2002); HADLEY CENTRE, THE GREENHOUSE EFFECT AND CLIMATE CHANGE 5 (1999). An additional 1.5 gigatons is released into the atmosphere from land-use changes, such as deforestation. Cement production contributes a small additional amount. *Id.*

14 Kevin Jardine, *Finger On The Carbon Pulse*, ECOLOGIST, Nov./Dec. 1994, at 220.

15 WORLDWATCH INSTITUTE, *supra* note 13, at 52; TOM M.L. WIGLEY, THE SCIENCE OF CLIMATE CHANGE: REPORT OF THE PEW CENTER ON CLIMATE CHANGE 5 (1999). Atmospheric concentrations of CO_2 have reached their highest levels in 160,000 years. United Nations Environment Programme (UNEP), GEO-2000, *The State of the Environment—Global Issues* (1999), *at* <http://www.unep.org/geo2000/english/0034.htm>.

16 "[M]ethane and nitrous oxide concentrations have increased by 145 and 15 percent respectively [since 1750]." Colin Warbrick & Dominic McGoldrick, *Global Warming and the Kyoto Protocol*, 47 INT'L & COMP. L.Q. 446, 447 (1998). "The primary natural source of [methane] is microbial decay of organic matter under anoxic conditions in wetlands. Anthropogenic sources, which in sum may be twice as great as natural sources, include rice cultivation, domestic ruminants, bacterial decay in landfills and sewage, leakage during the mining of fossil fuels, leakage from natural gas pipelines, and biomass burning." JAMES HANSEN ET AL., NASA GODDARD INSTITUTE FOR SPACE STUDIES, GLOBAL WARMING IN THE 21ST CENTURY: AN ALTERNATIVE SCENARIO 4 (2000), available at <http://www.giss.nasa.gov/gpol/cites/2000.html> (last visited Nov. 10, 2000). Significant sources of nitrous oxide include nitrogen-based fertilizers, the clearing of land, biomass burning, and fossil fuel combustion. UNEP, *supra* note 15.

17 Guy Brasseur, *Global Warming and Ozone Depletion: Certainties and Uncertainties, in* GLOBAL WARMING AND THE CHALLENGE OF INTERNATIONAL COOPERATION: AN INTERDISCIPLINARY ASSESSMENT 29–30 (Gary C. Bryner ed., 1994). Overall, CO_2 accounts for 65 percent of the total radiative forcing resulting from anthropogenically released greenhouse gases, methane contributes an addi-

Increases in the concentration of greenhouse gases reduces the efficiency with which the Earth's surface radiates to space. It results in an increased absorption of the outgoing infrared radiation by the atmosphere, with this radiation re-emitted at higher altitudes and lower temperatures.[18] This resulting change in net radiative energy, which is termed "radiative forcing," tends to warm the lower atmosphere and the Earth's surface.[19]

The latest assessment by the Intergovernmental Panel on Climate Change (IPCC)[20] concluded that rising concentrations of greenhouse gases are the primary cause[21] for the increase in average global temperatures of about 0.6 degrees Celsius in the past century.[22] In the Northern

---

tional 19 percent, chlorofluorocarbons, 10 percent, and nitrous oxide about 6 percent. GRAEME APLIN, GLOBAL ENVIRONMENTAL CRISES 222 (2d ed. 1999).

[18]   Intergovernmental Panel on Climate Change, Working Group I, *Technical Summary*, at 24 (2001).

[19]   *Id.*

[20]   The IPCC, comprised of 2,500 climate scientists from throughout the world, was established by the United Nations in 1988 to gather information and coordinate research related to climate change, to evaluate proposals for reducing greenhouse gas emissions, and to assess the viability of response mechanisms. G.A. Res. 43/53, U.N. GAOR, 2d Comm., 43d Sess., Supp. No. 49, at 133, U.N. Doc. A/43/49 (1989); David Lewis Feldman, *Iterative Functionalism and Climate Management Organizations: From Intergovernmental Panel on Climate Change to Intergovernmental Negotiating Committee, in* INTERNATIONAL ORGANIZATIONS & ENVIRONMENTAL POLICY 1195–96 (Robert V. Bartlett *et al.* eds., 1995).

[21]   Intergovernmental Panel on Climate Change, Working Group I, *Summary for Policymakers*, at 1 (2001) (hereafter *Summary for Policymakers*). The IPCC concluded:

> The warming over the last 40 years due to anthropogenic greenhouse gases can be identified despite uncertainties in forcing due to anthropogenic sulphate aerosol and natural factors (volcanoes and solar irradiance). The anthropogenic sulphate aerosol forcing, while uncertain, is negative over this period and therefore cannot explain the warming. Changes in natural forcing during most of this period are also estimated to be negative and are unlikely to explain the warming.

Id. at 6.

[22]   UNITED KINGDOM DEP'T OF THE ENV'T, TRANSPORT AND THE REGIONS, CLI-

Hemisphere, "the increase in temperature in the 20th century is likely to have been the largest of any century during the past 1000 years."[23]

According to the latest report by the Intergovernmental Panel on Climate Change, projected increases in atmospheric greenhouse gases over the next century could elevate temperatures on Earth by 1.4–5.8°C by the year 2100,[24] with the trend accelerating thereafter.[25] While this

---

MATE CHANGE AND ITS IMPACTS 9 (1999). "20th Century global mean temperature is at least as warm as any other century since at least 1400AD." IPCC, *Contribution of Working Group I to the IPCC Second Assessment Report*, IPCC-XI/Doc. 3, at SPM.2 (1995). Warming has accelerated in the last 25 years, more than doubling that of the 20th century average. William K. Stevens, *1999 Continues Warming Trend Around Globe*, N.Y. TIMES, Dec. 19, 1999, at 1.

[23]   *Summary for Policymakers, supra* 21, at 1.

[24]   Intergovernmental Panel on Climate Change, *Third Assessment Report of Working Group I*, Executive Summary, at 8 (2001). The IPCC, comprised of 2,500 climate scientists from throughout the world, was established by the United Nations in 1988 to gather information and coordinate research related to climate change, to evaluate proposals for reducing greenhouse gas emissions, and to assess the viability of response mechanisms. G.A. Res. 43/53, U.N. GAOR, 2d Comm., 43d Sess., Supp. No. 49, at 133, U.N. Doc. A/43/49 (1989); David Lewis Feldman, *Iterative Functionalism and Climate Management Organizations: From Intergovernmental Panel on Climate Change to Intergovernmental Negotiating Committee, in* INTERNATIONAL ORGANIZATIONS & ENVIRONMENTAL POLICY 1195–96 (Robert V. Bartlett *et al.* eds., 1995).

The IPCC's most recent estimate of anticipated temperature increases is substantially higher than that in its last major assessment report in 1997, where it projected a temperature increase of between 1.0 to 3.4°C by 2100. The highest estimates primarily reflect lower projections for sulphur dioxide emissions over the next century. Sulphur dioxide exerts a cooling effect on the atmosphere by deflecting incoming solar radiation. *Id.*; WIGLEY, *supra* note 15, at 21.

Moreover, a new study by the United Kingdom's Hadley Centre indicates that the carbon absorption capabilities of vegetation and soil, which are now responsible for sopping up fifty percent of carbon emissions, may start to decline with rising temperatures. As a consequence, the Centre now projects that concentrations of carbon dioxide could rise to 1000 ppm, resulting in temperature increases of 8 degrees Celsius by the end of the century. Peter M. Cox, *et al.*, *Acceleration of Global Warming Due to Carbon-Cycle Feedbacks in a Coupled Climate Model*, 408 NATURE 184, 186 (2000).

[25]   IPCC, THE IPCC ASSESSMENT OF KNOWLEDGE RELEVANT TO ARTICLE 2 OF

may seem like a slight shift in temperatures, it would "very likely be without precedent during at least the last 10,000 years."[26] A comparison with past changes of this magnitude demonstrates the possible implications:

> The last time it was three degrees [Fahrenheit] warmer than now was 100,000 years ago. Then, Central Europe had a climate like Africa's. And just three degrees separate today from the other climatic extreme, the last ice age of 10,000 years ago. Then, half of Europe lay under ice, and the sea level was 390 feet lower than it is today. A bitter north wind nipped at the ears of the polar bears living atop the frozen Baltic . . . Since the end of the last ice age, average global temperatures have never fluctuated by more than one degree.[27]

2.2. Cetaceans and Climate Change

In assessing the possible impacts of climate change on cetaceans, it must be emphasized at the outset that our ability to assess future impacts at the regional level, which is critical for ascertaining the possible ramifications for many cetacean species,[28] remains limited.[29] Climate researchers use computer models, derived from weather forecasting, to represent the

---

THE UNITED NATIONS FRAMEWORK CONVENTION ON CLIMATE CHANGE: A SYNTHESIS REPORT § 3.2 (1995) (draft).

[26] IPCC, *supra* note 18, at 8. Such a change could be ten to 50 times as fast as the natural average rate of temperature change since the last glaciation. MARGRET M.I. VAN VUUREN & MAARTEN KAPPELLE, DUTCH NATIONAL RESEARCH PROGRAMME ON GLOBAL AIR POLLUTION AND CLIMATE CHANGE, BIODIVERSITY AND GLOBAL CHANGE 14 (1998).

[27] *The Calamitous Cost of a Hotter World*, WORLD PRESS REV., July, 1995, at 15.

[28] INTERNATIONAL WHALING COMMISSION, REPORT OF THE IWC WORKSHOP ON CLIMATE CHANGE AND CETACEANS 2 (1996) [hereafter IWC REPORT].

[29] *Climate Change 2001, supra* note 5, at 587; William C.G. Burns, *The Impact of Climate Change on Pacific Island Developing Countries in the 21st Century, in* CLIMATE CHANGE IN THE SOUTH PACIFIC: IMPACTS AND RESPONSE IN AUSTRALIA, NEW ZEALAND, AND SMALL ISLAND STATES 234 (Alexander Gillespie & William C.G. Burns eds., 2000).

Earth's energy and water cycles and to predict how enhanced levels of greenhouse gases will affect the Earth's climate. The most sophisticated of these models, general circulation models (GCMs), use a three dimensional grid overlaying the surface of the Earth with grid points a few hundred kilometers per side, within which cells are stacked about 20 layers deep.[30]

Vertical layers of the model represent levels in the atmosphere and depths in the ocean, dividing the surface of the planet into a series of horizontal boxes separated by lines similar to latitudes and longitudes.[31] Within each grid point, a series of equations are run on a super-computer, producing simulations of key climatic components, including wind, air-pressure, temperature, humidity, ice coverage and land surface processes.[32] Climate models are usually run for several simulated decades, with the derived results compared to actual statistics on climatic indicia over this period . . . , such as mean temperatures and precipitation. The models are then run with changes in external forcing, such as projected increases in atmospheric greenhouse gas concentrations, over a series of decades or centuries. "The differences between the two climates provide an estimate of the consequent climate change due to changes in that forcing factor."[33]

However, as Solman and Nunez recently observed, computer models remain crude instruments for regional climate projections:

[General circulation models] have difficulty in reproducing regional climate patterns, and large discrepancies are found among

---

[30]   Hadley Centre for Climate Prediction and Research, *Regional Climate*, available at <http://www.meto.gov.uk/sec5CR_div.bak/Brochure/regn_pre.html>. *See also* IPCC, AN INTRODUCTION TO SIMPLE CLIMATE MODELS USED IN THE IPCC SECOND ASSESSMENT REPORT 10 (1997).

[31]   Kevin J. Hennessy, *CSIRO Climate Change Output*, available at <http://www.dar.csiro.au/pub/programs/climod/impacts/data.htm>.

[32]   Eric J. Barron, *Climate Models: How Reliable Are Their Predictions?*, CONSEQUENCES 17, 18 (Aug. 1995).

[33]   M.E. Schlesinger, *Model Projections of CO_2-Induced Equilibrium Climate Change, in* CLIMATE CHANGE AND SEA LEVEL CHANGE 171 (R.A. Warrick *et al.* eds., 1993).

models. In many regions of the world, the distribution of significant surface variables, such as temperature and rainfall, are often
influenced by the local effects of topography and other thermal
contrasts, and the coarse spatial resolution of the GCMs can not
resolve these effects.[34]

Climate researchers have developed several strategies to conduct
regional assessments. Nested models seek to simulate regional climates
by the application of limited area models nested in a GCM.[35] In recent
years, some of these models have yielded high correlations between
regional climate predictions and observed climatic phenomena, including precipitation, thermal inertia of water bodies and temperature.[36]
Downscaling by statistical means, or deriving statistical relationships
between observed local climatic variables and large-scale variables, has
also proved successful in linking large-scale spatial averages of precipitation and surface temperature to local precipitation and temperature-
time series.[37]

With the caveat that regional climate assessments remain speculative,
recent research indicates that cetaceans may be seriously threatened by
projected warming in the next century. In the Antarctic, where 90 percent

---

[34]   Silvina A. Solman & Mario N. Nunez, *Local Estimates of Global Climate
Change: A Statistical Downscaling Approach*, 19 INT'L J. CLIMATOLOGY 835,
835–36 (1999). *See also* HADLEY CENTRE, *supra* note 13, at 14.

[35]   K. YA. KONDRATYEV & A.P. CRACKNELL, OBSERVING GLOBAL CLIMATE
CHANGE 381 (1998); HADLEY CENTRE, *supra* note 13, at 14.

[36]   KONDRATYEV & CRACKNELL, *supra* note 35, at 383–84; *see also* Norman
Miller, *Climatically Sensitive California: Past, Present, and Future Climate, in*
POTENTIAL IMPACTS OF CLIMATE CHANGE AND VARIABILITY FOR THE CALIFORNIA
REGION, REPORT TO THE UNITED STATES GLOBAL CHANGE RESEARCH PROGRAM
NATIONAL ASSESSMENT 25–26 (1998).

[37]   Solman & Nunez, *supra* note 34, at 836; *see also* Thomas R. Karl, *The U.S.
National Climate Change Assessment: Do The Models Project a Useful Picture of
Regional Climate?*, Hearings before the Subcommittee on Oversight &
Investigations, Committee on Energy and Commerce, July 22, 2002, <http://ener-
gycommerce.house.gov/107/hearings/07252002Hearing676/Karl1142.htm>;
Hartmut Grassl, *Status and Improvements of Coupled General Circulation Models*,
288 SCI. 1991, 1994 (2000) (statistical downscaling used to effectively simulate
meteorological variables in Scandinavian mountain area).

of the world's great whales feed,[38] temperatures in some areas have risen four–five degrees Celsius in the last 50 years,[39] substantially more than the world average during that period.[40] While the lack of long time-series and natural climatic variability in the region makes it impossible to attribute definitively the region's warming to climate change,[41] recent modeling by the Hadley Centre for Climate Prediction and Research provides some evidence for such a link.[42]

---

[38]   *See* Gerard Baker, *Japan Threatens to Quit Whaling Commission*, FIN. TIMES, May 28, 1994, at 4. Three species of whales reside year-round in the ice pack: bottlenose, minke, and killer whales, while sperm, humpback, blue, fin, sei, and some minke stocks migrate to the Southern Ocean during the Antarctic winter. *See also* Paul Lincoln Stoller, *Protecting the White Continent: Is the Antarctic Protocol Mere Words or Real Action?*, 12 ARIZ. J. INT'L & COMP. L. 335, 336 & n.5 (1995).

[39]   K. Reid & J.P. Croxall, *Environmental Response of Upper Trophic-Level Predators Reveals a System Change in an Antarctic Marine Ecosystem*, 268 PROC. ROYAL SOC'Y LONDON B. 377, 377 (2000). *See also* John Turner *et al.*, *Climate Change (Communication Rising): Recent Temperature Trends in the Antarctic*, 418 SCI. 291–92 (2002) (warming in some portions of the western side of the Antarctic Peninsula have increased 1.09°C per decade between 1951–2000, but overall trends on continent remain speculative); Raymond C. Smith, *Marine Ecosystem Sensitivity to Climate Change*, 49 BIOSCI. 393, 395 (1999). Mid-winter surface air temperatures in the western Antarctic have risen 5.5°C over the 1941–1991 period. Sherwood Willing Wise, *The Antarctic Ice Sheet: Rise and Demise?*, 15(9) J. LAND USE & ENVTL. L. 383, 384 (2000); R.C. Smith *et al.*, *Surface Air Temperature Variations in the Western Antarctic Peninsula Region, in* 70 FOUNDATIONS FOR ECOLOGICAL RESEARCH WEST OF THE ANTARCTIC PENINSULA 20 (R.M. Ross *et al.* eds., 1996).

[40]   IPCC, CONTRIBUTION OF WORKING GROUP I TO THE IPCC SECOND ASSESSMENT REPORT, IPCC-XI/Doc. 3, at SPM. 20 (1995). *See also* Grover Foley, *The Threat of Rising Seas*, 29 ECOLOGIST 76, 78 (1999) ("Antarctica appears to be warming faster than anywhere else on the planet. . . .").

[41]   SCAR Global Change Programme, *A Summary of Change in the Antarctic*, available at <http://www.antcrc.utas.edu.au/scar/newsletter2/2summary.html> (last modified Sep. 20, 2000). *See also* Andrew Clark & Eugene Murphy, *A Long-Term Fast Ice Record from the South Orkney Islands*, 1 GLOBAL CHANGE RES. 1 (1996); Sean Ryan, *Global Warming*, SUNDAY TIMES, Mar. 26, 1995.

[42]   British Antarctic Survey, *supra* note 39 ("Measurements made over the Antarctic Peninsula and the Falkland Islands show that the level of peak electron concentration in the ionosphere F-region (at about 300 km altitude) has fallen by about 8km over 38 years . . . . While the lower atmosphere warms in response to increasing concentrations of greenhouse gasses, the upper atmosphere cools.

Recent research forecasts that a doubling of greenhouse gases from pre-industrial times could reduce sea ice in the Southern Hemisphere by more than 40 percent in the next century.[43] This may have several adverse effects on the abundance of the zooplankton species krill (*Euphausiacea*),[44] the primary prey species for whales in the Southern Hemisphere.[45] First, a diminution in sea ice may lead to a decline in the

―――――――――――

Theoretical studies indicate that the observed fall in the height of the F-region is compatible with expected temperature changes in the thermosphere"). However, annual mean temperature increases of 2 degrees Celsius over the past 50 years on the Antarctic Peninsula are not consistent with predictions of climate models. *Id.*

[43]   IWC REPORT, *supra* note 28, at 3. However, some researchers argue that warming either may have very little effect on ice sheets in the Antarctic, or may even portend an increase in volume, at least for the next century or two, due to an increase in snowfall caused by higher evaporation. University of Tasmania, Antarctic Cooperative Research Centre, *Polar Ice Sheets, Climate and Sea-Level Rise*, Feb. 10, 2000, available at <http://www.antcrc.utas.edu.au/antcrc/about/Position_Statement_2.html> (last visited Nov. 26, 2000). *See also* C. J. van der Veen, *Polar Ice Sheets and Global Sea Level: How Well Can We Predict the Future?*, 32 Global & Planetary Change 165, 188 (2002); .British Antarctic Survey, *supra* note 39; David Schneider, *The Rising Seas*, Mar. 19, 1997, SCI. AM. 96, 114; C.L. Hulbe, *Recent Changes to Antarctic Peninsula Ice Shelves: What Lessons Have Been Learned,?* 1 NATURAL SCI. (Apr. 11, 1997), available at <http://naturalscience.com/ns/articles/01-06/ns_clh.html>.

However, Bamber *et al.*, suggest that models of the draining of discharge from the Antarctic Ice Sheet as a consequence of warming may have underestimated ice stream flow rates, "implying that parts of the interior of Antarctica and probably former ice sheets can respond more rapidly to climate forcing than model simulations might suggest." Jonathan L. Bamber *et al.*, *Widespread Complex Flow in the Interior of the Antarctic Ice Sheet*, 287 SCI. 1249 (2000).

[44]   "Krill" is a general term that encompasses about 85 species of ocean crustaceans in the group called *euphausiids*. Five species of krill are found in the Antarctic, the most abundant being *Euphausia superba*, which grow up to about six centimeters and live between five to ten years. Australian Antarctic Division, *Krill: Magicians of the Southern Ocean*, <http://www.antdiv.gov.au/resources/more_res/krill.html> (last visited Nov. 26, 2000).

[45]   Christophe Barbraud & Henri Weimerskirch, *Emperor Penguins and Climate Change*, 411 NATURE 183, 185 (2001); O. BALASHOV. & B. HARE, POLAR MELTDOWN: THE CHANGING CLIMATE IN ANTARCTICA: A REPORT FOR GREENPEACE INTERNATIONAL (1997); David Helvarg, *On Thin Ice*, SIERRA, Nov./Dec. 1999, at 40. Blue whales may consume as much as four tons of krill per day. Hulbe, *supra*

productivity of algae, the primary source of food for krill during the winter.[46] Second, a reduction in sea ice could deny krill larvae critical protection from predators. Cetacean species that migrate long distances might also have to alter the timing and order of migration to follow the ice front, adversely affecting their energetics, such as feeding and reproductive biology.[47]

Finally, sea ice decline could result in the proliferation of the pelagic tunicate *Salpa Thompsoni*, one of the most abundant macrozooplankton species in the ice-free and seasonal pack-ice zone of the Southern Ocean.[48] Salps persist in low numbers under sub-optimal conditions but can rapidly proliferate when sea ice recedes and phytoplankton becomes more readily available during early spring.[49]

---

note 43. Krill are the major biomass component of the epipelagic marine ecosystem in the Seasonal Pack-Ice Zone and parts of the Ice-free and the high-Antarctic Zone, comprising approximately 500 million tons of biomass. Charles Arthur, *Global Warming Poses New Threat to Whales' Survival*, INDEPENDENT, June 26, 1997, at 3. Krill support an array of species in the region, including penguins, fur seals, and seabirds, such as the albatross. Duncan M. Cunningham & Philip J. Moors, *The Decline of Rockhopper Penguins Eudyptes chrysocome at Campbell Island, Soutern Ocean and the Influence of Rising Sea Temperatures*, 94 EMU 27, 34 (1994); Debora MacKenzie, *In for the Krill*, NEW SCI., June 5, 1999, at 26; Andrew Brierley, *Kingdom of the Krill*, NEW SCI., Apr. 17, 1999, at 36, 39–40.

[46]   World Wide Fund for Nature, *Climate Change: Parks at Risk*, available at <http://www.panda.org/climate/parks/dr_I_park9.htm> (last visited Nov. 26, 2000). Brine channels on the underside of sea ice connect to underlying water and nutrients that are important for the growth of algal species. Alfred Wegener Institute for Polar & Marine Research, *Microstructure*, <http://www.awi-bremerhaven.de/Eistour/mikrostruktur-e.html> (last visited Nov. 26, 2000).

[47]   Thomas Karl, *The Arctic and the Antarctic, in* IPCC, THE REGIONAL IMPACTS OF CLIMATE CHANGE: AN ASSESSMENT OF VULNERABILITY (Robert T. Watson *et al.* eds., 1997), at 98.

[48]   Arthur, *supra* note 45, at 3.

[49]   *See* K.H. KOCK & V. SIEGEL, INTERNATIONAL WHALING COMMISSION, TEMPORAL VARIATIONS IN SEA-ICE DYNAMICS AND KRILL ABUNDANCE IN THE ANTARCTIC PENINSULA REGION—IMPLICATIONS FOR THE KRILL-DOMINATED FOOD WEB 3 (1996); V. Siegel & V. Loeb, *Recruitment of Antarctic Krill 'Euphausia superba' and Possible Causes for its Variability*, 123 MAR. ECO. PROGRESS. SER. 45, 54 (1995).

This salp proliferation could prove disastrous for krill populations in the region. Salps could act as strong competitors of krill for food prior to the onset of phytoplankton blooms in the spring. This increased competition for food can stunt krill gonadal development, resulting in a reduction in recruitment the following year.[50] Moreover, dense salp blooms can interfere with krill reproduction and kill off their larvae.[51]

Warming and possible shifts in wind patterns could also affect the distribution and characteristics of polynyas in the Antarctic region.[52] Polynyas are areas of open waters in the polar ice pack, formed by a combination of currents, tides, upwellings and winds.[53] While snow and ice reflect most of the sun's incident energy, dark polynya water absorbs it, resulting in nutrient upwelling and profuse blooms of phytoplankton.[54] Cetacean species that rely on ice edges for phytoplankton foraging might be adversely affected by reductions in the areal extent and latitudinal shift of ice-edge habitats.[55] For example, recent research indicates that the calving of ice shelves in the Antarctic as a consequence of climate change is blocking the movement of sea ice offshore, substantially reducing areas of open waters.[56] After the calving of the huge B-15 iceberg in 2000, phytoplankton production in portions of the Ross Sea declined by between 32 and 95 percent.[57]

---

[50]   Siegel & Loeb, *supra* note 49, at 54.

[51]   Arthur, *supra* note 45, at 3. Krill also face other serious threats, including loss of prey species and direct damage from ozone depletion. Colin Woodard, *Food-Chain Alarm from a Low-Ozone Zone*, CHRISTIAN SCI. MONITOR, Dec. 11, 1998, at 8. Perhaps in the future, Krill will face overexploitation by commercial fishing concerns. *See infra* note 119.

[52]   IWC REPORT, *supra* note 28, at 10.

[53]   *See Anomalous Sea Ice Conditions in the Cosmonaut Sea During 1999*, <http://www.atmosp.physics.utoronto.ca/ANTARCTIC/cosmo_1999.html> (last visited Sept. 24, 2000); Fred Breummer, *Northern Oases*, 114 CANADIAN GEO. 1, 54 (1994).

[54]   Laura Cheshire, *Phytoplankton and Polynyas*, available at <http.nasadaacs.eos.nasa.gov/yearbooks/95/polynya.html> (last modified Sept. 15, 2000).

[55]   IWC REPORT, *supra* note 28, at 10.

[56]   Kevin R. Arrigo, *Ecological Impact of a Large Antarctic Iceberg*, 29 GEOPHYSICAL RES. LETTERS X-1, X-2 (2002) (advanced copy edition).

[57]   *Id.* at X3–X4.

The populations of several baleen whale species in the Antarctic, including blue and humpback, were decimated in the past by commercial whaling operations,[58] and blue whales may never recover.[59] Reductions in food supplies as a consequence of warming could further diminish the carrying capacity of whales in the Antarctic and push these species closer to extinction in the next century.[60] Also, the latest abundance estimates of minke whales in the Southern Hemisphere by the Scientific Committee of the IWC reveal that stocks may have plummeted in the past 15 years from 760,000 to only about 360,000 currently.[61] In assessing the possible causes for this decline, Dr. Sidney Holt of the International Ocean Institute concluded that warming was "the likeliest hypothesis" for the crash.[62] While the IWC's Scientific Committee emphasized that the abundance estimate decline may be attributable to problems associated with the latest survey,[63] there is certainly

---

[58]   By the early 1960s, aided by new technology, including explosive harpoons and stronger vessels, whalers drove blue and humpback whales to the point of commercial extinction. JAMES C.F. WANG, HANDBOOK ON OCEAN POLITICS & LAW 152 (1992). Blue whale populations have plummeted from a pre-exploitation level of 200,000 to as few as 500 in the Southern Hemisphere, and humpbacks have declined from 120,000 to approximately 10,000. WORLD WIDE FUND FOR NATURE, WANTED ALIVE! 1 (1998); John Carey, *Embattled Behemoths; Whales*, 25 INT'L WILDLIFE 4 (1995).

[59]   *See Whaling 1989/1990*, 23 ORYX 184 (1989).

[60]   World Wide Fund for Nature, *Blue Whales: The Largest Animal Ever to Live on Earth*, (2001), <http://www.panda.org/climate/summit2001/bluewhale2.pdf>, World Wide Fund for Nature, *Protected Areas at Risk* (1997), <http://www.panda.org/climate/pubs/parks/dr_i_park9.htm>. A portent of the danger to whale populations of declining krill biomass may be the substantial drop in Adelie penguin populations in recent years, which some researchers attribute in part to declines in krill populations. KOCK & SIEGEL, *supra* note 49, at 1; Greenpeace, *Antarctic Warming—Early Signs of Global Climate Change* (1995), <http://www.greenpeace.org/search/shtml>.

[61]   Report of the Science Committee, International Whaling Commission, IWC/53/4, at 35 (2001); Geoffrey Lean & Robert Mendick, *Whale Population Devastated by Warming*, The Independent, July 29, 2001, <http://news.independent.co.uk/uk/environment/story.jsp?story=85923>.

[62]   Lean, *supra* note 61.

[63]   Report of the Scientific Committee, *supra* note 61, at 39. The Committee

cause for concern given the fact that minke whales would be targeted should the commercial moratorium on whaling be lifted in the future.

Climate change may also have grave implications for cetaceans in the Arctic. Temperatures in the region have increased at several times the global rate over the past century,[64] with sea-ice thickness declining more than 40 percent, from 10.2 feet to 5.9 feet, since 1958,[65] and sea-ice areal extent declining between 3.0 and 4.5 percent per decade in the past 20 years.[66]

It should be emphasized that it is difficult to establish a causal link between melting ice and anthropogenic climate change, as this phenomena could also be attributed to other factors, such as changes in precipitation and snow cover, or advective processes accompanying the North Atlantic Oscillation in the late 1980s and 1990s.[67] One recent

---

concluded that the decline in abundance estimates may be real or related to "changes in the proportion of the population that is present in the survey region at the time of the survey" or "changes in the survey process over the course of the surveys that compromise the comparability of estimates across years." *Id.*

[64]   *See* Greenpeace, *The Threat of Climate Change to Arctic Wildlife* (1997), <http://www.greenpeace.org/~comms/97/arctic/library/biodiversity/wildlife.html>. "The polar amplification of warming in the Arctic is attributed to the positive albedo feedback of snow and sea-ice." IWC REPORT, *supra* note 28, at 13. *See also* MICHAEL E. MASS & RAYMOND S. BRADLEY, AMERICAN GEOPHYSICAL UNION, NORTHERN HEMISPHERE TEMPERATURES DURING THE PAST MILLENNIUM: INFERENCES, UNCERTAINTIES, AND LIMITATIONS (1999).

[65]   Sydney Levitus, *Anthropogenic Warming of Earth's Climate System*, 292 SCI. 267, 269 (2001); *Research Predicts Summer Doom for Northern Icecap*, N.Y. TIMES, July 11, 2000, at D2; D.A. Rothrock *et al.*, *Thinning of the Arctic Sea-Ice Cover*, 26 GEOPHYSICAL RES. LETTERS 3469, 3471 (1999).

[66]   *See* Ola M. Johannessen *et al.*, *Satellite Evidence for an Arctic Sea Ice Cover in Transformation*, 286 SCI. 1937, Oct. 1, 1999; Kenneth Blackman, *Global Warming Worries Indigenous People*, INTER PRESS SERVICE, Aug. 13, 1998, *available at* LEXIS, World Library.

[67]   Rothrock, *supra* note 65, at 3471. *See also* Kristin Leutwyler, *Icelandic Weather System May Explain Melting Arctic Ice*, Sci. Am., Oct. 5, 2001, <http://www.sciam.com/news/100301/1.html>; M.C. Serreze *et al.*, *Observational Evidence of Recent Change in the Northern High-Latitude Environment*, 46 CLIMATIC CHANGE 159, 170 (2000).

study, however, compared satellite and surface observations with two existing computer models and concluded that there was less than a 0.1 percent chance that ice shrinkage is attributable to natural cycles.[68]

In its most recent regional assessment, the Intergovernmental Panel on Climate Change concluded that projected warming trends in the Arctic over the next century could result in a further 50 percent decline of sea ice.[69] Richard Moritz, Director of the Surface Heat Budget for the Arctic Ocean (SHEBA) project goes further in a recent assessment, predicting that the Arctic's year-round icepack could totally disappear in 50 years.[70]

Further losses of sea ice over the next century could have adverse impacts on cetaceans in the region. While no single species dominates the Arctic food chain, as does krill in the Antarctic,[71] sea ice decline associated with warming could result in the diminution of phytoplankton populations.[72] This could lead to "knock-on effects" throughout the Arctic food chain, meaning declines in the stocks of several key prey

---

[68]   Konstantin Y. Vinnikov *et al.*, *Global Warming and Northern Hemisphere Sea Ice Extent*, 286 SCI. 1936 (1999). *See also* D.K. Perovich *et al.*, *Year on Ice Gives Climate Insights*, EOS, TRANSACTIONS AM. GEOPHYSICAL UNIT, Oct. 12, 1999, at 481

[69]   Karl, *supra* note 47, at 93. However, the IPCC cautioned that the inadequacy of regional polar models render such projections highly speculative. Indeed, some researchers argue that sea ice levels would not be substantially changed under doubled CO_2 conditions. *Id.*

[70]   Mariana Gosnell, *Meltdown? Sea Ice May Be Thawing, Which Could Mean Disruption of Life at Earth's Polar Ends*, INT'L WILDLIFE, July–Aug. 1998, at 12. *See* Lars H. Smedsrud & Tore Furevik, *Towards an Ice-Free Arctic?*, 2 CICERONE (Feb. 2000), available at <http://www.cicero.uio.no/cicerone/00/2/en/smedsrud.pdf>. The Arctic Ocean could be ice-free during summer months by the end of this century.

[71]   IWC REPORT, *supra* note 28, at 14.

[72]   ENVIRONMENTAL INVESTIGATION AGENCY, WHALES IN A CHANGING OCEAN 6 (1994).

The freshening of high latitude seas from freshwater inputs and melt water could be expected to increase the period of halothermal stratification, increase the depth of the halocline and exacerbate the saline concentration gradient. The longer period of stratification is predicted to disadvantage larger phytoplankton such as diatoms

species of cetaceans, such as copepods and plankton-feeding fish, including Arctic cod, a key prey species for narwhal and beluga whales.[73] Some cetacean species in the region, such as fin and bowheads, have demonstrated adaptability in feeding behavior and may be able to shift to other prey species.[74] However, other species—or stocks of such species, including narwhals and belugas—might be seriously affected by the loss of ice-dependent prey species.[75]

Polynyas are important spring feeding and breeding grounds for marine mammals in the Arctic, as well as overwintering sites for white and possibly bowhead whales.[76] Warming and the attendant ice melt might result in greater stratification of the water column and decreased nutrient supplies, limiting the growth of phytoplankton populations that are a critical link in the cetacean food chain in the region.[77]

---

and increase the number of smaller species. This will lengthen the food chain between primary producers and larger consumer species, effectively reducing the biomass of the latter. . . . The greater depth of the halocline means that all phytoplankton will spend a greater amount of time in sub-optimal illumination, while the strengthened concentration gradient will make it more difficult for nutrients to enter the surface layer from below. Both of these factors are expected to reduce overall productivity.

Id.

[73]   *Id. See also* IWC REPORT, *supra* note 28, at 14; Cynthia T. Tynan & Douglas P. DeMaster, *Observations and Predictions of Arctic Climatic Change: Potential Effects on Marine Mammals*, ARCTIC, Dec. 1997, available at LEXIS, World Library.

[74]   B. Würsig & J. Orega-Ortiz, *Global Climate Change and Marine Mammals*, Proceedings of the Thirteenth Annual Conference of the European Cetacean Society, Valencia, Spain, Apr. 5–8, 1999, at 352; IWC REPORT, *supra* note 28, at 14.

[75]   *Id. See also* Executive Secretary, Convention on the Conservation of Migratory Species of Wild Animals, *Climate Change and Migratory Species*, Tenth Meeting of the Scientific Council, May 2–4, 2000, at sec. 16. Changes in thermohaline circulation and the intensification of coastal upwelling as a consequence of warming may also adversely affect the abundance of cephalopod species. *Id.* at 13.

[76]   M. Holst & I. Stirling, *A Note on Sightings of Bowhead Whales in the North Water Polynya, North Baffin Bay, May–June 1998*, 1 J. CETACEAN RES. MGMT. 153, 153 (1999).

[77]   Tynan & DeMaster, *supra* note 73.

Projected reductions in sea ice area could also open up the Northwest Passage. This could expose cetaceans to increased ship traffic and dangers associated with mineral exploitation, as well as bycatch threats should new fishing areas appear in the region.[78] Collisions with vessels pose a serious threat to many cetacean stocks throughout the world.[79] Vessel noise may also disrupt cetacean migration patterns,[80] increase mortality through stress,[81] result in hearing loss,[82] and interfere with communications, which may result in strandings.[83] Mineral exploitation could threaten cetaceans through pollution,[84] noise,[85] and in the case of

------

[78]   IWC REPORT, *supra* note 28, at 13; Smedsrud & Furevik, *supra* note 70; *Research Predicts Summer Doom for Northern Icecap, supra* note 62.

[79]   Masami Fujiwara & Hal Caswell, *Demography of the Endangered North Atlantic Right Whale*, 414 NATURE 537, 539 (2001); David Laist, *Collisions Between Ships and Whales*, 17(1) MARINE MAMMAL SCI. 35, 36 (2001); WORLD WIDE FUND FOR NATURE, *supra* note 58, at 1 (1998).

[80]   Kim E.W. Shelden & David J. Rugh, *Bowhead Status Report*, National Marine Mammal Laboratory 17 (1998), available at <http://nmm101.afsc. noaa.gov/CetaceanAssessment/bowhead/bmsos.htm>; Alexander Gillespie, *Whale-Watching and the Precautionary Principle: The Difficulties of the New Zealand Domestic Response*, 17 N.Z. U.L. REV. 254, 261–62 (1997); James E. Scarff, *The International Management of Whales, Dolphins, and Porpoises: An Interdisciplinary Assessment*, 6 ENVTL. L.Q. 326, 416 (1977).

[81]   David Harrison, *Noise Drives Whales Crazy*, OBSERVER, May 31, 1998, available at LEXIS, World Library.

[82]   *See* Natural Resources Defense Council, *Sounding the Depths*, <http:// www.nrdc.org/wildlife/sound/sdinx.asp> (last visited Sept. 24, 2000).

[83]   *Id. See also* Judith D. Hutchinson, *Fisheries Interactions: The Harbour Porpoise—a Review, in* THE CONSERVATION OF WHALES AND DOLPHINS 154–55 (Mark P. Simmons & Judith D. Hutchinson eds., 1996).

[84]   Bernd Würsig, *Cetaceans and Oil: Ecologic Perspectives, in* SEA MAMMALS AND OIL: CONFRONTING THE RISK 129–65 (Joseph R. Geraci & David J. St. Aubin eds., 1990); Thomas Land, *Co-Ordinated Action is Key to Black Sea Pollution Strategy*, LLOYDS LIST, Jan. 2, 1998, available at LEXIS, World Library.

[85]   Mark P. Simmonds & Susan J. Mayer, *An Evaluation of Environmental and Other Factors in Some Recent Marine Mammal Mortalities in Europe: Implications for Conservation and Management*, 5 ENVTL. REV. 89, 96 (1997) (Seismic testing has resulted in displacement of whales in Gulf of Mexico and New Zealand). *See*

oil and gas exploration, water dispersal during the drilling phase.[86]

In other regions of the world, warming may also alter ocean upwelling patterns, fostering increased blooms of dinoflagellates, many of which produce brevitoxins.[87] Dinoflagellate blooms have been associated with the deaths of marine species throughout the world, including cetaceans in the Mediterranean.[88] The warming of tropical waters may also contribute to epizootics, such as the one that killed thousands of striped dolphins in the Mediterranean in the early 1990s,[89] and augment the spread of marine disease agents and parasites.[90] Anticipated reductions in river flow associated with climate change could also increase the concentrations of pollutants in coastal areas in the region.[91]

Warming trends are also likely to raise ocean surface water temperatures to above 26 degrees Celsius in the next century.[92] This tempera-

---

also SWISS COALITION FOR THE PROTECTION OF WHALES, POLAR EXPOSURE: ENVIRONMENTAL THREATS TO ARCTIC MARINE LIFE AND COMMUNITIES 13 (1997).

[86]   *See id.*

[87]   ENVIRONMENTAL INVESTIGATION AGENCY, *supra* note 72, at 26.

[88]   William C.G. Burns, *The Agreement on the Conservation of Cetaceans of the Black Sea, Mediterranean Sea and Contiguous Atlantic Area (ACCOBAMS): A Regional Response to the Threats Facing Cetaceans*, 1 J. INT'L WILDLIFE L. & POL'Y 113, 116 (1998); Fred Pearce, *Dead in the Water: Attempts to Save the Grossly Polluted Mediterranean Seen as Doomed as the Sea Itself*, NEW SCIENTIST, Feb. 4, 1998; Joby Warrick, *Dead Dolphins and Toxin Fish: Scientists Hunt Down Seaborne Saboteurs*, INT'L HERALD TRIB. (Neuilly-sur-Seine, France), Sept. 24, 1997, at 6.

[89]   Burns, *supra* note 88, at 115; Alex Aguilar, *Population Biology, Conservation Threats and Status of Mediterranean Striped Dolphins (Stenella Coeruleoalba)*, 2(1) J. CETACEAN RESEARCH & MGMT. 17, 22–23 (2000); Seamus Kennedy, *Infectious Disease of Cetacean Populations, in* THE CONSERVATION OF WHALES & DOLPHINS, *supra* note 83, at 344–45.

[90]   IWC REPORT, *supra* note 28, at 16.

[91]   Ann Milner Roberts, *Climatic Catastrophe in the Mediterranean*, FINANCIAL TIMES BUSINESS REP., Nov. 20, 1997.

[92]   NASA Goddard Institute for Space Studies, *How Will the Frequency of Hurricanes Be Affected by Climate Change?*, (1999), available at <http://www.

ture increase could result in a greater exchange of energy and add momentum to the vertical exchange processes critical to the development of tropical typhoons and cyclones.[93] As a consequence, some researchers predict that the occurrence of tropical typhoons and cyclones could increase by as much as 50 to 60 percent,[94] and their intensity by 10 to 20 percent.[95] Increased precipitation associated with such storms could result in more land pollutants running into coastal waterways inhabited by whales,[96] as well as the introduction of river-borne contaminants into Arctic waters.[97] Elevated levels of atmospheric carbon dioxide could also increase seawater acidity, potentially raising the concentration of heavy metals in ocean ecosystems, and thus exacerbating the toxic effect of these substances on cetaceans.[98]

---

giss.nasa.gov/research/intro/druyan.02/>; Thomas R. Karl *et al.*, *The Coming Climate*, SCI. AM. (1997), available at <http://www.sciam.com/0597/issue/0597karl.html>.

[93]   *See* Leonard Doyle, *Insurers Refuse to Cover Global Warming Risks*, INDE-PENDENT, May 8, 1992, at 11.

[94]   NASA, *supra* note 92; R.J. Haarsman, *Tropical Disturbances in a GCM*, 8 CLIMATE DYNAMICS 247 (1993).

[95]   Thomas R. Knutson, Robert E. Tuleya & Yoshio Kurihara, *Simulated Increase of Hurricane Intensities in a CO$_2$-Warmed Climate*, 279 SCI. 1018, 1018 (1998). Not all climatologists agree that warming will result in increases in the incidence or intensity of storms. *See* Bette Hileman, *Climate Observations Substantiate Global Warming Models*, CHEM. & ENG. NEWS, Nov. 27, 1995, at <http://pubs.acs.org/hotartcl/cenear/951127/pgl.html>; G.J. Holland, *The Maximum Intensity of Tropical Cyclones*, 54 J. ATMOSPHERIC SCI. 2519 (1995).

[96]   Arthur, *supra* note 45.

[97]   Tynan & DeMaster, *supra* note 73. As the International Whaling Commission has noted, pollution is one of the gravest threats facing cetaceans. IWC RESOLUTION, *supra* note 4. *See also* Letizia Marsili & Silvano Focardi, *Organochlorine Levels in Subcutaneous Blubber Biopsies of Fin Whales (Balaenoptera physalus) and Striped Dolphins (Stenella coeruleoalba) from the Mediterranean Sea*, 92 ENVTL. POLLU-TION 1 (1995); ALLISON MOTLUK, *Deadlier than the Harpoon*, NEW SCI., July 1, 1995; William C.G. Burns, *The International Whaling Commission and the Regulations of the Consumptive and Non-Consumptive Uses of Small Cetaceans: The Critical Agenda for the 1990s*, 13 WIS. INT'L L.J. 105, 119 (1994).

[98]   *See* Catherine Dold, *Toxic Agents Found to Be Killing Off Whales*, N.Y. TIMES, June 16, 1992. For an analysis of land-based toxic contaminants that

Recent research suggests that warming could also adversely affect the status of some species of fish that may be prey species for cetaceans or induce their migration,[99] resulting in distributional shifts of species that may have unpredictable effects, contribute further to anoxic conditions in the region and substantially raise sea surface temperatures.[100] Recent anomalous increases (2–3° C) of summer temperatures and deepening of the thermocline in coastal areas of the Western Mediterranean may provide a portent. This has resulted in massive mortality of benthic fauna (such as sponges and gorgonians) inhabiting hard substrates.[101]

Finally, flood control efforts in coastal areas necessitated by rising sea levels could threaten the habitat of species that live in shallow waters, including the highly endangered vaquita, the two susu species on the Indian subcontinent and the baiji in China.[102]

3.   THE IWC AND CLIMATE CHANGE

The ICRW was entered into 54 years ago by 15 nations "in the face of precipitous declines in the stocks of most important whale species"[103]

---

threaten species in the Arctic, *see* Arctic Council, *Regional Programme of Action for the Protection of the Arctic Marine Environment from Land-Based Activities*, <http://arctic-council.usgs.gov/99-0376-eng.pdf> (last visited Sept. 23, 2000).

[99]   Bernd Würsig, Randall R. Reeves & J.G. Ortega-Ortiz, *Global Climate Change and Marine Mammals, in* MARINE MAMMALS BIOLOGY & CONSERVATION, at 592–99 (Peter G.H. Evans & Juan Antonio Raga eds., 2001).

[100]   INTERGOVERNMENTAL PANEL ON CLIMATE CHANGE, CLIMATE CHANGE 2001: IMPACTS, ADAPTATION, AND VULNERABILITY, Contribution of the Working Group II to the Third Assessment Report of the Intergovernmental Panel on Climate Change, *Europe*, sec. 13.2.3.2.2 (2001); Roberta Danovaro *et al.*, *Deep-sea Ecosystem Response to Climate Change: The Eastern Mediterranean Case Study*, 16(9) TRENDS ECO. & EVOLUTION 505, 505 (2001).

[101]   Danovaro, *supra* note 100, at 505; J.C. Romano *et al.*, *Anomalie termique dans les eaux du golfe de Marseille durant l'été 1999. Une explication partielle de la mortalité d'invertébrés fixés?*, 323 C.R. ACAD. SCI. 415–27 (2000) (mortality linked to stability ocean temperature increase over course of few months).

[102]   Würsig, *supra* note 74, at 353.

[103]   William C.G. Burns, *The International Whaling Commission and the Future*

to "establish a system of international regulation for the whale fisheries to ensure proper and effective conservation and development of whale stocks."[104] For the first 35 years of its existence, the IWC focused almost exclusively, and for the most part unsuccessfully,[105] on establishing catch quotas for the commercial whaling industry. However, at the IWC's 38th Meeting, the Commission's Scientific Committee acknowledged the need to assess the impact of human influences other than direct exploitation, including environmental changes.[106] At the IWC's 44th Meeting, the parties decided that the Scientific Committee should establish a regular agenda item to address environmental change issues.[107]

In 1996, the Scientific Committee convened a workshop on climate change and cetaceans.[108] While observing that assessment of the possible impacts of climate change on cetaceans was "severely limited" by the constraints of climate models, the workshop concluded that "concerns about the ability of at least some cetacean populations to adapt to future conditions are justified."[109] It called on the IWC to encourage its members to join international efforts to reduce greenhouse gas emissions.[110] Additionally, the Scientific Committee invited scientists with

---

of Cetaceans: Problems and Prospects, 8 COLO. J. INT'L ENVTL. L. & POL'Y 31, 33 (1997).

[104]   Whaling Convention, *supra* note 3, pmbl.

[105]   "The first few decades of whale management under the IWC can be described as an 'era of "quota whaling,"' . . . during which the recommendations of the Commission's Scientific Committee were often ignored by pro-whaling nations eager to hunt as many whales as possible." (citations omitted). Sarah Suhre, *Misguided Morality: The Repercussions of the International Whaling Commission's Shift from a Policy of Regulation to One of Preservation*, 12 GEO. INT'L ENVTL. L. REV. 305, 309 (1999); Burns, *supra* note 103, at 35.

[106]   INTERNATIONAL WHALING COMMISSION, THIRTY-SEVENTH REPORT OF THE INTERNATIONAL WHALING COMMISSION 151 (1986).

[107]   *Resolution on the Need for Research on the Environmental and Whale Stocks in the Antarctic Region*, 43 REP. INT'L WHALING COMMISSION 39–40 (1992).

[108]   IWC REPORT, *supra* note 28.

[109]   *Id.* at 22.

[110]   *Id.*

expertise in the field to attend future Committee meetings and recommended that a future workshop be convened to review progress.[111]

In the same year, the IWC endorsed the Scientific Committee's establishment of a Standing Working Group on Environmental Concerns (SWGEC) to assess the effects of environmental change on cetaceans in addition to the Committee's proposal for increased cooperation with other organizations working on environmental change issues.[112] At the 49th Meeting, the IWC endorsed the recommendations of the climate change workshop, as well as those from a meeting on pollution issues, and called on the Scientific Committee to produce detailed scientific proposals for future work on environmental concerns. It also encouraged party states to carry out relevant non-lethal research and called upon members to provide additional funds to support the work of the Scientific Committee and SWGEC.[113]

At the 50th Meeting, the IWC commended the body's Scientific Committee for its two ongoing initiatives on the impacts of pollutants

---

[111]   *Id.*

[112]   48th Meeting of the International Whaling Commission, *Resolution on Environmental Change and Cetaceans*, IWC/48/44, at 1 (1996). In the context of climate change issues, these other organizations include Global Ocean Ecosystem Dynamics (GLOBEC), a program adopted by UNESCO's International Geosphere-Biosphere Program to "advance our understanding of the structure and function of the global ocean ecosystem," *see* GLOBEC International, <http://www.ibss.iuf.net/links/globec/globec1.html> (last visited Nov. 26, 2000); The Convention on the Conservation of Antarctic Marine Living Resources (CCAMLR), 19 I.L.M. 841 (1980); the Scientific Committee on Antarctic Research/Antarctic Pack Ice Seals; the South Channel Ocean Productivity Experiment; and the Palmer Long Term Ecological Program of the National Science Foundation's Office of Polar Programs. IWC REPORT, *supra* note 28, at 3–4, 23; 50th Meeting of the International Whaling Commission, *Report of the Standing Working Group on Environmental Concerns*, IWC 50/4, Annex H, at 3 (1998). The IWC at its 50th Meeting also encouraged Japan to coordinate its Whale Research Programme under a Special Permit in the Antarctic with the SWGEC. 50th Meeting of the International Whaling Commission, *Resolution on Coordinating and Planning for Environmental Research in the Antarctic*, IWC Resolution 1998–7 (1998).

[113]   *See* ANNUAL REPORT OF THE INTERNATIONAL WHALING COMMISSION, RESOLUTION ON ENVIRONMENTAL CHANGE AND CETACEANS, IWC Resolution 1998–5, Appendix 6, 44 (1999).

and chemical contaminants and baleen whale habitat and prey studies related to climate change and identification of physical and biological habitat degradation and Arctic issues.[114] It also directed the Scientific Committee to accord high priority to implementing the research initiatives of the SWGEC and to produce costed proposals for non-lethal research.[115] Furthermore, the IWC addressed the critical issue of funding for such initiatives, allocating approximately US$170,000 from the Commission's reserves to fund environmental research in the eight priority areas identified by the Scientific Committee.[116] Additionally, the parties agreed to consider at the 51st meeting the establishment of a dedicated Environmental Research Fund and the attendance of invited participants with relevant expertise at future meetings of the Scientific Committee.[117] Finally, the parties agreed to establish a regular Commission agenda item for environmental concerns to facilitate reporting by the Scientific Committee on its progress in this context and reporting to the parties on national and regional initiatives.[118]

At the 51st Meeting, the IWC noted that the SWGEC had agreed to focus on one or two priority topics at each meeting to maximize its effectiveness.[119] The Scientific Committee endorsed SWGEC's decision to prioritize two programs in 2000: the Southern Ocean Whale and Ecosystem Research Programme (SOWER 2000) and POLLUTION 2000+.[120]

---

[114]   IWC RESOLUTION, *supra* note 4.

[115]   *Id.*

[116]   50th Meeting of the International Whaling Commission, *Resolution for the Funding of Work on Environmental Concerns*, IWC Resolution 1998–6. For a list of the eight priorities cited by the Scientific Committee, see *supra* note 4 and accompanying text.

[117]   *Resolution for the Funding of Work on Environmental Concerns, supra* note 116.

[118]   *Id.*

[119]   51st Meeting of the International Whaling Commission, *Resolution for the Funding of High Priority Scientific Research*, IWC Resolution 1999–5 (1999).

[120]   *Id.* The IWC also agreed to a feasibility study on fin and minke whales off West Greenland, and to accord priority to research in this context in 2000/2001 and subsequent years. *Id.*

The IWC decided to provide approximately US$214,000 for core funding of environmental research programs in 1999/2000.[121] However, it noted that the SOWER 2000 and POLLUTION 2000+ programs would cost more than US$510,000 in the first year alone,[122] and called upon parties to the IWC, other governments, international organizations and other bodies to provide supplemental funding for the programs.[123] At the 52nd Meeting of the Parties in 2000, the IWC observed once again that the Scientific Committee's available funding for environmental initiatives was insufficient to facilitate implementation or development of these programs.[124]

The SOWER 2000 research program should yield data relevant to the possible impacts of climate change on cetaceans. In cooperation with the Commission for the Conservation of Antarctic Marine Living Resources,[125]

---

[121]   *Id.*

[122]   *See Final Press Release, 1999 Annual Meeting, St. George's, Grenada*, available at <http://ourworld.compuserve.com/homepages/iwcoffice/Press99.htm> (last visited Sept. 25, 2000).

[123]   *Resolution for the Funding of High Priority Scientific Research, supra* note 119.

[124]   International Whaling Commission, Resolution 2000–7 (2000).

[125]   Convention on the Conservation of Antarctic Marine Living Resources, May 20, 1980, 33 U.S.T. 3476, 1329 U.N.T.S. 48, available at <http://www.eelink. net/~asilwildlife/aa.html> (The full text of the CCAMLR, and other Antarctic agreements, is available on the American Society of International Law—Wildlife Interest Group's website) [hereafter CCAMLR]. CCAMLR applies to the Antarctic marine living resources of the area south of 60 degrees south latitude and to the Antarctic marine living resources of the area between that latitude and the Antarctic Convergence which form part of the Antarctic marine ecosystem. *Id.* art. I(1). Antarctic marine resources are populations of fin fish, mollusks, crustaceans and all other species of living organisms, including birds, found south of the Antarctic convergence. *Id.* at art. I(2). The Convention seeks to prevent the decrease of any harvested population to levels below those which ensure its stable recruitment. *Id.* at art. II (3)(a). Parties to the Convention pledge to not engage in activities that will contravene the purposes of the agreement. *Id.* at art. III. The Commission was established under the Convention to ensure achievement of the Convention's objectives by, *inter alia*, facilitating research and studies of Antarctic marine living resources; identifying conservation needs; and formulating and adopting conservation measures. *Id.* at art. IX. The Commission and the Convention's Scientific Committee

and the Southern Ocean Global Ocean Ecosystems Dynamics (GLOBEC) program,[126] the SWGEC will conduct an international survey program with two major components: abundance estimates of minke whales and other baleen whales, and an assessment of the status of Southern Hemisphere blue whales.[127] The IWC hopes that this research will facil-

---

are required under the Convention to develop cooperative working relationships, as appropriate, with inter-governmental and non-governmental organizations which could contribute to their work including the International Whaling Commission. *Id.* at art. XXIII(3).

IWC observers are also onboard research vessels participating in the CCAMLR's 2000 Krill Synoptic Survey, which seeks to improve estimates of the pre-exploitation biomass of krill, a critical parameter for establishing the sustainable yield of the Southern Ocean krill fishery. CCAMLR, *2000 Krill Synoptic Survey of Area 48*, <http:www.ccamlr.org/English/e_scientific_committee/e_sc_krill_surv.html> (last modified Aug. 11, 1999). The IWC observers will also conduct observations of whale abundance and distribution. Personal correspondence from Eugene Sabourenkov, Science Officer, CCAMLR Secretariat.

For additional information on the Commission, *see* the IWC website, <http://www. ccamlr.org/>; *see also* Stuart Kaye, *Legal Approaches to Polar Fisheries Regimes: A Comparative Analysis of the Convention for the Conservation of Antarctic Marine Living Resources and the Bering Sea Doughnut Hole Convention*, 26 CAL. W. INT'L L.J. 75 (1995).

[126]   The Global Ocean Ecosystems Dynamics (GLOBEC) was adopted by the International Geosphere-Biosphere Programme, to "advance our understanding of the structure and functioning of the global ocean ecosystem, its major subsystems, and its response to physical forcing so that a capability can be developed to forecast the response of the marine ecosystem to global change." GLOBEC is co-sponsored by the Scientific Committee on Oceanic Research (SCOR) and the Intergovernmental Oceanographic Commission of UNESCO. Intergovernmental Oceanographic Commission, *GLOBEC Open Science Meeting*, <http://ioc.unesco. org/iyo/activities/conferences/globec.htm> (last visited Nov. 12, 2000). Southern Ocean GLOBEC is one of GLOBEC's major research programs. Its major research activities will begin over the next two years and will focus on the impact of physical forces on population dynamics and predator-prey interactions between key species in the region, with special emphasis on the overwintering strategies of zooplankton and top predators. Researchers hope this research will advance the understanding of Southern ocean ecosystems and enhance the ability to monitor and predict climate change impacts. U.S. GLOBEC, <http://www.pml.ac.uk/globec> (last updated Oct. 29, 2000); *see also* IWC REPORT, *supra* note 28, at 4.

[127]   International Whaling Commission, *Report of the Scientific Committee*,

itate mapping of cetacean distribution and abundance in relation to krill distribution in the Antarctic and possible changes in cetacean foraging behavior in response to changes in krill abundance and distribution.[128]

4. CETACEANS AND CLIMATE CHANGE: PROSPECTS IN THE NEXT CENTURY

4.1. The Institutional Role of the IWC in Protecting Cetaceans from Climate Change

It is difficult to be sanguine about the prospects for the IWC to effectively address the threats that cetacean species may face from climate change. First, it is doubtful whether the IWC possesses, or will be able to cobble together, the financial resources necessary to conduct meaningful climate research. Cetacean research is extremely expensive because many species are highly migratory and rarely come near land.[129]

Annex Z, *Report of the ad hoc Working Group on Future SOWER Planning, reprinted in* 1 J. CETACEAN RES. & MGMT. 263–66 (Supp. 1999).

The overall long-term objective of the SOWER program is to:
> Define how spatial and temporal variability in the physical (e.g. sea surface temperature, salinity, mixed layer depth, upwelling, extent of ice cover) and biological (e.g. prey availability) environment influence cetacean species in order to determine those processes in the marine ecosystem which best predict long-term changes in cetacean distribution, abundance, stock structure, extent and timing of migrations and fitness.

Scientific Committee, International Whaling Commission, Annex H, *Report of the Standing Working Group on Environmental Concerns, reprinted in* 2 J. CETACEAN RES. & MGMT. 217 (2000). "A specific objective of the programme is to 'relate distribution, abundance and biomass of baleen whale species to the same for krill in a large area in a single season.'" International Whaling Commission, Scientific Committee, *Report of the SOWER 2000 Workshop*, Annex E, at 32 (2000).

[128] International Whaling Commission, *Report of the Scientific Committee*, Annex H, Appendix 3, *Observer's Report of the Meeting of the Southern Ocean GLOBEC Planning Group*, (1999), *reprinted in* 1 J. CETACEAN RES. & MGT. 204 (Supp. 1999).

[129] *See* Joseph P. Rosati, *Enforcement Questions of the International Whaling Commission: Are Exclusive Economic Zones the Solution?*, 14 CAL. W. INT'L L.J. 114, 124 n.101 (1984); Scarff, *supra* note 80, at 333.

As a consequence of the cost-prohibitive nature of such research, "there are few cases where whale or dolphin populations have been studied for long enough to determine their overall status, let alone identify the key environmental factors which control populations."[130] Cetacean research in the context of climate change will be particularly costly. Climate modeling is a very expensive proposition and research in the context of cetaceans will necessitate extensive modeling. For example, in the Arctic, modeling will be required to ascertain an imposing suite of relevant indicia, including regional ice dynamics, winds, mesoscale features and mechanisms of nutrient resupply.[131]

Yet the parties to the IWC allocated less than US$200,000 at the 50th Meeting to address the impact of *eight* major environmental threats to cetaceans.[132] As indicated above, the parties at the 51st Meeting did provide additional funding for the 1999/2000 research program, which includes a climate change component, but they acknowledged a US$300,000 shortfall for the first year of the program alone.

It is unlikely that the parties will be forthcoming with substantial additional funding for environmental research. As Burke observed, the "IWC is . . . given little or no capacity of its own to increase knowledge and understanding of whales . . . It must rely on member states and on private groups, neither of which can be presumed to do objective science or to interpret conditions without bias."[133] Given the bitter rancor

---

130   Paul Thompson & Sue Mayer, *Defining Future Research Needs for Cetacean Conservation, in* THE CONSERVATION OF WHALES & DOLPHINS, *supra* note 83, at 412. *See also* Koen van Waerebeek *et al.*, *Spatial and Temporal Distribution of the Minke Whale, Balaenoptera acutorostrata (Lacépède, 1804, in the Southern Northeast Atlantic Ocean and the Mediterranean Sea, with Reference to Stock Identity*, 1 J. CETACEAN RES. & MGMT. 223 (1999); Animal Welfare Institute, *Debunking the RMP: The Case for Practical Reality over Abstract Theory* 14–17 (1993).

131   Tynan & DeMaster, *supra* note 73, at 20.

132   *See supra* note 116 and accompanying text.

133   WILLIAM T. BURKE, THE NEW INTERNATIONAL LAW OF FISHERIES 292 (1994). *See also* Steinar Andresen, *The Whaling Regime, in* SCIENCE & POLITICS IN INTERNATIONAL ENVIRONMENTAL REGIMES 52 (Steiner Andresen *et al.*, 2000); Gregory Rose & George Paleokrassis, *Compliance with International Environmental Obligations: A Case Study of the International Whaling Commission, in*

that characterizes IWC deliberations in this era, primarily over whether the commercial moratorium on whaling imposed in the 1980s should be lifted,[134] it is difficult to believe that the parties will bolster the Secretariat's autonomy by providing it with a substantial new source of funding.[135]

IMPROVING COMPLIANCE WITH INTERNATIONAL ENVIRONMENTAL LAW 148, 156 (James Cameron, Jacob Werksman & Peter Roderick eds., 1996).

[134]   Japan and Norway and their allies in the IWC contend that the moratorium on commercial whaling should be lifted on the grounds that sustainable harvesting of at least one species, minke whales, is now tenable. Parties that oppose lifting the moratorium base their position on ethical or moral grounds or question the sustainability of the harvest. *See* William C.G. Burns, *The Forty-Ninth Meeting of the International Whaling Commission: Charting the Future of Cetaceans in the Twenty-First Century*, 8 COLO. J. INT'L ENVTL. L. & POL'Y 64, 67 (1997); Kristen Fletcher, *The 49th Annual Meeting of the International Whaling Commission: Prelude to the Next Fifty Years*, 1 J. INT'L WILDLIFE L. & POL'Y 134, 134 (1998). As Michael Canny, the Chairman of the International Whaling Commission, recently concluded, there is increasing concern "that the inability of the IWC to reach a consensus on fundamental questions . . . will lead to a breakup of the IWC with detrimental effects on the conservation of whales." Michael Canny, *Opening Statement of the Government of Ireland*, IWC/49/OS/Ireland (1997). *See also* William Aron *et al.*, *The Whaling Issue*, 24 MARINE POL'Y 179, 179 (2000) (IWC "verges on extinction").

[135]   Indeed, Norway has expressly declared its misgivings about IWC efforts to assess environmental impact on cetaceans, with the exception of the impact of pollution. At the 51st Meeting, Norway's delegate argued that the IWC should concentrate its limited resources on monitoring of and research related to abundance and distribution of cetaceans, changes in biological parameters and the effects of pollution. The Norwegian delegate also issued the veiled threat that if the Scientific Committee were to accord less priority to advice on whaling issues, Norway might be compelled to seek advice on its whaling activities from another international body, such as the North Atlantic Marine Mammal Commission (NAMMCO), an intergovernmental organization established under the Agreement on Cooperation in Research, Conservation and Management of Marine Mammals in the North Atlantic, <http://www.eelink.net/~asilwildlife/nam.html>, by the Faroe Islands, Greenland, Iceland and Norway in 1992 to conduct scientific study, conservation and management of marine mammals in the North Atlantic region. *See* Professor Lars Walløe, *IWC should focus on central issues, not on general environmental topics*, <http://www.highnorth.no/Library/Policies/National/lw-IWC-99.htm> (last visited Sept. 25, 2000). For an overview of NAMMCO, *see* David D. Caron, *The International Whaling Commission and the Atlantic Marine Mammal Commission:*

Even assuming, *arguendo*, that the IWC will be able to conduct adequate research on its own or in cooperation with other agencies, its ability to protect cetaceans from climate change may be extremely limited. Should the moratorium on commercial whaling be lifted in the future, quotas will be determined under the Revised Management Scheme (RMS). The RMS is a mechanism for estimating the abundance of discrete species and sustainable catch limits, as well as to establish methods to ensure such limits are adhered to. It consists of the Revised Management Procedure, "a framework to assess the viability of exploiting discrete stocks of cetaceans to facilitate the establishment . . . of sustainable harvesting quotas for said stocks," and several other components, including an inspection and observation scheme to deter cheating.[136] The IWC should be encouraged to incorporate possible climate

---

the Institutional Risks of Coercion in Consensual Structures, 89 AM. J. INT'L L. 154, 164–65 (1995). Some fear that NAMMCO will ultimately become "an option for those in the North Atlantic region that decide to withdraw from the IWC or, more likely, to opt out of particular obligations." *Id.* at 165.

[136]   William C.G. Burns, *The International Whaling Commission and the Future of Cetaceans: Problems and Prospects*, 8 COLO. J. INT'L ENVTL. L. & POL'Y 33, 54–56 (1997); *Resolution of Provisions for Completing the Revised Management scheme Proposed by US, UK, Netherlands, Denmark, Norway*, IWC/48/42, Agenda Item 11.4.2 (1996). Under the RMS, catches are not to be permitted on stocks that are below 54 percent of the estimated carrying capacity. International Whaling Commission, *Whale Population Estimates*, <http://ourworld.compuserve.com/homepages/iwcoffice/Estimate.htm> (last visited Nov. 26, 2000). Current scientific research on the RMP centers largely on simulation testing of possible application to specific species and ocean areas. Testing began with North Atlantic and Southern Hemisphere minke whales and has now moved on to North Pacific minke and Bryde's whales. International Whaling Commission, *Report of the Scientific Committee, Annex D. Report of the Sub-Committee on the Revised Management Procedure*, 1 J. CETACEAN RES. & MGMT. 263–66 (Supp. 1999).

The IWC has accepted and endorsed the Revised Management Procedure; however, several outstanding issues remain before the IWC will consider lifting the commercial moratorium, including the specification of the inspection and observer system and "arrangements to ensure that total catches over time are within limits set under the RMS." International Whaling Commission, *Final Press Release*, 1998 Annual Meeting, May 20, 1998.

At its 50th Meeting, the parties to the IWC passed a resolution agreeing that any catch limits established under the RMS "shall be calculated by deducting all human-change impacts into the RMS framework. Unfortunately, this will do little to protect the cetacean species most vulnerable to climate change because the depleted status of most species would preclude the setting of catch quotas for more than one or two species in the near future.[137] Thus, if the IWC is going to protect vulnerable species from climate change, this protection will have to occur outside the IWC's framework for the establishment of harvesting quotas.

The parties could also vote to expand the boundaries of existing sanctuaries established by the IWC in the Pacific sector of the Southern Ocean, the Southern Hemisphere, and the Indian Ocean,[138] or to create new sanctuaries. This could provide additional protection for cetacean species that may be threatened by climate change by precluding direct exploitation. However, because only a few species are likely to be subject to commercial whaling in the future, the lifting of the moratorium would not benefit the majority of the species most threatened by climate change.

---

induced mortalities that are known or can be reasonably estimated, other than commercial catches, from the total allowable removal." International Whaling Commission, *Resolution on Total Catches Over Time*, IWC Resolution 1998–2.

[137]   The IWC is currently conducting simulation trials for only two species of whales for the purposes of establishing quotas under the RMP, minke and Bryde's whales. International Whaling Commission—Scientific Committee, *Report of the Sub-Committee on the Revised Management Procedure*, Annex D, 2 J. CETACEAN RES. & MGMT. 79, 85–91 (Supp. 2000).

[138]   Article V(1)(c) of the Whaling Convention, *supra* note 4, permits the parties to establish "open and closed waters, including the designation of sanctuary areas." The IWC designated most of the Pacific sector of the Southern Ocean as a sanctuary at the outset of the ICRW, banning the catching of baleen whales. In 1979, it established the Indian Ocean Sanctuary, prohibiting commercial whaling in "the waters of the Northern Hemisphere from the coast of Africa to 100°E, including the Red and Arabian Seas and the Gulf of Oman; and the waters of the Southern Hemisphere in the sector from 20°E to 130°E, with the Southern boundary set at 55°S." Whaling Convention, Schedule, sec. III(7)(a), as amended at the 51st Annual Meeting (1999) <http://ourworld.compuserve.com/homepages/iwcoffice/Schedule. htm>. Ninety percent of the world's whales feed in the Southern Ocean Sanctuary. Gerard Baker, *Japan Threatens to Quit Whaling Commission*, FIN. TIMES, May 28, 1994, at 4. For a history of the establishment of IWC sanctuaries, *see* Cassandra Phillips, *Conservation in Practice: Agreements, Regulations, Sanctuaries and Action Plans, in* THE CONSERVATION OF WHALES & DOLPHINS 460–63 (Mark P. Simmonds & Judith D. Hutchinson eds., 1996).

"Interplay management" "refers to deliberate efforts by participants in tributary or recipient regimes to prevent, encourage, or shape the way one regime affects problem solving under another."[139] The IWC may ultimately find that its most effective tool for protecting cetaceans from climate change lies in such management in the form of advocacy of their protection in other forums.

For example, as outlined above, Antarctic warming may result in diminution of krill stocks, seriously threatening cetaceans in the region.[140] Thus, it may be incumbent upon the IWC to lobby the Convention on the Conservation of Antarctic Marine Living Resources, the primary body that manages marine resources in the region,[141] to limit commercial harvesting of krill in the future.[142] CCAMLR's effort to model how harvesting of a prey species might impact predators dependent on that species "has yet to provide reliable quantitative results."[143] The IWC should encourage additional research in this context and perhaps assist in developing these models in the context of cetaceans.

---

[139]   Olav Schram Stokke, *The Interplay of International Regimes: Putting Effectiveness Theory to Work*, Fridjof Nansen Institute, FNI Rep. 14/2001, at 17, <http://www.fni.no/ca/01-14-oss.pdf>, (last visited Aug. 10, 2002).

[140]   *See supra* notes 44–50 and accompanying text.

[141]   *See supra* note 125 and accompanying text.

[142]   Up to this point, efforts to commercially exploit krill in the Antarctic have been minimal, with only approximately 100,000 tons being harvested annually. Reid & Croxall, *supra* note 39, at 383. However, an American agri-business concern is gearing up for a much larger harvest to supply the aquaculture industry and as a protein supplement for human food. Moreover, it is anticipated that fishers from the United Kingdom, South Africa, Russia and the United States will soon join those from five states currently targeting krill in the region. Beth C. Clark & Alan D. Hemmings, *Problems and Prospects for the Convention on the Conservation of Antarctic Marine Living Resources Twenty Years On*, 4 J. Int'l Wildlife Law & Pol'y 47, 59 (2001). *See also Ozone Hole Killing Antarctic Krill Stocks, Scientists Warn*, Deutsche Press-Agentur, Feb. 8, 1999 available at LEXIS, World Library. "[K]rill may take over as the major issue facing CCAMLR . . ." The need for precaution in establishing harvesting quotas for krill is reinforced by high uncertainty about trends in predator demand. Reid & Croxall, *supra* note 39, at 383.

[143]   Clark & Hemmings, *supra* note 142, at 61; Graeme Parkes, *Precautionary Fisheries Management: The CCAMLR Approach*, 24 Marine Pol'y 83, 86 (2000).

The IWC might take its cue from the current initiative by the Agreement on Small Cetaceans of the Baltic and North Seas (ASCOBANS),[144] a regional treaty established under the Convention on the Conservation of Migratory Species of Wild Animals (CMS),[145] to influence European policy on marine mammal bycatch in fisheries operations. Citing unsustainable bycatch of harbor porpoises and other cetaceans in the North and Celtic Seas,[146] ASCOBANS is pressing the European Commission to restrict marine mammal bycatch in fishing nets to less than 1.7 percent annually.[147]

Similarly, the IWC should press the Intergovernmental Panel on Climate Change, the primary scientific research body informing the decisionmaking of the United Nations Framework Convention on Climate Change (UNFCCC),[148] to incorporate cetacean data in its assessment reports. The IPCC assessments to date have not addressed this specific issue; however, the IPCC is currently conducting an assessment of climate change impacts on biodiversity for the Convention on Biological Diversity.[149] However, as discussed in the next section of this chapter, this input will only make a difference if the UNFCCC has the institutional will to meaningfully address climate change during this century.

---

[144]   The Agreement on the Conservation of Small Cetaceans of the Baltic and North Seas (ASCOBANS) entered into force in 1994 and is reprinted in II THE MARINE MAMMAL COMMISSION COMPENDIUM OF SELECTED TREATIES, INTERNATIONAL AGREEMENTS, AND OTHER RELEVANT DOCUMENTS ON MARINE RESOURCES, WILDLIFE AND THE ENVIRONMENT 1612 (1994). For additional information on the treaty, see Robin Churchill's chapter in this volume.

[145]   19 I.L.M. 15 (1980).

[146]   UNEP, *Bycatch Limits Needed to Conserve Europe's Dolphins and Porpoises*, <http://www.unep.org/documents> (Apr. 14, 2001).

[147]   *Id.*; 3rd Session of the Meeting of Parties, ASCOBANS, Annex 9c, Resolution 3, *Incidental Take of Cetaceans* (2000).

[148]   UNCED, Framework Convention on Climate Change, opened for signature, June 4, 1992, *reprinted in* 31 I.L.M. 849 (1992) (hereafter *UNFCCC*). *See also* Wayne A. Morrissey, *Global Climate Change: Adequacy of Commitments Under the U.N. Framework Convention and Berlin Mandate*, Congressional Research Service Report for Congress (Oct. 25, 1996), <http://www.cnie.org/nle/clim-14.html> (last visited Nov. 12, 2000).

[149]   Convention on Migratory Species, *Report of the Tenth Meeting of the CMS Scientific Council*, May 2–4, 2001, at 20.

4.2.   The Institutional Role of the United Nations Framework Convention on Climate Change

Even if the research initiatives of the IWC and other organizations improve our understanding of the impact of climate change on cetacean species, this impact can be averted only if nations demonstrate the resolve to substantially reduce greenhouse gas emissions. The primary international instrument to achieve this objective is the United Nations Framework Convention on Climate Change (UNFCCC),[150] which entered into force in 1994 and has been ratified by 186 countries.[151] The overarching objective of UNFCCC is to "achieve . . . stabilization of greenhouse gas concentrations in the atmosphere at a level that would prevent dangerous anthropogenic interference with the climate system."[152]

Reflecting the Convention's emphasis on common but differentiated responsibility,[153] the Convention requires developed country parties to "take the lead in combating climate change and the adverse effects thereof."[154] Article 4(2) requires developed country parties and other parties included in Annex I[155] to "adopt national policies and take cor-

---

[150]   *Supra* note 148.

[151]   International Institute for Sustainable Development, *UNFCCC COP-6 Part II Highlights*, 12(170) Earth Negotiations Bulletin (2001), <http://www.iisd.ca/link-ages/vol12/enb12170e.html>. For the negotiating history leading up to the UNFCCC, *see* Elizabeth P. Barratt-Brown *et al.*, *A Forum for Action on Global Warming: The UN Framework Convention on Climate Change*, 4 COLO. J. INT'L ENVTL. L. & POL'Y 103, 106–09 (1993).

[152]   UNFCCC, *supra* note 148, at art. 2.

[153]   *Id.*, at arts. 3(1) & 4(1); Paul G. Harris, *Common But Differentiated Responsibility: The Kyoto Protocol and United States Policy*, 7 N.Y.U. ENVTL. L.J. 27, 27 (1999). The principle of common but differentiated responsibility was also adopted in Principle 7 of the Rio Declaration on Environment and Development, requiring nations to share responsibility for confronting environmental problems, but taking into account nations' different contributions to these problems and capacity to confront them. *See* Rio Convention on Environment and Development, Principle 7, U.N. Doc. A/CONF.151/5/Rev. 1, 31 I.L.M. 874 (1992)

[154]   UNFCCC, *supra* note 148, at art. 3(1).

[155]   Annex I of the UNFCCC is comprised of country parties that were members of the Organization for Economic Cooperation and Development at the time

responding measures on the mitigation of climate change, by limiting [their] anthropogenic emissions of greenhouse gases and protecting and enhancing [their] greenhouse gas sinks and reservoirs."[156]

Unfortunately, the record of the UNFCCC's Annex I parties has been disheartening. Initially, the major greenhouse gas emitting states agreed to "aim" to reduce their greenhouse gas emissions to 1990 levels by 2000.[157] All industrialized nations flouted this pledge, leading the Organization for Economic Cooperation and Development to conclude that emissions from industrialized nations could rise between 11 to 24 percent in the next 15 years.[158]

At the First Conference of the Parties to the UNFCCC held in Berlin in 1995,[159] the parties concluded that their existing commitments were inadequate on three grounds. First, most Annex I nations were not on track to meet their initial aim by 2000. Second, the UNFCCC contained no provision for controlling greenhouse emissions beyond 2000. Third, the parties acknowledged that stabilization of emissions at 1990 levels would be insufficient to stabilize atmospheric greenhouse gas concentrations. Consequently, in a decision referred to as the "Berlin Mandate," the parties established a process to strengthen UNFCCC commitments through adoption of a protocol or other legal instrument, with the goal

---

of adoption of the treaty, some Eastern European nations, and some nations that were part of the former Soviet Union.

[156]   UNFCCC, *supra* note 148, at art. 4(2)(a).

[157]   *Id*. at art. 4(2).

[158]   Bas Arts, *New Arrangements in Climate Policy*, 52 CHANGE 1, 2 (2000) ("[T]he industrialized nations have increased emissions by an average of about 10% above 1990 levels"). *See also EU Moves on Emissions, Warms CO_2 Pollution Rising*, Reuters News Service, (Mar. 9, 2000) <http://www.planetark. org/dailynewsstory.cfm?newsid=5917>; *Cabinet Okays Plan to Fight Global Warming* (1999), <http://www.theglobeandmail.com/gam/Environment/19991215/UWARMN. html>.

[159]   Under the UNFCCC, the parties established a Conference of the Parties to regularly review and promote implementation of the treaty. UNFCCC, *supra* note 148, at art. 7(1)(2). The Conference of the Parties is to be held annually unless otherwise decided by the parties. *Id*. at art. 7(4).

of establishing quantified emissions limitation and reduction objectives for the period past 2000.[160]

At the Third Conference of the Parties of the UNFCCC, held in Kyoto, Japan, in 1997, the parties adopted the Kyoto Protocol,[161] under which industrialized nations agree to reduce their collective emissions of six greenhouse gases[162] by at least 5 percent below 1990 levels by 2008 to 2012.[163] However, the United States is unlikely to ratify the Protocol given the fierce opposition of several powerful sectors in the U.S., including organized labor, fossil fuel producers, influential members of the Senate,[164] and now the Bush

---

[160]   UNFCCC, Conference of the Parties, 1st Sess., UN Doc. FCCC/CP/ 1999/7/Add.1, Decision 1/CP.1, at 4–6 (June 6, 1999). The Conference of the Parties agreed to establish an "Ad Hoc Group on the Berlin Mandate" (AGBM) to, *inter alia*, "set quantified limitation and reduction, objectives within specified time-frames, such as 2005, 2010 and 2020, for [Annex I parties] anthropogenic emissions by sources and removals by sinks of greenhouse gases not controlled by the Montreal Protocol. . . ." *Id.*

[161]   Kyoto Protocol to the United Nations Framework Convention on Climate Change, Dec. 10, 1997, FCCC/CP/1997/L.7/Add. 1, 37 I.L.M. 22.

[162]   The six greenhouse gases regulated under the Kyoto Protocol are: carbon dioxide, methane, nitrous oxide, hydrofluorocarbons, perfluorocarbons, and sulphur hexafluoride. *Id.* at Annex A.

[163]   *Id.* art. 3(1). "Individual States' commitments to reductions are differentiated with a view to meeting the 5 percent overall target; the European Community and all its member States are committed to 8 percent reductions, the United States to 7 percent and Japan and Canada to 6 percent. New Zealand, the Russian Federation and Ukraine will stabilise emissions at 1990 levels, whilst some States negotiated an actual increase in emissions." Peter G.G. Davies, *Global Warming and the Kyoto Protocol*, 47 INT'L & COMP. L.Q. 446, 453 (1998).

[164]   Andrew C. Revkin, *Senators Doubt Progress on Global Warming Plan*, N.Y. TIMES, Sept. 29, 2000, at <http://www.nytimes.com/2000/09/29/science/29CLIM. html>; Gretchen Vogel & Andrew Lawler, *Hot Year, But Cool Response in Congress*, 280 SCI. 1684, 1684 (1998). Opponents to the Protocol argue that it will have serious adverse impacts on the economies of developed nations, including substantial reductions in economic growth and increased unemployment. *See* Gregg VanHelmond, *Squandering the Surplus: $11 Billion on the Unratified Kyoto Protocol*, HERITAGE FOUNDATION BACKGROUNDER, No. 132 (Sept. 17, 1999), available at <http://www.heritage.org/library/backgrounder/bg1322.html>; Margo Thorning, *The Impact of the Kyoto Protocol on U.S. Economic Growth and*

administration.[165] This would severely undercut the treaty's effectiveness, as the U.S. is responsible for approximately one-quarter of greenhouse gas emissions.[166] Additionally, several other industrialized nations, including Japan and Russia, have been able to extract concessions in the ensuing negotiations to implement the Protocol that are likely to substantially diminish emissions reductions over the next decade.[167]

Projected Budget Surpluses, American Council for Capital Formation (Mar. 25, 1999), available at <http://www.accf.org/Mar99test.htm>; Consumer Alert, *'Cooler Heads' Members Say Kyoto Protocol Will Have Devastating Effects on Consumers, Seniors, the Poor, Small Business* (Mar. 16, 1998), <http://www.consumeralert. org/issues/enviro/MarchPR.htm>. Moreover, opponents to the Kyoto Protocol contend that it places developed nations at an economic disadvantage vis-à-vis developing nations who are not required to make emissions reductions commitments under the treaty. *See* U.S. Senate Resolution 2019, 105th Congress, 1st Sess. (1997).

[165]   In a letter sent to Senators Hagel, Helms, Craig and Roberts on March 13, 2001, President Bush expressed the Administration's opposition to the Protocol on the grounds that it exempted developing nations from obligations and would seriously harm the U.S. economy. Suraje Dessai, *the Climate Regime from The Hague to Marrakech: Saving or Sinking the Kyoto Protocol?*, Tyndall Centre Working Paper No. 12, at 5 (Dec. 2001). *See also Bush Move on Climate "Regrettable'—IPCC*, Planet Ark, Mar. 30, 2001; <http://www.planetark.org/dailynewsstory.cfm? newsid=10304&newsdate=30-Mar-2001>.

[166]   Bharat H. Desai, *Institutionalizing the Kyoto Climate Accord*, 29 ENVTL. POL'Y & L. 159, 161 (1999); *see also* Alex Barnum, *Can World Unite, Halt Climate Threat?*, S.F. CHRON., Nov. 28, 1997, at A21. U.S. refusal to adopt the Protocol might result in European nations balking also. *Risky Business*, GLOBAL CHANGE, Oct. 1998, at 2. This could doom the agreement because it requires ratification by 55 nations, including Annex I nations accounting for a least 55 percent of total carbon dioxide emissions in 1990. Kyoto Protocol, *supra* note 157, at art. 25(1). Up to this point, the Protocol has been adopted by only 34 nations, only one of which is an Annex I party. United Nations Framework Convention on Climate Change Secretariat, *Kyoto Protocol, Status of Ratification*, <http://www.unfccc.de/resource/ kpstats.pdf>.

[167]   The compromises made by the parties to the UNFCCC for implementation of the Kyoto Protocol's provisions may result in a net *increase* of 9 percent in greenhouse gas emissions over the next decade. Mustafa H. Babiker *et al.*, *The Evolution of a Climate Regime: Kyoto to Marrakech*, MIT Joint Program on the Science and Policy of Global Change, Rep. No. 82, at 13 (2002). Japan's Central Environmental Council also recently recommended that industries be permitted for now to imple-

Many industrialized nations also continue to demonstrate very little commitment to reducing emissions. For example, U.S. emissions rose 2.5 percent in 1999 and 3.1 percent in 2000, well above its average 1.3 percent annual growth rate over the last decade.[168] Global emissions reached a new high in 2001, the eighth annual record since 1990.[169] A recent report concluded that if present trends continue, emissions in the developed world could increase 40 percent over 1990 levels by 2010.[170]

Moreover, because many greenhouse gases persist in the atmosphere for decades, "their radiative forcing—their tendency to warm Earth—persists for periods that are long compared with human life spans."[171]

---

ment measures to reduce greenhouse gas emissions on a voluntary basis, likely scuppering prospects for complying with the Kyoto Protocol. *Toothless Global Warming Bill*, Japan Times Online, May 1, 2002, <http://www.japantimes.co.jp/cgi-bin/getarticle.pl5?ed20020501a1.htm>; Mick Corliss, *Emissions Trading Plan Put on Back Burner*, Japan Times Online, Jan. 18, 2002; <http://www.japantimes.co.jp/cgi-bin/getarticle.pl5?nn20020118b2.htm>; Alex Kirby, *Japan Cools on Climate Pact*, BBC News, Jan. 3, 2002, <http://news.bbc.co.uk/hi/english/sci/tech/newsid_1740000/1740677.stm>.

[168]   *Carbon Dioxide Emissions Up 3.1 Percent in 2000*, Reuters, Nov. 12, 2001, <http://enn.com/news/wire-stories/2001/11/11122001/reu_carbon_45543.asp>. While preliminary data indicates that U.S. carbon dioxide emissions declined in 2001, <http://www.eia.doe.gov/oiaf/1605/flash/flash.html>, the first such decline since 1991, this is likely primarily attributable to warmer weather and the economic downturn, Environment News Service, July 2, 2002, <http://ens-news.com/ens/jul2002/2002-07-02-09.asp#anchor1>.

[169]   WORLDWATCH INSTITUTE, *supra* note 13, at 52.

[170]   Breffni O'Rourke, *Europe: Meeting Kyoto Pollution Cuts Will Be Difficult*, Radio Free Europe/Radio Liberty (2000), <http://www.rferl.org/nca/features/2000/10/05102000185206.asp>. *See also* Colin Macilwain, *Emissions Targets 'Unrealistic' Says US Climate Change Body*, 406 NATURE 333 (2000); *U.S. 'Unlikely to Meet its Targets*,' BBC News (Sept. 21, 2000), <http://news.bbc.co.uk/hi/english/sci/tech/newsid_934000/934194.stm>. One encouraging item has been the decline of European Union emissions by 1.8 percent over the past decade. WORLDWATCH INSTITUTE, *supra* note 13, at 52. However, continuing increases in energy demand, limited prospects for increasing the share of non-fossil fuels to total energy supply, and the phasing out of nuclear energy by OECD nations cloud the prospects for compliance with Kyoto. J.W. Sun, *The Kyoto Negotiations on Climate Change—An Arithmetic Perspective*, 30 ENERGY POL'Y 83, 84 (2002).

[171]   Bette Hileman, *Climate Observations Substantiate Global Warming Models*,

When coupled with the fact that developing countries were excluded from reduction commitments,[172] it is likely that full implementation of the Kyoto Protocol will only reduce warming by one-twentieth of one degree by 2050[173] and delay doubling of atmospheric concentrations of carbon dioxide from pre-industrial levels by less than a decade.[174] Thus, many potential impacts of climate change, including impacts on cetacean species, may be inevitable. Moreover, averting the climate change impacts beyond the middle of the next century will require far more dramatic reductions in greenhouse gas emissions. The IPCC has estimated that it would be necessary to reduce greenhouse emissions by more than 60 percent to stabilize atmospheric concentrations at 1990 levels.[175] Given how difficult it was to secure agreement to the much more modest reductions contemplated under the Kyoto Protocol, one must be skeptical about the prospects for nations agreeing to such dramatic cutbacks in the future. Recent research indicates that a commitment to increased energy efficiency, energy conservation, and renewable energy sources could reduce emissions by Annex I states to 1990 levels over the next few decades.[176] However, major greenhouse emitters, most

---

CHEMICAL & ENGINEERING NEWS (Nov. 27, 1995), <http://pubs.acs.org/hotartcl/cenear/951127/pgl.html>; *see also* Thomas R. Karl *et al.*, *The Coming Climate*, May 1, 1997, SCI. AM. 78, 80 ("as much as 40 percent of [carbon dioxide] tends to remain in the atmosphere for centuries"); Silvia Kusidio, *Climatic Changes Are No Longer Preventable, Warn Experts*, DEUTSCHE PRESSE-AGENTUR, Mar. 22, 1995 *available at* LEXIS, News file ("Even a worldwide stabilization of the emissions would not prevent a rise in the greenhouse gases . . . for the next 200 years . . .").

[172]   *See supra* notes 153–56 and accompanying text.

[173]   Martin Parry *et al.*, *Buenos Aires and Kyoto Targets Do Little to Reduce Climate Change Impacts*, 8 GLOBAL ENVTL. CHANGE 285, 285 (1998). *See also* Brian C. O'Neill & Michael Oppenheimer, *Dangerous Climate Impacts and the Kyoto Protocol*, 296 SCI. 1971, 1971 (2002) (Kyoto Protocol would "reduce warming only marginally").

[174]   Roger Jones, *Climate Change in the South Pacific*, 35 TIEMPO 17, 20 (2000).

[175]   *See* World Meteorological Organization/United Nations Environment Programme Intergovernmental Panel on Climate Change, *Climate Change: The IPCC Scientific Assessment* xxxvi (J.T. Houghton, G.J. Jenkins & J.J. Ephrams eds., 1990).

[176]   Committee for the National Institute for the Environment, *Energy Efficiency: Budget, Climate Change, and Electricity Restructuring Issues II*, available

notably the U.S., appear to lack the resolve to commit the necessary financial resources to develop such technologies.

Finally, as indicated earlier, the UNFCCC currently only binds developed countries and economies in transition to the reduction of greenhouse gas emissions.[177] However, given the tremendous projected increases in greenhouse gas emissions in developing countries over the next century,[178] the future effectiveness of the UNFCCC is contingent on engaging these nations in the regime's mission.[179] It is far from certain that developing nations will commit themselves to substantial emission reductions. There is great trepidation among developing countries about possible economic impacts and a sense of unfairness, given the tremendous disparity in per capita emissions between industrialized and developing nations.[180]

---

at <http://cnie.org/nle/eng-48a.html>; Steven Bernow *et al.*, *America's Global Warming Solutions*, A Study for the World Wildlife Fund and the Energy Foundation (Oct. 6, 1999), <http://www.tellus.org/energy/publications/solution. pdf>; William K. Stevens, *Price of Global Warming?*, N.Y. TIMES, Oct. 10, 1995, at B6.

[177]   UN Framework Convention on Climate Change, *supra* note 148, at art. 4(2) & Annex I; Kyoto Protocol, *supra* note 154, at art. 3. Article 10 of the Kyoto Protocol did affirm the existing commitments imposed on all parties under Article 4(1) of the UNFCCC. These commitments include national emissions reporting to the Conference of the Parties and formulation and implementation of programs containing measures to mitigate climate change.

[178]   China's carbon dioxide emissions alone will probably exceed that of the entire OECD by the middle of this century. *See* Francis Cairncross, *Global Warming Won't Cost the Earth*, INDEPENDENT, Mar. 28, 1995, at 13; *see also* Kim Ji-Soo, *Seoul Resists Pressure to Commit to Reducing CO_2 Emissions*, KOREA HERALD, Dec. 8, 1999 at 3. South Korea, with the 11th highest emissions of greenhouse gases in the world, projected that its emissions will rise seventy-six percent from its 1998 levels by 2020). "[B]y 2100, carbon-dioxide emissions from developing countries will probably be more than the rich world's output." *Global Warming and Cooling Enthusiasm*, ECONOMIST, Apr. 1, 1995, at 33.

[179]   Clare Breidenich *et al.*, *The Kyoto Protocol to the United Nations Framework Convention on Climate Change*, 92 AM. J. INT'L L. 315, 331 (1998); *Kyoto Protocol: The Unfinished Agenda*, 39 CLIMATIC CHANGE (1998) at 9.

[180]   Anita Margrethe Halvorssen, *Climate Change Treaties—New Developments at the Buenos Aires Convention*, 1998 COLO. J. INT'L ENVTL. L & POL'Y 1, 1 (1998);

5. CONCLUSION

The IWC's recognition of the need to address environmental change issues, including the possible impacts of climate change on cetaceans, is laudable. However, its limited research resources and the speculative future of the UNFCCC likely means that cetaceans will face increasing threats from climate change in the next century. It remains to be seen whether many species of cetaceans that were driven to the brink of extinction by harvesting can now survive the onslaught of environmental change, including the specter of global warming. If the IWC is to make a serious commitment to addressing the possible impacts of environmental change on species under its regulation, the parties to the agreement must substantially increase funding for research and use their influence in other forums to effectuate policies that will protect the interests of cetaceans. If the parties fail to do so, the IWC's ultimate legacy may be that it saved whales from extinction by commercial harvesting but failed them in their time of greatest need.

Davies, *supra* note 163, at 457. "20 percent of the world's population is responsible for 63 percent of carbon dioxide emissions, while another 20 percent is responsible for only 2 percent of these emissions." Robert Engelman, *Population, Consumption and Equity*, TIEMPO, Dec. 1998, at 5.

Developing countries repulsed an effort at the Third Conference of the Parties in Kyoto to establish emission limitation objectives for wealthier developing states. *Id.* At the Fourth Meeting of the Conference of the Parties, in Buenos Aires, Argentina became the first developing country to commit itself to take on an emissions reduction target for the initial commitment period under Kyoto of 2008 to 2012. *Address by the President of the Republic of Argentina*, Report of the Conference of the Parties on Its Fourth Session, United Nations Framework Convention on Climate Change, Conference of the Parties, 4th Sess., U.N. Doc.FCCC/CP/1998/16, Annex I, at 35 (1999). However, it is unclear how substantive this commitment will be, or if other developing nations will follow suit. Susan Fletcher, *Global Climate Change Treaty: The Kyoto Protocol*, Congressional Research Report for Congress, 98–2, available at <http://www.cnie.org/nle/clim-3.html>.

Part V

The Ecosystem Role of Cetaceans

Chapter 13
Whales—The New Scapegoat for Overfishing

Michael Donoghue[1]

1. INTRODUCTION

In recent years, the advocates of whaling have advanced a new rationale for culling whale populations: they are depriving humanity of food. This has the added attraction of providing a scapegoat for the overfishing of marine resources that has resulted in dramatic declines in several important commercial fish stocks, particularly in the Northern Hemisphere (*e.g.*, Newfoundland cod, North Sea haddock).[2]

The argument has been promulgated in the popular media in a variety of ways: in pamphlets distributed at international meetings,[3] through paid advertisements,[4] and even on the editorial pages of several newspapers.[5]

---

[1] Relationship Manager (Marine), Department of Conservation, P.O. Box 10–420, Wellington, New Zealand, donoghue@pop.ihug.co.nz. The author is particularly grateful to Dr. Jock Young, for making available his carefully researched contribution on the *Potential for Impact of Large Whales on Commercial Fishing in the South Pacific Ocean*. (March 1999, CSIRO Marine Research, Hobart).

[2] *See* C. Deere, *Net Gains: Linking Fisheries Management, International Trade and Sustainable Development*, 19–20, International Union for the Conservation of Nature (1999).

[3] *See* material distributed at the 51st and 52nd Annual Meetings of the International Whaling Commission (IWC) by groups such as The Riches of the Sea and All-Japan Seamen's Union

[4] ZIZ television channel ran an advertisement on 5 September 2000 in St Kitts, stating that whale populations are threatening fish stocks, but the IWC wants to prevent all whales from being caught.

[5] *See* THE DOMINION, Wellington, New Zealand Dec. 1, 2000.

It is argued that recovering populations of whales will threaten global food security by distorting the marine food chain. A pamphlet distributed at the 51st Annual Meeting of the International Whaling Commission (IWC) in May 1999 bore the title: *Whales compete with fishermen for limited resources—Whales consume approximately six times as much fish as the world's fisheries catch.*[6]

An oft-quoted scientific source for this assertion is a paper co-authored by the Director of the Cetacean Research Institute in Tokyo,[7] which was originally distributed at the Committee on Fisheries (COFI) meeting of the Food and Agriculture Organisation (FAO) in October 1998. The paper has not been published in a peer-reviewed journal, but was tabled at the COFI meeting as a "For Information Document."

The paper by Ohsumi and Tamura is entitled *Estimation of total food consumption by cetaceans in the world's oceans,* and has been used as the scientific rationale to underpin a diplomatic offensive whose intent is to have whales regarded by the international community as greedy and dangerous competitors with humans. No new research was carried out in the production of the paper. The authors simply searched the literature for a selection of some of the estimates available for whale populations, and listed them. They then took three different estimates (mostly published over 20 years ago) of the daily energy requirements for whales, and then converted them into values for daily prey consumption. The population estimate was multiplied by the estimated prey consumption to calculate how much whales eat.

While this argument may appear to have some superficial logic, its scientific basis is questionable. Several members of the IWC's Scientific Committee have pointed out that some of the estimates of whale populations used in the paper were outdated and/or highly speculative. It was

---

6   Pamphlet distributed by The Institute of Cetacean Research, Tokyo Suisan Bldg, 4–18 Toyomi-cho, Chuo-ku, Tokyo, and The Riches of the Sea, 8–3, Higashi Nihonbashi 2-chome, Chuo-ku, Tokyo.

7   *Estimation of total food consumption by cetaceans in the world's oceans,* T.Tamura & S. Ohsumi.

For Information document tabled at meeting of the IWC Scientific Committee, IWC 51, May 1999.

also noted that many whales, particularly in the Southern Hemisphere, rarely, if ever, eat fish, and that the use of the word "food" in the paper's title was therefore incorrect and prejudicial. There are also considerable differences of opinion on the validity of using laboratory-based research on ecological energetics to estimate rates of prey consumption.

This chapter provides a brief history of the recent development of the argument that recovering whale populations pose a threat to global food security, and an assessment of its scientific validity.

2.   OF FISH AND WHALES

The new offensive on whales as predators of humans' resources can be traced back to the "scientific whaling" programs undertaken by Norway and Iceland in the late 1980s and early 1990s. When the IWC voted in 1982 to establish a moratorium on commercial whaling, to take effect from 1986, both countries had active commercial whaling operations in the North Atlantic. Norway objected to the moratorium decision under the provisions of Article 5 of the International Convention for the Regulation of Whaling,[8] and consequently was not bound by it. Iceland did not object within the required 90 days, but invoked Article 8 of the Convention, which permits contracting governments to issue to their nationals special permits to conduct scientific research.[9] The only stipulations are that the results of the research must be reported to the IWC, and that all whales taken shall be processed and dealt with according to the directions of the contracting government issuing the permit.

Soon after the moratorium came into effect, the Icelandic government announced that it was beginning a four-year research program, beginning in 1986. According to the Icelandic Minister of Fisheries,[10] one of

---

[8]   International Convention for the Regulation of Whaling, Dec. 2, 1946, 62 Stat. 1716, T.I.A.S No. 19\849, 161 U.N.T.S. 361.

[9]   *Id*. at art. VIII. "[A]ny Contracting government may grant to any of its nationals a special permit authorizing that national to kill, take, and treat whales for purposes of scientific research subject to such restrictions as to number and subject to such other conditions as the Contracting Government thinks fit."

[10]   Statement to 40th Annual Meeting of the IWC by H. Asgrimsson; 40 REP. INT'L WHALING COMMISSION 15 (1990).

the principal aims of the program was "to study the role of whales in the marine ecosystem."

During the four years of the Icelandic research program, 292 fin whales and 90 sei whales were killed. At this time, most of the members of the IWC believed that special permits should be issued only when there was no other way of gathering information that was essential for the management and conservation of whales, and not to assess fisheries' interactions or to aid ecosystem management. In response to Iceland issuing its special permit, the IWC passed Resolutions in 1986 (Resolution on Special Permits for Scientific Research) and 1987 (Resolution on Scientific Research Programs) that called on contracting governments to issue special permits only "under exceptional circumstances, to meet critically important research needs."

Following the establishment of the moratorium on commercial whaling, Norway temporarily ceased its whaling activities, but in 1988 hunts for minke whales resumed, under the authority of a special permit issued by the Norwegian Ministry of Fisheries. Drawing on and embellishing the Icelandic precedent, Norway stated in 1990 that their special permit catch was an integral part of:

> . . . studies related to energetics to provide information for inclusion into a multi-species model . . . to provide the scientific basis for proper management of marine mammal populations, taking into account their relationships with other marine species.[11]

With the gradual shift towards a rationale for scientific whaling that was more focused on fisheries' interactions with whales than with management of whale stocks, debate within the IWC on the merits of Norway's special permit program became more intense. Norway explained at the 41st Annual Meeting that:

> The interactions between minke whales, seals and other living marine resources is a question of vital importance to Norway where the dependence on the total ecological balance in the seas off Norway is essential in the long term. For Norway, the multi-species research approach is a critical research need. . . .[12]

---

[11]   *Id.*

[12]   *Id.*

This position laid the foundation for two arguments that would be used frequently in the future by whaling countries, namely that whales distort marine ecosystems, and that research into fisheries' interactions is a critical research need for the IWC.

In 1990, the Commission passed three separate Resolutions on proposals for special permit whaling by Iceland, Norway, and Japan.[13] Despite a series of resolutions from successive IWC meetings calling for a cessation of their special permit program, however, Norway continued to kill minke whales for research purposes over several years, until full-scale commercial whaling resumed in 1995.

Catches by Norway 1988–1995

Year	Scientific	Commercial	Total
1988	29	0	29
1989	17	0	17
1990	5	0	5
1991	0	0	0
1992	95	0	95
1993	69	157	226
1994	77	202	279
1995	0	218	218

The most recent use of special permits to investigate the interactions between whales and fisheries has been by Japan. From 1994–1999, 100 minke whales were killed annually in the North Pacific under a special permit program known as the Japanese Research Program in the North Pacific (JARPN). The principal objective of the research, according to the government of Japan, was clarification of the stock structure of North Pacific minke whales. At the 52nd Annual Meeting of the IWC in 2000, however, Japan proposed a significant expansion of their research program, to be known as JARPN II:

The priority in this second phase will be on feeding ecology, involving the studies on prey consumption by cetaceans, prey pref-

---

13   40 REP. INT'L WHALING COMMISSION 35–36 (1990)

erence of cetaceans and ecosystem models. Bryde's and sperm whales will be included as part of this research.[14]

The inclusion of Bryde's and sperm whales in the research take, and the acknowledgement that the main purpose of the special permit was to investigate prey consumption (which most current IWC members do not believe to be a critically important research need) sparked a strong reaction. A Resolution was passed by a large majority at the July 2000 meeting of the IWC, which urged Japan not to proceed with the lethal take of whales. This Resolution also:

> . . . affirms that gathering information on interactions between whales and prey species is not a critically important issue which justifies the killing of whales for research purposes.[15]

Japan nevertheless went ahead with the lethal research, prompting a strong diplomatic response. In August, an unprecedented demarche, involving 15 countries, took place in Tokyo, urging the government of Japan to refrain from killing whales in the North Pacific.[16]

Because both sperm and Bryde's whales are protected under the provisions of the U.S. Endangered Species Act,[17] U.S. Secretary of Commerce, Norman Mineta, certified to President Clinton that Japan's actions were contrary to U.S. law.[18] Therefore, Japan was prohibited from access to pelagic fish resources within the U.S. Exclusive Economic

---

[14]   Government of Japan—Research plan for Cetacean Studies in the Western North Pacific under Special Permit (JARPN II) IWC/SC/52/O1.

[15]   IWC/52/36, Resolution on whaling under special permit in the North Pacific Ocean.

[16]   *Fifteen Nations Ask Japan to Stop Research Whaling*, REUTERS, Aug. 21, 2000.

[17]   16 U.S.C. § 1531 (1973), available at <http://eelink.net/~asilwildlife/docs. html>.

[18]   The Japanese were certified under the Fishermen's Protective Act of 1967 (the "Pelly Amendment"), 22 U.S.C. § 1978. Under this statute, when the Secretary of Commerce or Interior finds that nationals of a foreign country are "engaging in trade or taking which diminishes the effectiveness of any international program for endangered or threatened species," the President may bar the importation of products from that nation. For a copy of Mineta's certification letter to President Clinton, *see* <http://www.noaa.gov/whales/minetaletter.htm>.

Zone. In his statement of September 13, 2000, President Clinton stated:

> Following Secretary Mineta's certification that Japan is undermining international whaling protections with its expanded whaling program, I am today directing that Japan be denied future access to fishing rights in U.S. waters, and directing members of my Cabinet to consider additional steps we might take, including possible trade sanctions.[19]

Given the degree of international tension that has been generated by lethal research programs investigating the feeding habits of large whales, it is reasonable to examine what new information has been obtained. A number of papers have been published on the stomach contents of harpooned minke whales in the North Atlantic and North Pacific Oceans, demonstrating that they feed on a range of pelagic fish species.[20] Data collected in these programs have been incorporated into the construction of multi-species models, particularly in Iceland and Norway. Papers on multi-species management have been published in the journals of the IWC and the North Atlantic Marine Mammal Commission (NAMMCO), as well as those of fisheries' organizations such as the International Council for the Exploration of the Seas (ICES). Many of the publications attempt to estimate the impacts of cetaceans on fish stocks.[21]

The complexity surrounding trophic (feeding) interactions makes predicting outcomes for even the simplest trophic web very difficult, however, and has only been demonstrated for simplified freshwater systems. Models of multi-species interactions have shown that expected outcomes are rarely met. This is because each trophic level is affected by a mix of competition, predation and environment, the importance of each vary-

---

[19]   Statement of the President, September 13, 2000. Distributed by the Office of International Information Programs, U.S. Department of State, <http://usinfo.state.gov>.

[20]   *See* T. Haug *et al.*, *Spatial and temporal variations in northeast Atlantic minke whale feeding habits*, in WHALES, SEALS, FISH AND MAN—PROCEEDINGS OF THE INTERNATIONAL SYMPOSIUM ON THE BIOLOGY OF MARINE MAMMALS IN THE NORTHEAST ATLANTIC at 225–39 (A.S. Blix, L. Walloe & O. Ulltang eds., 1995).

[21]   *See* O. Flaaten. & K. Stollery, *The economic costs of biological predation. Theory and application to the case of the Northeast Atlantic minke whale's consumption of fish*, 8 ENVTL. & RESOURCE ECON. 75–95 (1996).

ing with the species concerned. Potential predatory interactions in even simple food webs make predictions highly uncertain. More than 15 years ago, Beverton concluded that, even when food chains were dependent on a simple two-stage linkage, the outcomes were unpredictable.[22] Most food chains are far more complex. It is not surprising that there is little scientific consensus on the reliability of multi-species fisheries' models to predict the impact of cetaceans on fish stocks.

Returning to the original proposition of the Ohsumi/Tamura paper, what scientific evidence is available to assess whether whales consume three to six times as much fish as is taken by the global fishing fleet?

A distinction needs to be made between the two types of whales—baleen whales (*Mysticeti*—12 species) and toothed whales (*Odontoceti*—67 species). Most of the large whale species, with the notable exception of the sperm whale, are baleen whales. Instead of teeth, they have a thick fringe of stiff hair-like material (baleen) that hangs from the upper jaw. Their basic feeding method is to engulf large quantities of water, close the mouth and lift the tongue to force the water through the curtain of baleen. Organisms are trapped in the baleen as the water is filtered through, and are swallowed by the whale.

Apart from sperm whales, most of the species that have been commercially hunted over the past 200 years have been baleen whales. Generally speaking, with one notable exception, the eastern North Pacific gray whale, baleen whale populations are significantly (or severely) depleted. In the Southern Hemisphere, the total biomass of baleen whales was estimated in 1988 to be between 4 percent and 8 percent of their biomass in 1900.[23] Even if fish were their main prey, such greatly reduced populations are unlikely to have any significant adverse impact on fish stocks.

The likelihood that the remnant populations of most baleen whale species could have a significant impact on commercial fish resources in the Southern Hemisphere is made even more improbable by a consideration of their life history. With the exception of the Bryde's whale,

---

[22]   *Id.*

[23]   S. Katona & H. Whitehead, *Are Cetacea ecologically important?*, 26 OCEANOGR. MAR. BIOL. ANN. REV. 553–68 (1988).

whose distribution is between tropical and temperate waters, baleen whales in the Southern Hemisphere undertake annual migrations between breeding grounds in lower latitudes and feeding grounds in the Antarctic Ocean. Their principal prey is krill (*Euphausia superba*), a shrimp-like crustacean found only in the high latitudes of the Southern Ocean, where it is found in enormous quantities and is the dominant organism of the Antarctic marine ecosystem. The daily intake of blue and fin whales during their time on the feeding grounds has been estimated at ~ 4 percent of body weight per day. However, for the remainder of the year these whales eat only one-tenth of their summer intake. With the marked seasonality in food supply, daily intake averaged over the year amounts to 1.2–1.6 percent of body weight per day.[24] This pattern appears to be true also for sei and humpback whales.[25]

Furthermore, the food webs that support baleen whales and commercial fisheries are usually quite different, especially in the Southern Hemisphere. For example, the blue whale feeds mainly on krill in summer in the Southern Ocean, returning to warmer northern waters to breed. Tropical tunas, on the other hand, feed in the much less productive waters of the tropics. At times the groups overlap both in space and time, but because of their different feeding strategies they do not compete for the same food. A major review of baleen whale food found that Bryde's whales—the only true resident of the South Pacific and Coral Sea—ate planktonic crustaceans (mainly species of *Euphausia* and *Thyssanoessa*) rather than small fish.[26] The proposition that baleen whales have any significant (or indeed discernible) impact on commercial fish catches in the Southern Hemisphere is thus not supported by the scientific evidence.

In the North Atlantic and North Pacific Oceans, however, the feeding ecology of baleen whales is different. Humpback, fin, sei, and minke whales have all been observed to feed on a range of prey species, includ-

---

[24]   C. Lockyer, *Review of baleen whale (Mysticeti) reproduction and implications for management*, REP. INT'L WHALING COMMISSION (Special Issue 6) 27–50.

[25]   C. Lockyer, *Estimates of growth and energy budget for the sperm whale, Physeter catadon, in* 3 MAMMALS IN THE SEA, 5 FAO Fisheries Series, at 489–504.

[26]   A. Kawamura, *Review of food of baleen whales*, 32 SCI. REP. WHALES RES. INST. 155–97 (1980).

ing planktonic crustaceans, small pelagic fish, and squid. Stomach contents of minke whales taken from coastal Norwegian waters under special permit included commercial fish species such as cod, herring, capelin and pollock,[27] although they also eat krill (*Meganyctiphanes* spp.).[28] In the North Pacific Ocean, data collected during the JARPN program show that minke whales feed on a range of small fish such as Pacific saury and Japanese saury, as well as some planktonic euphausiids (*Euphausia pacifica*).[29]

With regard to toothed whales, only one species, the sperm whale, has been a major target of commercial whaling. Hundreds of thousands of sperm whales have been killed over the past 250 years, and all the evidence from the thousands of stomach contents that have been examined shows that they feed predominantly at great depths on mesopelagic fishes and cephalopods. Sperm whales also feed on large, deep ocean cephalopods including the giant squid (*Architeuthis* spp.), which underlines their close relationship not with surface waters of most commercial fisheries, but with the deep ocean. The inclusion on the government of Japan's special permit of up to ten sperm whales to be taken in the JARPN II program was therefore especially puzzling. Despite the examination of thousands of stomach contents from sperm whales taken during the era of commercial whaling, there has never been a serious suggestion that they are significant consumers of any commercial fish stocks.

While the presence of fish in the stomachs of whales taken in "scientific" or commercial whaling operations demonstrates that fish are a part of their diet, because the whale is killed in the sampling process, it does not provide any indication of how often whales feed on fish and how often they feed on planktonic crustaceans. Nor does it clarify what proportion of the fish eaten by whales are commercial species and what proportion is of no commercial value. For example, although minke

---

[27]   E.S. Nordøy & A.S. Blix, *Diet of Minke whales in the Northeastern Atlantic*, 42 REP. INT'L WHALING COMMISSION 393–98 (1992).

[28]   *Id.*

[29]   T. Tamura, Y. Fujise & K. Shimazaki, *Diet of minke whales Balaenoptera acutorostrata in the north-western part of the North Pacific in summer, 1994 and 1995*, 64 FISHERIES SCI. 71–76 (1998).

whales in the Southern Hemisphere feed predominantly on krill, their secondary choice of prey appears to be lanternfish (family *Myctophidae*).[30] Although these small deep-water fishes have the greatest biomass of any vertebrate family, they are rarely found in concentrations suitable for fishing.

Regardless of how much these animals eat, it is far from clear that this is having an impact on other "top predators" such as commercial fishers. The greatest difficulty in predicting or assessing the impact of marine mammals on fisheries is in determining (first qualitatively, then quantitatively) the food webs that underpin the top predators (whales and commercial fishing interests). Trites *et al.*[31] provided a simplified model (Figure 1) that described the various pathways leading to commercial fisheries and marine mammals. The pathways leading to the top predators are quite separate and join only at the lower feeding levels. It is this last point that is perhaps the most critical, as it shows that marine mammals and fisheries usually occupy different trophic niches. That is, the food webs that support them are usually quite different.

Although seals, whales, dolphins, and sea birds all consume prey that form part of commercial fish stocks, fish take by far the greatest share of the available food source. Bax[32] reported that in six widely separated areas the ratio of consumption by fish to that taken by commercial fishers ranged from 2:1 to as high as 35:1. Furthermore, the study by Bax suggested that although marine mammal consumption of fish was higher than that of seabirds, it was always much lower than that of predatory fish, which were the targets of commercial fishers. Similarly, in the eastern Central Pacific Ocean, the amount of fish eaten by predatory fish is reportedly an order of magnitude higher than that eaten by marine mammals.[33]

---

[30]   A. Kawamura, *A review of baleen whale feeding in the Southern Ocean*, 44 REP. INT'L WHALING COMMISSION 261–71 (1994).

[31]   A.W. Trites, V. Christensen & D. Pauly, *Competition between fisheries and marine mammals for prey and primary production in the Pacific Ocean*, 22 J. N.W. ATLANTIC FISH. SCI. 22, 173–87 (1997).

[32]   N.J. Bax, *A comparison of the fish biomass flow to fish, fisheries and mammals on six marine ecosystems*, 193 ICES MAR. SCI. SYMPOSIA 217–24 (1991).

[33]   V. Christensen, *Managing fisheries involving predator and prey species*, 6 REV. FISH BIOLOGY FISH. 1–26 (1996).

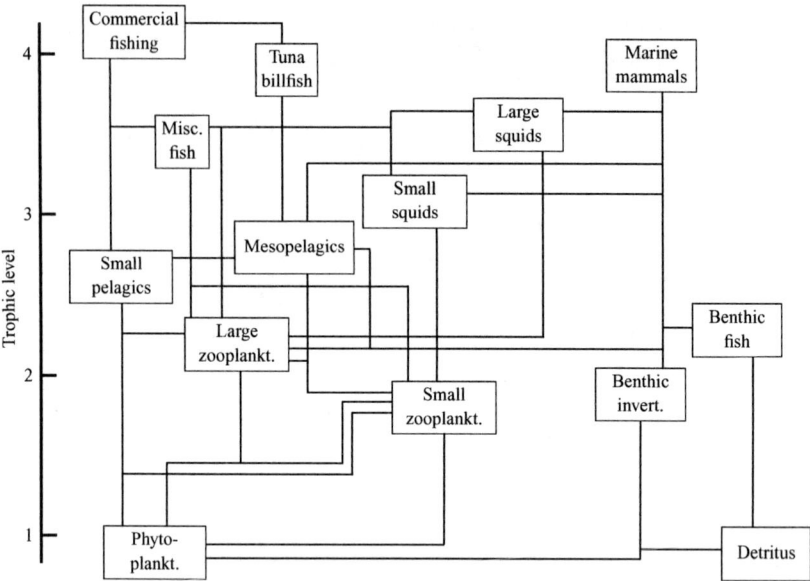

Figure 1 Schematic representation of a simplified food chain leading to marine mammals and fisheries.

Furthermore, it is the nature of the predation that is important: differences between the size of fish taken by fisheries and those taken through natural predation indicate quite different impacts on mortality. Pauly and Christensen[34] estimated that 25 to 35 percent of primary production was necessary to account for the recorded global marine fish catches and discards. However, the same data indicated that the biomass lost to predation by fish is between three to 25 times larger than the commercial catch. This incongruous relationship is possible only because natural predation removes more prey from lower feeding levels in the marine food web than fishing does. They concluded that trophic position and life-history stage are as important as species in assessing the effects of predation in marine systems. One study[35] estimated that there was only a 35 percent overlap in the fish caught by fisheries and those

---

[34]   D. Pauly & V. Christensen, *Primary production required to sustain global fisheries*, 374 NATURE 255–57 (1995).

[35]   Trites *et al.*, *supra* note 31.

preyed upon by marine mammals in the Pacific Ocean. Furthermore, the overlap was not whale predation but rather dolphins and seals.

In recent years, it has become increasingly evident that conflicts occasionally arise between commercial longline fisheries and certain species of small toothed whales which remove or damage fish that have been captured in fishing operations before the line is retrieved or as it is hauled aboard. Killer whales, pilot whales, and false killer whales have all been implicated in this behavior, in fisheries as far apart as Alaska and Fiji.

Attempts at deterring whales from approaching and depredating longlines have been unsuccessful thus far, and interactions appear to be increasing in both scope and scale. This clearly influences some fishers' attitudes toward cetaceans and may lead to retaliatory measures. However, the extent of this problem has yet to be quantified, and the problem of depredation by cetaceans appears to be of considerably less significance than the taking of hooked fish by sharks. In 1999, the Indian Ocean Tuna Commission (IOTC) released a survey it had conducted on the depredation of longlines by marine mammals. The survey concluded that sharks and marine mammals were poaching as many as 20–30 percent of the tuna caught on some longlines in portions of the Indian Ocean.[36] However, it also cautioned that, "in general, little information is available. Depredation of longline caught fish appears to be less serious in other oceans and attains a maximum in the tropical western Indian Ocean."[37] The Commission encouraged the Scientific Committee to carry out a five-year research plan on the depredation by marine mammals and sharks of tunas caught on longlines.[38]

Overall, the scientific literature suggests that the impacts of whales on commercial fish stocks can be summarized as follows:

---

[36]   *See* Appendix XII, *Proposal For Research on Predation by Marine Mammals and Sharks on Tunas Caught by the Longline Fishery in the Indian Ocean*, Report of the Indian Ocean Tuna Commission. Report of the Fourth Session, IOTC/S/04/99/R[E], Dec. 13–16, 1999.

[37]   *Id*. at IOTC/S/04/99/R[E], at para. 34.

[38]   *Id. at* para. 35. *See also Proposal for Research on Predation, supra* note 36, at app. XII.

- Most populations of great whales remain at historically low levels;
- Baleen whales in the Southern Hemisphere do not compete with any commercial fisheries;
- Baleen whales in the Northern Hemisphere consume some commercial fish species in some areas;
- The total consumption of commercial fish species by baleen whales is unknown, but is at the least an order of magnitude less than the consumption by other commercial fish species;
- Sperm whales mainly feed on non-commercial species of deep-water cephalopods;
- Some species of small toothed whales are known to bite hooked fish off commercial longlines.

Despite the apparently strong scientific evidence to the contrary, Japan vigorously contended at the 51st Annual Meeting of the IWC that whales are consuming up to six times the world's fish catch, prompting the following intervention by the New Zealand Commissioner, Jim McLay:

One of the references in the Japanese paper is a publication by Trites, Christensen, and Pauly,[39] (which, unlike the Ohsumi/Tamura paper, has been peer-reviewed and published in a reputable journal, the Journal of Northwest Atlantic Fisheries Science). They state that:

A large fraction (>60%) of the food caught by marine mammals consists of deep sea squids and very small deep sea fishes not harvestable by humans, thus limiting the extent of direct competition between fisheries and marine mammals. Moreover, the most important consumers of commercially exploited fish are other predatory fish, not marine mammals.

And later, they concluded:

The excessive build up and overcapitalization in the North Pacific's fishing leads unavoidably to overfishing and potentially threatens marine mammals with food web competition. . . . It is clear that . . . the North Pacific fisheries cannot continue to expand as they have previously.

---

[39]   Trites *et al.*, *supra* note 31.

None of these conclusions will surprise those familiar with the current state of the world's fisheries.

It is misleading to blame cetaceans for declining fish catches. The fundamental problem facing the world's fisheries resource is the legacy of decades of over-fishing and the gross over-capacity of fishing effort.

We should all recall that, in past times, there were far more great whales than today and considerably more fish, and that both have been over-exploited. If others insist on running their unsubstantiated views to the contrary, we for one intend to confront and rebut those arguments.[40]

Japan has continued, however, to promote the proposition that whales pose a serious threat to the world's fish stocks. On November 17, 2000, the Mainichi Shimbun published an interview with Dr Seiji Ohsumi, Director of the Institute for Cetacean Research, from which the following is extracted:

Question: *Mr. Ohsumi, you have emphasized, "Whales are heavy eaters."*

Answer: *According to our calculations about how much food whales eat, it was found that the annual intake of food by whales in waters around the world is an estimated 280 million to 500 million tons. The world catch of fish is approximately 84 million tons (in 1994), so this means whales consume three to six times as much marine resources as humans do.*

Because whales pose a threat to fisheries resources, there is the need to resume whaling also for the protection of the fishing industry. Until recently, the question of "what and how much whales are eating" has not been taken up as a subject for discussion, but we find it now necessary to deal with the issue.

The nub of the debate is neatly encapsulated in these two paragraphs. Whaling nations seek to justify a return to commercial whaling as part

---

[40]   The intervention was made at the 51st Meeting of the IWC, on May 26, 1999.

of what they promote as a marine ecosystem management plan. In the meantime, whaling vessels and crews are kept in employment by carrying out "scientific whaling." Additionally, whales can be blamed for causing the decline in fish catches that has in reality been brought about by a series of human actions such as the huge over-capacity in the world's fishing fleets, degradation of coastal waters in populated areas through pollution and waste discharge, and global climate change. Expect this argument to run and run.

Appendix

International Convention for the Regulation of Whaling Washington, December 2, 1946

The Governments whose duly authorised representatives have subscribed hereto,

- Recognizing the interest of the nations of the world in safeguarding for future generations the great natural resources represented by the whale stocks;
- Considering that the history of whaling has seen over-fishing of one area after another and of one species of whale after another to such a degree that it is essential to protect all species of whales from further over-fishing;
- Recognizing that the whale stocks are susceptible of natural increases if whaling is properly regulated, and that increases in the size of whale stocks will permit increases in the number of whales which may be captured without endangering these natural resources;
- Recognizing that it is in the common interest to achieve the optimum level of whale stocks as rapidly as possible without causing widespread economic and nutritional distress;
- Recognizing that in the course of achieving these objectives, whaling operations should be confined to those species best able to sustain exploitation in order to give an interval for recovery to certain species of whales now depleted in numbers;
- Desiring to establish a system of international regulation for the whale fisheries to ensure proper and effective conservation and development of whale stocks on the basis of the principles embodied in the provisions of the International Agreement for the Regulation of Whaling, signed in London on 8th June, 1937, and the protocols to that Agreement signed in London on 24th June, 1938, and 26th November, 1945; and

- • Having decided to conclude a convention to provide for the proper conservation of whale stocks and thus make possible the orderly development of the whaling industry;

Have agreed as follows:—

Article I

1. This Convention includes the Schedule attached thereto which forms an integral part thereof. All references to "Convention" shall be understood as including the said Schedule either in its present terms or as amended in accordance with the provisions of Article V.

2. This Convention applies to factory ships, land stations, and whale catchers under the jurisdiction of the Contracting Governments and to all waters in which whaling is prosecuted by such factory ships, land stations, and whale catchers.

Article II

As used in this Convention:—

1. "Factory ship" means a ship in which or on which whales are treated either wholly or in part;

2. "Land station" means a factory on the land at which whales are treated whether wholly or in part;

3. "Whale catcher" means a ship used for the purpose of hunting, taking, towing, holding on to, or scouting for whales;

4. "Contracting Government" means any Government which has deposited an instrument of ratification or has given notice of adherence to this Convention.

Article III

1. The Contracting Governments agree to establish an International Whaling Commission, hereinafter referred to as the Commission, to be composed of one member from each Contracting Government. Each member shall have one vote and may be accompanied by one or more experts and advisers.

2. The Commission shall elect from its own members a Chairman and Vice-Chairman and shall determine its own Rules of Procedure. Decisions of the Commission shall be taken by a simple majority of those members voting except that a three-fourths majority of those members voting shall be required for action in pursuance of Article V. The Rules of Procedure may provide for decisions otherwise than at meetings of the Commission.

3. The Commission may appoint its own Secretary and staff.

4. The Commission may set up, from among its own members and experts or advisers, such committees as it considers desirable to perform such functions as it may authorize.

5. The expenses of each member of the Commission and of his experts and advisers shall be determined by his own Government.

6. Recognizing that specialized agencies related to the United Nations will be concerned with the conservation and development of whale fisheries and the products arising therefrom and desiring to avoid duplication of functions, the Contracting Governments will consult among themselves within two years after the coming into force of this Convention to decide whether the Commission shall be brought within the framework of a specialized agency related to the United Nations.

7. In the meantime the Government of the United Kingdom of Great Britain and Northern Ireland shall arrange, in consultation with the other Contracting Governments, to convene the first meeting of the Commission, and shall initiate the consultation referred to in paragraph 6 above.

8. Subsequent meetings of the Commission shall be convened as the Commission may determine.

Article IV

1. The Commission may either in collaboration with or through independent agencies of the Contracting Governments or other public or private agencies, establishments, or organizations, or independently

   (a) encourage, recommend, or if necessary, organize studies and investigations relating to whales and whaling;

  (b) collect and analyze statistical information concerning the current condition and trend of the whale stocks and the effects of whaling activities thereon;

  (c) study, appraise, and disseminate information concerning methods of maintaining and increasing the populations of whale stocks.

2. The Commission shall arrange for the publication of reports of its activities, and it may publish independently or in collaboration with the International Bureau for Whaling Statistics at Sandefjord in Norway and other organizations and agencies such reports as it deems appropriate, as well as statistical, scientific, and other pertinent information relating to whales and whaling.

Article V

1. The Commission may amend from time to time the provisions of the Schedule by adopting regulations with respect to the conservation and utilization of whale resources, fixing *(a)* protected and unprotected species; *(b)* open and closed seasons; *(c)* open and closed waters, including the designation of sanctuary areas; *(d)* size limits for each species; *(e)* time, methods, and intensity of whaling (including the maximum catch of whales to be taken in any one season); *(f)* types and specifications of gear and apparatus and appliances which may be used; *(g)* methods of measurement; and *(h)* catch returns and other statistical and biological records.

2. These amendments of the Schedule *(a)* shall be such as are necessary to carry out the objectives and purposes of this Convention and to provide for the conservation, development, and optimum utilization of the whale resources; *(b)* shall be based on scientific findings; *(c)* shall not involve restrictions on the number or nationality of factory ships or land stations, nor allocate specific quotas to any factory or ship or land station or to any group of factory ships or land stations; and *(d)* shall take into consideration the interests of the consumers of whale products and the whaling industry.

3. Each of such amendments shall become effective with respect to the Contracting Governments ninety days following notification of the amendment by the Commission to each of the Contracting Governments, except that *(a)* if any Government presents to the Commission

objection to any amendment prior to the expiration of this ninety-day period, the amendment shall not become effective with respect to any of the Governments for an additional ninety days; *(b)* thereupon, any other Contracting Government may present objection to the amendment at any time prior to the expiration of the additional ninety-day period, or before the expiration of thirty days from the date of receipt of the last objection received during such additional ninety-day period, whichever date shall be the later; and *(c)* thereafter, the amendment shall become effective with respect to all Contracting Governments which have not presented objection but shall not become effective with respect to any Government which has so objected until such date as the objection is withdrawn. The Commission shall notify each Contracting Government immediately upon receipt of each objection and withdrawal and each Contracting Government shall acknowledge receipt of all notifications of amendments, objections, and withdrawals.

4. No amendments shall become effective before 1st July, 1949.

Article VI

The Commission may from time to time make recommendations to any or all Contracting Governments on any matters which relate to whales or whaling and to the objectives and purposes of this Convention.

Article VII

The Contracting Government shall ensure prompt transmission to the International Bureau for Whaling Statistics at Sandefjord in Norway, or to such other body as the Commission may designate, of notifications and statistical and other information required by this Convention in such form and manner as may be prescribed by the Commission.

Article VIII

1. Notwithstanding anything contained in this Convention any Contracting Government may grant to any of its nationals a special permit authorizing that national to kill, take and treat whales for purposes of scientific research subject to such restrictions as to number and subject to such other conditions as the Contracting Government thinks fit, and the killing, taking, and treating of whales in accordance with the provisions

of this Article shall be exempt from the operation of this Convention. Each Contracting Government shall report at once to the Commission all such authorizations which it has granted. Each Contracting Government may at any time revoke any such special permit which it has granted.

2. Any whales taken under these special permits shall so far as practicable be processed and the proceeds shall be dealt with in accordance with directions issued by the Government by which the permit was granted.

3. Each Contracting Government shall transmit to such body as may be designated by the Commission, in so far as practicable, and at intervals of not more than one year, scientific information available to that Government with respect to whales and whaling, including the results of research conducted pursuant to paragraph 1 of this Article and to Article IV.

4. Recognizing that continuous collection and analysis of biological data in connection with the operations of factory ships and land stations are indispensable to sound and constructive management of the whale fisheries, the Contracting Governments will take all practicable measures to obtain such data.

Article IX

1. Each Contracting Government shall take appropriate measures to ensure the application of the provisions of this Convention and the punishment of infractions against the said provisions in operations carried out by persons or by vessels under its jurisdiction.

2. No bonus or other remuneration calculated with relation to the results of their work shall be paid to the gunners and crews of whale catchers in respect of any whales the taking of which is forbidden by this Convention.

3. Prosecution for infractions against or contraventions of this Convention shall be instituted by the Government having jurisdiction over the offence.

4. Each Contracting Government shall transmit to the Commission full details of each infraction of the provisions of this Convention by persons or vessels under the jurisdiction of that Government as reported by its inspectors. This information shall include a statement of measures taken for dealing with the infraction and of penalties imposed.

Article X

1. This Convention shall be ratified and the instruments of ratifications shall be deposited with the Government of the United States of America.

2. Any Government which has not signed this Convention may adhere thereto after it enters into force by a notification in writing to the Government of the United States of America.

3. The Government of the United States of America shall inform all other signatory Governments and all adhering Governments of all ratifications deposited and adherences received.

4. This Convention shall, when instruments of ratification have been deposited by at least six signatory Governments, which shall include the Governments of the Netherlands, Norway, the Union of Soviet Socialist Republics, the United Kingdom of Great Britain and Northern Ireland, and the United States of America, enter into force with respect to those Governments and shall enter into force with respect to each Government which subsequently ratifies or adheres on the date of the deposit of its instrument of ratification or the receipt of its notification of adherence.

5. The provisions of the Schedule shall not apply prior to 1st July, 1948. Amendments to the Schedule adopted pursuant to Article V shall not apply prior to 1st July, 1949.

Article XI

Any Contracting Government may withdraw from this Convention on 30th June, of any year by giving notice on or before 1st January, of the same year to the depository Government, which upon receipt of such a notice shall at once communicate it to the other Contracting Governments. Any other Contracting Government may, in like manner, within one month of the receipt of a copy of such a notice from the depository Government give notice of withdrawal, so that the Convention shall cease to be in force on 30th June, of the same year with respect to the Government giving such notice of withdrawal.

The Convention shall bear the date on which it is opened for signature and shall remain open for signature for a period of fourteen days thereafter.

In witness whereof the undersigned, being duly authorized, have signed this Convention.

Done in Washington this second day of December, 1946, in the English language, the original of which shall be deposited in the archives of the Government of the United States of America. The Government of the United States of America shall transmit certified copies thereof to all the other signatory and adhering Governments.

SIGNATORIES:	FOR CHILE: Augustín R. Edwards	FOR PERU: Carlos Rotalde
FOR AGENTINA: Oscar Ivanissevich José Manuel Moneta Guillermo Brown Pedro H. Bruno Videla	FOR DENMARK: Peter Friedrich Erichsen	FOR THE UNION OF SOVIET SOCIALIST REPUBLICS: Alexander S. Bogdanov Eugine I. Nikishin
FOR AUSTRALIA: Francis F. Anderson	FOR FRANCE: Francis Lacoste	FOR THE UNITED KINGDOM OF GREAT BRITAIN AND NORTHERN IRELAND: A.T.A. Dobson J. Thomson
FOR BRAZIL: Paulo Fróes da Cruz	FOR THE NETHER- LANDS: Guy Richardson Powles	FOR THE UNITED STATES OF AMERICA: Remington Kellogg Ira N. Gabrielson William E.S. Flory
FOR CANADA: H.H. Wrong H.A. Scott	FOR NEW ZEALAND: Birger Bergersen	FOR THE UNION OF SOUTH AFRICA: H.T. Andrews

Protocol to the International Convention for the Regulation of Whaling Signed at Washington under date of December 2, 1946

The Contracting Governments to the International Convention for the Regulation of Whaling signed at Washington under date of 2nd December, 1946 which Convention is hereinafter referred to as the 1946 Whaling Convention, desiring to extend the application of that Convention to helicopters and other aircraft and to include provisions on methods of inspection among those Schedule provisions which may be amended by the Commission, agree as follows:

Article I

Subparagraph 3 of the Article II of the 1946 Whaling Convention shall be amended to read as follows:

> "3. 'whale catcher' means a helicopter, or other aircraft, or a ship, used for the purpose of hunting, taking, killing, towing, holding on to, or scouting for whales."

Article II

Paragraph 1 of Article V of the 1946 Whaling Convention shall be amended by deleting the word "and" preceding clause (h), substituting a semicolon for the period at the end of the paragraph, and adding the following language: "and (i) methods of inspection".

Article III

1. This Protocol shall be open for signature and ratification or for adherence on behalf of any Contracting Government to the 1946 Whaling Convention.

2. This Protocol shall enter into force on the date upon which instruments of ratification have been deposited with, or written notifications

of adherence have been received by, the Government of the United States of America on behalf of all the Contracting Governments to the 1946 Whaling Convention.

3. The Government of the United States of America shall inform all Governments signatory or adhering to the 1946 Whaling Convention of all ratifications deposited and adherences received.

4. This Protocol shall bear the date on which it is opened for signature and shall remain open for signature for a period of fourteen days thereafter, following which period it shall be open for adherence.

IN WITNESS WHEREOF the undersigned, being duly authorized, have signed this Protocol.

DONE in Washington this nineteenth day of November, 1956, in the English Language, the original of which shall be deposited in the archives of the Government of the United States of America. The Government of the United States of America shall transmit certified copies thereof to all Governments signatory or adhering to the 1946 Whaling Convention.

SIGNATORIES:	FOR CHILE: Augustín R. Edwards	FOR PERU: Carlos Rotalde
FOR AGENTINA: Oscar Ivanissevich José Manuel Moneta Guillermo Brown Pedro H. Bruno Videla	FOR DENMARK: Peter Friedrich Erichsen	FOR THE UNION OF SOVIET SOCIALIST REPUBLICS: Alexander S. Bogdanov Eugine I. Nikishin
FOR AUSTRALIA: Francis F. Anderson	FOR FRANCE: Francis Lacoste	FOR THE UNITED KINGDOM OF GREAT BRITAIN AND NORTHERN IRELAND: A.T.A. Dobson J. Thomson
FOR BRAZIL: Paulo Fróes da Cruz	FOR THE NETHER-LANDS: Guy Richardson Powles	FOR THE UNITED STATES OF AMERICA: Remington Kellogg Ira N. Gabrielson William E.S. Flory
FOR CANADA: H.H. Wrong H.A. Scott	FOR NEW ZEALAND: Birger Bergersen	FOR THE UNION OF SOUTH AFRICA: H.T. Andrews

Agreement on the Conservation of Small Cetaceans of the Baltic and North Seas (ASCOBANS) (1991)

The Parties,

Recalling the general principles of conservation and sustainable use of natural resources, as reflected in the World Conservation Strategy of the International Union for the Conservation of Nature and Natural Resources, the United Nations Environment Programme, and the World Wide Fund for Nature, and in the report of the World Commission on Environment and Development,

Recognizing that small cetaceans are and should remain an integral part of marine ecosystems,

Aware that the population of harbour porpoises of the Baltic Sea has drastically decreased,

Concerned about the status of small cetaceans in the Baltic and North Seas,

Recognizing that by-catches, habitat deterioration and disturbance may adversely affect these populations,

Convinced that their vulnerable and largely unclear status merits immediate attention in order to improve it and to gather information as a basis for sound decisions on management and conservation,

Confident that activities for that purpose are best coordinated between the States concerned in order to increase efficiency and avoid duplicate work,

Aware of the importance of maintaining maritime activities such as fishing,

Recalling that under the Convention on the Conservation of Migratory Species of Wild Animals (Bonn 1979), Parties are encouraged to conclude agreements on wild animals which periodically cross national jurisdictional boundaries

Recalling also that under the provisions of the Convention on the Conservation of European wildlife and Natural Habitats (Berne 1979), all small cetaceans regularly present in the Baltic and North Seas are listed in its Appendix II as strictly protected species, and

Referring to the Memorandum of Understanding on Small Cetaceans in the North Sea signed by the Ministers present at the Third International Conference on the Protection of the North Sea, have agreed as follows:

1. Scope and interpretation

1.1. This agreement shall apply to all small cetaceans found within the area of the agreement.

1.2. For the purpose of this agreement:

(a) "Small cetaceans" means any species, subspecies or population of toothed whales Odontoceti, except the sperm whale *Physeter macrocephalus*;

(b) "Area of the agreement" means the marine environment of the Baltic and North Seas, as delimited to the north-east by the shores of the Gulfs of Bothnia and Finland; to the south-west by latitude 48° 30' N and longitude 5° W; to the north-west by longitude 5° W and a line drawn through the following points: latitude 60° N / longitude 5° W, latitude 61° N / longitude 4° W, and latitude 62° N / longitude 3° W; to the north by latitude 62° N; and including the Kattegat and the Sound and Belt passages but excluding the waters between Cape Wrath and St Anthony Head;

(c) "Bonn Convention" means the Convention on the Conservation of Migratory Species of Wild Animals (Bonn 1979);

(d) "Regional Economic Integration Organization" means an organization constituted by sovereign States which has competence in respect of the negotiation, conclusion and application of international agreements in matters covered by this agreement;

(e) "Party" means a range State or any Regional Economic Integration Organization for which this agreement is in force;

(f) "Range State" means any State, whether or not a Party to the agreement, that exercises jurisdiction over any part of the range of a species covered by this agreement, or a State whose flag vessels, outside national jurisdictional limits but within the area of the agreement, are engaged in operations adversely affecting small cetaceans;

(g) "Secretariat" means, unless the context otherwise indicates, the Secretariat to this agreement.

2. Purpose and basic arrangements

2.1. The Parties undertake to cooperate closely in order to achieve and maintain a favourable conservation status for small cetaceans.

2.2. In particular, each Party shall apply within the limits of its jurisdiction and in accordance with its international obligations, the conservation, research and management measures prescribed in the Annex.

2.3. Each Party shall designate a Coordinating Authority for activities under this agreement.

2.4. The Parties shall establish a Secretariat and an Advisory Committee not later than at their first Meeting.

2.5. A brief report shall be submitted by each Party to the Secretariat not later than 31 March each year, commencing with the first complete year after the entry into force of the agreement for that Party. The report shall cover progress made and difficulties experienced during the past calendar year in implementing the agreement.

2.6. The provisions of this agreement shall not affect the rights of a Party to take stricter measures for the conservation of small cetaceans.

3. The Coordinating Authority

3.1. The activities of each Party shall be coordinated and monitored through its Coordinating Authority which shall serve as the contact point for the Secretariat and the Advisory Committee in their work.

4. The Secretariat

4.1. The Secretariat shall, following instructions provided by the meetings of the Parties, promote and coordinate the activities undertaken in accordance with Article 6.1 of this agreement and shall, in close consultation with the Advisory Committee, provide advice and support to the Parties and their Coordinating Authorities.

4.2. In particular, the Secretariat shall: facilitate the exchange of information and assist with the coordination of monitoring and research among Parties and between the Parties and international organizations engaged in similar activities; organize meetings and notify Parties, the observers mentioned in Article 6.2.1 and the Advisory Committee; coordinate and circulate proposals for amendments to the agreement and its Annex; and present to the Coordinating Authorities, each year no later than 30 June, a summary of the Party reports submitted in accordance with Article 2.5, and a brief account of its own activities during the past calendar year, including a financial report.

4.3. The Secretariat shall present to each Meeting of the Parties a summary of, *inter alia*, progress made and difficulties encountered since the last Meeting of the Parties. A copy of this report shall be submitted to the Secretariat of the Bonn Convention for information to the Parties of that Convention.

4.4. The Secretariat shall be attached to a public institution of a Party or to an international body, and that institution or body shall be the employer of its staff.

5. The Advisory Committee

5.1. The Meeting of the Parties shall establish an Advisor Committee to provide expert advice and information to the Secretariat and the Parties on the conservation and management of small cetaceans and on other matters in relation to the running of the agreement, having regard to the need not to duplicate the work of other international bodies and the desirability of drawing on their expertise.

5.2. Each Party shall be entitled to appoint one member of the Advisory Committee.

5.3. The Advisory Committee shall elect a chairman and establish its own rules of procedure.

5.4. Each Committee member may be accompanied by advisers, and the Committee may invite other experts to attend its meetings. The Committee may establish working groups.

6. The meeting of the Parties

6.1. The Parties shall meet, at the invitation of the Bonn Convention Secretariat on behalf of any Party, within one year of the entry into force of this agreement, and thereafter, at the notification of the Secretariat, not less than once every three years to review the progress made and difficulties encountered in the implementation and operation of the agreement since the last Meeting, and to consider and decide upon:

(a) The latest Secretariat report;

(b) Matters relating to the Secretariat and the Advisory Committee;

(c) The establishment and review of financial arrangements and the adoption of a budget for the forthcoming three years;

(d) Any other item relevant to this agreement circulated among the Parties by a Party or by the Secretariat not later than 90 days before the Meeting, including proposals to amend the agreement and its Annex; and

(e) The time and venue of the next Meeting.

6.2.1. The following shall be entitled to send observers to the Meeting: the Depositary of this agreement, the secretariats of the Bonn Convention, the Convention on International Trade in Endangered Species of Wild Fauna and Flora, the Convention on the Conservation of European wildlife and Natural Habitats, the Convention for the Prevention of Marine Pollution by Dumping from Ships and Aircraft, the Convention for the Prevention of Marine Pollution from Landbased Sources, the Common Secretariat for the Cooperation on the Protection of the Wadden Sea, the International Whaling Commission, the North-East Atlantic Fisheries Commission, the International Baltic Sea Fisheries Commission, the Baltic Marine Environment Protection Commission, the International Council for the Exploration of the Sea, the International Union for the Conservation of Nature and Natural Resources, and all non-Party Range States and Regional Economic Integration Organizations bordering on the waters concerned.

6.2.2. Any other body qualified in cetacean conservation and management may apply to the Secretariat not less than 90 days in advance of the Meeting to be allowed to be represented by observers. The Secretariat shall communicate such applications to the Parties at least 60 days before the Meeting, and observers shall be entitled to be present unless that is opposed not less than 30 days before the Meeting by at least one third of the Parties.

6.3. Decisions at Meetings shall be taken by a simple majority among Parties present and voting, except that financial decisions and amendments to the agreement and its Annex shall require a three-quarters majority among those present and voting. Each Party shall have one vote. However, in matters within their competence, the European Economic Community shall exercise their voting rights with a number of votes equal to the number of their member States which are Parties to the agreement.

6.4. The Secretariat shall prepare and circulate a report of the Meeting to all Parties and observers within 90 days of the closure of the Meeting.

6.5. This agreement and its Annex may be amended at any Meeting of the Parties.

6.5.1. Proposals for amendments may be made by any Party.

6.5.2. The text of any proposed amendment and the reasons for it shall be communicated to the Secretariat at least 90 days before the opening of the Meeting. The Secretariat shall transmit copies forthwith to the Parties.

6.5.3. Amendments shall enter into force for those Parties which have accepted them 90 days after the deposit of the fifth instrument of acceptance of the amendment with the Depositary. Thereafter they shall enter into force for a Party 30 days after the date of deposit of its instrument of acceptance of the amendment with the Depositary.

7. Financing

7.1. The Parties agree to share the cost of the budget, with Regional Economic Integration Organizations contributing 2.5 per cent of the administrative costs and other Parties sharing the balance in accordance with the United Nations scale, but with a maximum of 25 per cent per Party.

7.2. The share of each Party in the cost of the Secretariat and any additional sum agreed for covering other common expenses shall be paid to the Government or international organization hosting the Secretariat, as soon as practicable after the end of March and in no case later than before the end of June each year.

7.3. The Secretariat shall prepare and keep financial accounts by calendar years.

8. Legal matters and formalities

8.1. This is an agreement within the meaning of the Bonn Convention, Article IV (4).

8.2. The provisions of this agreement shall in no way affect the rights and obligations of a Party deriving from any other existing treaty, convention, or agreement.

8.3. The Secretary-General of the United Nations shall assume the functions of Depositary of this agreement.

8.3.1. The Depositary shall notify all Signatories, all Regional Economic Integration Organizations and the Bonn Convention Secretariat of any signatures, deposit of instruments of ratification, acceptance, approval or accession, entry into force of the agreement, amendments, reservations and denunciations.

8.3.2. The Depositary shall send certified true copies of the agreement to all signatories, all non-signatory Range States, all Regional Economic Integration Organizations and the Bonn Convention Secretariat.

8.4. The agreement shall be open for signature at the United Nations Headquarters by 31 March 1992 and thereafter remain open for signature at the United Nations Headquarters by all Range States and Regional Economic Integration Organizations, until the date of entry into force of the agreement. They may express their consent to be bound by the agreement (a) by signature, not subject to ratification, acceptance or approval, or (b) if the agreement has been signed subject to ratification, acceptance or approval, by the deposit of an instrument of ratification, acceptance or approval. After the date of its entry into force, the agreement shall he open for accession by Range States and Regional Economic Integration Organizations.

8.5. The agreement shall enter into force 90 days after six Range States have expressed their consent to be bound by it in accordance with

Article 8.4. Thereafter, it shall enter into force for a State and Regional Economic Integration Organization on the 30th day after the date of signature, not subject to ratification, acceptance or approval, or of the deposit of an instrument of ratification, acceptance, approval or accession with the Depositary.

8.6.   The agreement and its Annex shall not be subject to general reservations. However, a Range State or Regional Economic Integration Organization may, on becoming a Party in accordance with Article 8.4 and 8.5, enter a specific reservation with regard to any particular species, subspecies or population of small cetaceans. Such reservations shall be communicated to the Depositary on signing or at the deposit of an instrument of ratification, acceptance, approval or accession.

8.7.   A Party may at any time denounce this agreement. Such denunciation shall be notified in writing to the Depositary and take effect one year after the receipt thereof.

In witness whereof the undersigned, being duly authorized thereto, have affixed their signatures to this agreement.

Done at New York on 17 March 1992, the English, French, German and Russian texts of the agreement being equally authentic.

ANNEX

Conservation and management plan

The following conservation, research, and management measures shall be applied, in conjunction with other competent international bodies, to the populations defined in Article l.l:

1. Habitat conservation and management

Work towards (a) the prevention of the release of substances which are a potential threat to the health of the animals, (b) the development, in the light of available data indicating unacceptable interaction, of modifications of fishing gear and fishing practices in order to reduce bycatches and to prevent fishing gear from getting adrift or being discarded at sea, (c) the effective regulation, to reduce the impact on the animals, of activities which seriously affect their food resources, and (d) the prevention of other significant disturbance, especially of an acoustic nature.

2. Surveys and research

Investigations, to be coordinated and shared in an efficient manner between the Parties and competent international organizations, shall be conducted in order to (a) *assess* the status and seasonal movements of the populations and stocks concerned, (b) locate areas of special importance to their survival, and (c) identify present and potential threats to the different species.

Studies under (a) should particularly include improvement of existing and development of new methods to establish stock identity and to estimate abundance, trends, population structure and dynamics, and migrations. Studies under (b) should focus on locating areas of special importance to breeding and feeding. Studies under (c) should include research on habitat requirements, feeding ecology, trophic relationships, dispersal, and sensory biology with special regard to effects of pollution, disturbance and interactions with fisheries, including work on methods to reduce such interactions. The studies should exclude the killing of animals and include the release in good health of animals captured for research.

3. Use of by-catches and strandings

Each Party shall endeavour to establish an efficient system for reporting and retrieving by-catches and stranded specimens and to carry out, in the framework of the studies mentioned above, full autopsies in order to collect tissues for further studies and to reveal possible causes of death and to document food composition. The information collected shall be made available in an international database.

4. Legislation

Without prejudice to the provisions of paragraph 2 above, the Parties shall endeavour to establish (a) the prohibition under national law, of the intentional taking and killing of small cetaceans where such regulations are not already in force, and (b) the obligation to release immediately any animals caught alive and in good health. Measures to enforce these regulations shall be worked out at the national level.

5. Information and education

Information shall be provided to the general public in order to ensure support for the aims of the agreement in general and to facilitate the reporting of sightings and strandings in particular; and to fishermen in order to facilitate and promote the reporting of by-catches and the delivery of dead specimens to the extent required for research under the agreement.

Agreement on the Conservation of Cetaceans of the Black Sea, Mediterranean Sea and Contiguous Atlantic Area (ACCOBAMS) (1996)

The Parties,

Recalling that the Convention on the Conservation of Migratory Species of Wild Animals, 1979, encourages international co-operative action to conserve migratory species;

Recalling further that the third meeting of the Conference of the Parties to the Convention, held in Geneva in September 1991, urged Range States to collaborate with a view to concluding, under the Convention's auspices, a multilateral agreement for the conservation of small cetaceans of the Mediterranean and Black Seas;

Recognizing that cetaceans are an integral part of the marine ecosystem which must be conserved for the benefit of present and future generations, and that their conservation is a common concern;

Recognizing the importance of integrating actions to conserve cetaceans with activities related to the socio-economic development of the Parties concerned by this Agreement, including maritime activities such as fishing and the free circulation of vessels in accordance with international law;

Aware that the conservation status of cetaceans can be adversely affected by factors such as degradation and disturbance of their habitats, pollution, reduction of food resources, use and abandonment of non-selective fishing gear, and by deliberate and incidental catches;

Convinced that the vulnerability of cetaceans to such threats warrants the implementation of specific conservation measures, where they do not already exist, by States or regional economic integration organizations that exercise sovereignty and/or jurisdiction over any part of their range, and by States, flag vessels of which are engaged outside national

421

jurisdictional limits in activities that may affect the conservation of cetaceans;

Stressing the need to promote and facilitate co-operation among States, regional economic integration organizations, intergovernmental organizations and the non governmental sector for the conservation of cetaceans of the Black Sea, Mediterranean Sea, the waters which interconnect these seas, and the contiguous Atlantic area;

Convinced that the conclusion of a multilateral agreement and its implementation through co-ordinated, concerted actions will contribute significantly to the conservation of cetaceans and their habitats in the most efficient manner, and will have ancillary benefits for other species;

Acknowledging that, despite past or ongoing scientific research, knowledge of the biology, ecology, and population dynamics of cetaceans is deficient, and that it is necessary to develop co-operation for research and monitoring of these species in order to fully implement conservation measures;

Acknowledging further that effective implementation of such an agreement will require that assistance be provided, in a spirit of solidarity, to some Range States for research, training, and monitoring of cetaceans and their habitats, as well as for the establishment or improvement of scientific and administrative institutions;

Recognizing the importance of other global and regional instruments of relevance to the conservation of cetaceans, signed by many Parties, such as the International Convention for the Regulation of Whaling, 1946; the Convention for the Protection of the Mediterranean Sea against Pollution, 1976, its related protocols and the Action Plan for the Conservation of Cetaceans in the Mediterranean Sea adopted under its auspices in 1991; the Convention on the Conservation of European Wildlife and Natural Habitats, 1979; the United Nations Convention on the Law of the Sea, 1982; the Convention on Biological Diversity, 1992; the Convention for the Protection of the Black Sea against Pollution, 1992; and the Global Plan of Action for the Conservation, Management and Utilization of Marine Mammals of the United Nations Environment Programme, adopted in 1984; as well as initiatives of *inter alia* the General Fisheries Council for Mediterranean, the International Commission for Scientific Exploration of the Mediterranean, and the International Commission for the Conservation of Atlantic Tunas,

Have agreed as follows:

Article I

Scope, Definitions and Interpretation

1. a)   The geographic scope of this Agreement, hereinafter referred to as the "Agreement area", is constituted by all the maritime waters of the Black Sea and theMediterranean and their gulfs and seas, and the internal waters connected to or interconnecting these maritime waters, and of the Atlantic area contiguous to the Mediterranean Sea west of the Straits of Gibraltar. For the purpose of this Agreement:

   — the Black Sea is bounded to the southwest by the line joining Capes Kelaga and Dalyan (Turkey);
   — the Mediterranean Sea is bounded to the east by the southern limits of the Straits of the Dardanelles between the lighthouses of Mehmetcik and Kumkale (Turkey) and to the west by the meridian passing through Cape Spartel lighthouse, at the entrance to the Strait of Gibraltar; and
   — the contiguous Atlantic area west of the Strait of Gibraltar is bounded to the east by the meridian passing through Cape Spartel lighthouse and to the west by the line joining the lighthouses of Cape St. Vicente (Portugal) and Casablanca (Morocco).

   b)   Nothing in this Agreement nor any act adopted on the basis of this Agreement shall prejudice the rights and obligations, the present and future claims or legal views of any State relating to the law of the sea or to the Montreux Convention of 20 July 1936 (*Convention concernant le régime des détroits*), in particular the nature and the extent of marine areas, the delimitation of marine areas between States with opposite or adjacent coasts, freedom of navigation on the high seas, the right and the modalities of passage through straits used for international navigation and the right of innocent passage in territorial seas, as well as the nature and extent of the jurisdiction of the coastal State, the flag State and the port State.

c)   No act or activity undertaken on the basis of this Agreement shall constitute grounds for claiming, contending or disputing any claim to national sovereignty or jurisdiction.

2. This Agreement applies to all cetaceans that have a range which lies entirely or partly within the Agreement area or that accidentally or occasionally frequent the Agreement area, an indicative list of which is contained in Annex 1 to this Agreement.

3. For the purpose of this Agreement:

a)   "Cetaceans" means animals, including individuals, of those species, subspecies or populations of *Odontoceti* or *Mysticeti*;

b)   "Convention" means the Convention on the Conservation of Migratory Species of Wild Animals, 1979;

c)   "Secretariat of the Convention" means the body established under Article IX of the Convention;

d)   "Agreement secretariat" means the body established under Article III, paragraph 7, of this Agreement;

e)   "Scientific Committee" means the body established under Article III, paragraph 7, of this Agreement;

f)   "Range" means all areas of water that a cetacean inhabits, stays in temporarily, or crosses at any time on its normal migration route within the Agreement area.

g)   "Range State" means any State that exercises sovereignty and/or jurisdiction over any part of the range of a cetacean population covered by this Agreement, or a State, flag vessels of which are engaged in activities in the Agreement area which may affect the conservation of cetaceans;

h)   "Regional economic integration organization" means an organization constituted by sovereign States which has competence in respect of the negotiation, conclusion and application of international agreements in matters covered by this Agreement;

i)   "Party" means a Range State or a regional economic integration organization for which this Agreement is in force;

j)   "Subregion", depending on the particular context, means either the region comprising the coastal States of Black Sea or the region comprising the coastal States of the Mediterranean Sea and the contiguous Atlantic area; any reference in the Agreement to the States of a particular subregion shall be taken to mean the

States which have any part of their territorial waters within that subregion, and States, flag vessels of which are engaged in activities which may affect the conservation of cetaceans in that subregion; and

k)   "Habitat" means any area in the range of cetaceans where they are temporarily or permanently resident, in particular, feeding areas, calving or breeding grounds, and migration routes.

In addition, the terms defined in Article I, subparagraphs 1 a) to e), and i) of the Convention shall have the same meaning, *mutatis mutandis*, in this Agreement.

4. This Agreement is an agreement within the meaning of Article IV, paragraph 4, of the Convention.

5. The annexes to this Agreement form an integral part thereof, and any reference to the Agreement includes a reference to its annexes.

Article II

Purpose and Conservation Measures

1. Parties shall take co-ordinated measures to achieve and maintain a favourable conservation status for cetaceans. To this end, Parties shall prohibit and take all necessary measures to eliminate, where this is not already done, any deliberate taking of cetaceans and shall co-operate to create and maintain a network of specially protected areas to conserve cetaceans.

2. Any Party may grant an exception to the prohibition set out in the preceding paragraph only in emergency situations as provided for in Annex 2, paragraph 6, or, after having obtained the advice of the Scientific Committee, for the purpose of non-lethal *in situ* research aimed at maintaining a favourable conservation status for cetaceans. The Party concerned shall immediately inform the Bureau and the Scientific Committee, through the Agreement secretariat, of any such exception that has been granted. The Agreement secretariat shall inform all Parties of the exception without delay by the most appropriate means.

3. In addition, Parties shall apply, within the limits of their sovereignty and/or jurisdiction and in accordance with their international obligations, the conservation, research and management measures

prescribed in Annex 2 to this Agreement, which shall address the following matters:

a)   adoption and enforcement of national legislation;
b)   assessment and management of human-cetacean interactions;
c)   habitat protection;
d)   research and monitoring;
e)   capacity building, collection and dissemination of information, training and education; and
f)   responses to emergency situations.

Measures concerning fisheries activities shall be applied in all waters under their sovereignty and/or jurisdiction and outside these waters in respect of any vessel under their flag or registered within their territory.

4. In implementing the measures prescribed above, the Parties shall apply the precautionary principle.

Article III

Meeting of the Parties

1. The Meeting of the Parties shall be the decision-making body of this Agreement.

2. The Depositary shall convene, in consultation with the Secretariat of the Convention, a session of the Meeting of the Parties to this Agreement not later than one year after the date of its entry into force. Thereafter, the Agreement secretariat shall convene, in consultation with the Secretariat of the Convention, ordinary sessions of the Meeting of the Parties at intervals of not more than three years, unless the Meeting of the Parties decides otherwise.

3. The Agreement secretariat shall convene an extraordinary session of the Meeting of the Parties on the written request of at least two thirds of the Parties.

4. The United Nations, its Specialized Agencies, the International Atomic Energy Agency, any State not a Party to this Agreement, secretariats of other global and regional conventions or agreements concerned *inter alia* with the conservation of cetaceans, and regional or subregional fisheries management organizations with competence for species found temporarily or permanently resident in the Agreement area may be rep-

resented by observers in sessions of the Meeting of the Parties. Any other agency or body technically qualified in the conservation of cetaceans may be represented at sessions of the Meeting of the Parties by observers, unless at least one third of the Parties present object. Once admitted to a session of the Meeting of the Parties, an observer shall continue to be entitled to participate in future sessions unless one third of the Parties object at least thirty days before the start of the session.

5. Only Parties have the right to vote. Each Party shall have one vote. Regional economic integration organizations which are Parties to this Agreement shall exercise, in matters within their competence, their right to vote with a number of votes equal to the number of their member States which are Parties to the Agreement. A regional economic integration organization shall not exercise its right to vote if its member States exercise theirs and vice versa.

6. All decisions of the Meeting of the Parties shall be adopted by consensus except as otherwise provided in Article X of this Agreement. However, if consensus cannot be achieved in respect of matters covered by the annexes to the Agreement, a decision may be adopted by a two thirds majority of the Parties present and voting. In the event of a vote, any Party may, within one hundred and fifty days, notify the Depositary in writing of its intention not to apply the said decision.

7. At its first session, the Meeting of the Parties shall:
    a)   adopt its rules of procedure;
    b)   establish an Agreement secretariat to perform the secretariat functions listed in Article IV of this Agreement;
    c)   designate in each subregion, within an existing institution, a Coordination unit to facilitate implementation of the measures prescribed in Annex 2 to this Agreement;
    d)   elect a Bureau as provided for in Article VI;
    e)   establish a Scientific Committee, as provided for in Article VII; and
    f)   decide on the format and content of Party reports on the implementation of the Agreement, as provided for in Article VIII.

8. At each of its ordinary sessions, the Meeting of the Parties shall:
    a)   review scientific assessments of the conservation status of cetaceans of the Agreement area and the habitats which are

important to their survival, as well as the factors which may affect them unfavourably;

b)   review the progress made and any difficulties encountered in the implementation of this Agreement on the basis of the reports of the Parties and of the Agreement secretariat;

c)   make recommendations to the Parties as it deems necessary or appropriate and adopt specific actions to improve the effectiveness of this Agreement;

d)   examine and decide upon any proposals to amend, as may be necessary, this Agreement;

e)   adopt a budget for the next financial period and decide upon any matters relating to the financial arrangements for this Agreement;

f)   review the arrangements for the Agreement secretariat, the Co-ordination units and the Scientific Committee;

g)   adopt a report for communication to the Parties to this Agreement and to the Conference of the Parties of the Convention;

h)   agree on the provisional time and venue of the next meeting; and

i)   deal with any other matter relating to implementation of this Agreement.

Article IV

Agreement Secretariat

1. Subject to the approval of the Conference of the Parties to the Convention, an Agreement secretariat shall be established within the Secretariat of the Convention. If the Secretariat of the Convention is unable, at any time, to provide this function, the Meeting of the Parties shall make alternative arrangements.

2. The functions of the Agreement secretariat shall be:

a)   to arrange and service the sessions of the Meeting of the Parties;

b)   to liaise with and facilitate co-operation between Parties and non-Party Range States, and international and national bodies whose activities are directly or indirectly relevant to the conservation of cetaceans in the Agreement area;

c)   to assist the Parties in the implementation of this Agreement, ensuring coherence between the subregions and with measures adopted pursuant to other international instruments in force;

d) to execute decisions addressed to it by the Meeting of the Parties;

e) to invite the attention of the Meeting of the Parties to any matter pertaining to this Agreement;

f) to provide to each ordinary session of the Meeting of the Parties a report on the work of the Agreement secretariat, the Co-ordination units, the Bureau, and the Scientific Committee, and on the implementation of the Agreement based on information provided by the Parties and other sources;

g) to administer the budget for this Agreement;

h) to provide information to the general public concerning this Agreement and its objectives; and

i) to perform any other function entrusted to it under this Agreement or by the Meeting of the Parties.

3. The Agreement secretariat, in consultation with the Scientific Committee and the Co-ordination units, shall facilitate the preparation of guidelines covering *inter alia:*

a) the reduction or elimination, as far as possible and for the purposes of this Agreement, of adverse human-cetacean interactions;

b) habitat protection and natural resource management methods as they relate to cetaceans;

c) emergency measures; and

d) rescue methods.

Article V

Co-ordination Units

1. The functions of the subregional Co-ordination units shall be:

a) to facilitate implementation in the respective subregions of the activities provided for in Annex 2 to this Agreement, in accordance with instructions of the Meeting of the Parties;

b) to collect and evaluate information that will further the objectives and implementation of the Agreement and provide for appropriate dissemination of such information; and

c) to service meetings of the Scientific Committee and to prepare a report for communication to the Meeting of the Parties through the Agreement secretariat.

The designation of the Co-ordination units and their functions shall be reviewed, as appropriate, at each session of the Meeting of the Parties.

2. Each Co-ordination unit, in consultation with the Scientific Committee and the Agreement secretariat, shall facilitate the preparation of a series of international reviews or publications, to be updated regularly, including:
  a) reports on the status and trends of populations, as well as gaps in scientific knowledge;
  b) a subregional directory of important areas for cetaceans; and
  c) a subregional directory of national authorities, research and rescue centres, scientists and non-governmental organizations concerned with cetaceans.

Article VI

Bureau

1. The Meeting of the Parties shall elect a Bureau consisting of the Chairperson and Vice-Chairpersons of the Meeting of the Parties, and shall adopt rules of procedure for the Bureau, as proposed by the Agreement secretariat. The Chairperson of the Scientific Committee shall be invited to participate as an observer in the meetings of the Bureau. Whenever necessary, the Agreement secretariat shall provide secretariat services.

2. The Bureau shall:
  a) provide general policy guidance and operational and financial direction to the Agreement secretariat and the Co-ordination units concerning the implementation and promotion of the Agreement;
  b) carry out, between sessions of the Meeting of the Parties, such interim activities on its behalf as may be necessary or assigned to it by the Meeting of the Parties; and
  c) represent the Parties *vis-à-vis* the Government(s) of the host country (or countries) of the Agreement secretariat and the Meeting of the Parties, the Depositary and other international organizations on matters relating to this Agreement and its secretariat.

3. At the request of its Chairperson, the Bureau shall normally meet once per annum at the invitation of the Agreement secretariat, which shall inform all Parties of the date, venue and agenda of such meetings.

4. The Bureau shall provide a report on its activities for each session of the Meeting of the Parties which will be circulated to all Parties in advance of the session by the Agreement secretariat.

Article VII

Scientific Committee

1. A Scientific Committee, comprising persons qualified as experts in cetacean conservation science, shall be established as an advisory body to the Meeting of the Parties. The Meeting of the Parties will entrust the functions of the Scientific Committee to an existing organization in the Agreement area that assures geographically-balanced representation.

2. Meetings of the Scientific Committee shall be convened by the Agreement secretariat at the request of the Meeting of the Parties.

3. The Scientific Committee shall:
    a) provide advice to the Meeting of the Parties on scientific and technical matters having a bearing on the implementation of the Agreement, and to individual Parties between sessions, as appropriate, through the Co-ordination unit of the subregion concerned;
    b) advise on the guidelines as provided for in Article IV, paragraph 3, assess the reviews prepared in accordance with Annex 2 to this Agreement and formulate recommendations to the Meeting of the Parties relating to their development, contents and implementation;
    c) conduct scientific assessments of the conservation status of cetacean populations;
    d) advise on the development and co-ordination of international research and monitoring programmes, and make recommendations to the Meeting of the Parties concerning further research to be carried out;
    e) facilitate the exchange of scientific information and of conservation techniques;
    f) prepare for each session of the Meeting of the Parties a report of its activities which shall be submitted to the Agreement secretariat not less than one hundred and twenty days before the

session of the Meeting of the Parties and circulated forthwith by the Agreement secretariat to all Parties;

g) render timely advice on the exceptions of which it has been informed pursuant to Article II, paragraph 2; and

h) carry out, as may be necessary, other tasks referred to it by the Meeting of the Parties.

4. The Scientific Committee, in consultation with the Bureau and the respective Co-ordination units, may establish working groups as may be necessary to deal with specific tasks. The Meeting of the Parties shall agree a fixed budget allocation for this purpose.

Article VIII

Communication and Reporting

Each Party shall:

a) designate a focal point for this Agreement, and shall communicate without delay the focal point's name, address and telecommunication numbers to the Agreement secretariat, for prompt circulation to the other Parties and to the Co-ordination units; and

b) prepare for each ordinary session of the Meeting of the Parties, beginning with the second session, a report on its implementation of the Agreement with particular reference to the conservation measures and scientific research and monitoring it has undertaken. The format of such reports shall be determined by the first session of the Meeting of the Parties and reviewed as may be necessary at any subsequent session. Each report shall be submitted to the Agreement secretariat not less than one hundred and twenty days before the opening of the session of the Meeting of the Parties for which it has been prepared, and copies shall be circulated forthwith to the other Parties by the Agreement secretariat.

Article IX

Financial Arrangements

1. The scale of contributions to the budget of this Agreement shall be determined by the Meeting of the Parties at its first session. No regional economic integration organization shall be required to contribute more than 2.5 per cent of the administrative costs.

2. Decisions relating to the budget and any changes to the scale of contributions that may be found necessary shall be adopted by the Meeting of the Parties by consensus.

3. The Meeting of the Parties may establish a supplementary conservation fund from voluntary contributions of Parties or from any other source in order to increase the funds available for monitoring, research, training and projects relating to the conservation of cetaceans.

4. Parties are also encouraged to provide technical and financial support on a bilateral or multilateral basis to assist Range States which are developing countries or countries with economies in transition to implement the provisions of this Agreement.

5. The Agreement secretariat shall undertake periodically a review of potential mechanisms for providing additional resources, including funds and technical assistance, for the implementation of this Agreement, and shall report its findings to the Meeting of the Parties.

Article X

Amendment of the Agreement

1. This Agreement may be amended at any ordinary or extraordinary session of the Meeting of the Parties.

2. Proposals for amendments to the Agreement may be made by any Party. The text of any proposed amendment and the reasons for it shall be communicated to the Agreement secretariat not less than one hundred and fifty days before the opening of the session. The Agreement secretariat shall transmit copies forthwith to the Parties. Any comments on the text by the Parties shall be communicated to the Agreement secretariat not less than sixty days before the opening of the session. The Secretariat shall communicate to the Parties, as soon as possible after the last day for submission of comments, all comments submitted by that day.

3. Any additional annex or any amendment to the Agreement other than an amendment to its annexes shall be adopted by a two thirds majority of the Parties present and voting and shall enter into force for those Parties which have accepted it on the thirtieth day after the date on which two thirds of the Parties to the Agreement at the date of the

adoption of the additional annex or amendment have deposited their instruments of acceptance with the Depositary. For any Party that deposits an instrument of acceptance after the date on which two thirds of the Parties have deposited their instruments of acceptance, the additional annex or amendment shall enter into force on the thirtieth day after the date on which it deposits its instrument of acceptance.

4. Any amendment to an annex to the Agreement shall be adopted by a two thirds majority of the Parties present and voting and shall enter into force for all Parties on the one hundred and fiftieth day after the date of its adoption by the Meeting of the Parties, except for Parties that have entered a reservation in accordance with paragraph 5 of this Article.

5. During the period of one hundred and fifty days provided for in paragraph 4 of this Article, any Party may by written notification to the Depositary enter a reservation with respect to an amendment to an annex to the Agreement. Such reservation may be withdrawn by written notification to the Depositary, and thereupon the amendment shall enter into force for that Party on the thirtieth day after the date of withdrawal of the reservation.

Article XI

Effect of this Agreement on Legislation and International Conventions

1. The provisions of this Agreement shall not affect the right of any Party to maintain or adopt more stringent measures for the conservation of cetaceans and their habitats, nor the rights or obligations of any Party deriving from any existing treaty, convention or agreement to which it is a party, except where the exercise of those rights and obligations would threaten the conservation of cetaceans.

2. Parties shall implement this Agreement consistently with their rights and obligations arising under the law of the sea.

Article XII

Settlement of Disputes

1. Any dispute which may arise between two or more Parties with respect to the interpretation or application of the provisions of this Agreement shall be subject to negotiation between the Parties involved

in the dispute, or to mediation or conciliation by a third party if this is acceptable to the Parties concerned.

2. If the dispute cannot be resolved in accordance with paragraph 1 of this Article, the Parties may by mutual consent submit the dispute to arbitration or judicial settlement. The Parties submitting the dispute shall be bound by the arbitral or judicial decision.

Article XIII

Signature, Ratification, Acceptance, Approval or Accession

1. This Agreement shall be open for signature by any Range State, whether or not areas under its jurisdiction lie within the Agreement area, or regional economic integration organization, at least one member of which is a Range State, either by:

   a)   signature without reservation in respect of ratification, acceptance or approval; or

   b)   signature with reservation in respect of ratification, acceptance or approval, followed by ratification, acceptance or approval.

2. This Agreement shall remain open for signature at Monaco until the date of its entry into force.

3. This Agreement shall be open for accession by any Range State or regional economic integration organization mentioned in paragraph 1, above, on and after the date of entry into force of the Agreement.

4. Instruments of ratification, acceptance, approval or accession shall be deposited with the Depositary.

Article XIV

Entry into Force

1. This Agreement shall enter into force on the first day of the third month following the date on which at least seven coastal States of the Agreement area or regional economic integration organizations, comprising at least two from the subregion of the Black Sea and at least five from the subregion of the Mediterranean Sea and contiguous Atlantic area, have signed without reservation in respect of ratification, acceptance or approval, or have deposited their instruments of ratification, acceptance or approval in accordance with Article XIII of this Agreement.

2. For any Range State or regional economic integration organization which has:
   a) signed without reservation in respect of ratification, acceptance, or approval;
   b) ratified, accepted, or approved; or
   c) acceded to this Agreement after the date on which the number of Range States and regional economic integration organizations necessary to enable entry into force have signed it without reservation or have ratified, accepted or approved it, this Agreement shall enter into force on the first day of the third month following the signature without reservation, or deposit, by that State or organization, of its instrument of ratification, acceptance, approval or accession.

Article XV

Reservations

The provisions of this Agreement shall not be subject to general reservations. However, a specific reservation may be entered by any State in respect of a specifically delimited part of its internal waters, on signature without reservation in respect of ratification, acceptance or approval or, as the case may be, on the deposit of its instrument of ratification, acceptance, approval or accession. Such a reservation may be withdrawn at any time by the State which had entered it by notification in writing to the Depositary; the State concerned shall not be bound by the application of the Agreement to the waters which are the object of the reservation until thirty days after the date on which the reservation has been withdrawn.

Article XVI

Denunciation

Any Party may denounce this Agreement at any time by written notification to the Depositary. The denunciation shall take effect twelve months after the date on which the Depositary has received the notification.

Article XVII

Depositary

1. The original of this Agreement, in the Arabic, English, French, Russian and Spanish languages, each version being equally authentic,

shall be deposited with the Government of the Principality of Monaco, which shall be the Depositary. The Depositary shall transmit certified copies of the Agreement to all States and regional economic integration organizations referred to in Article XIII, paragraph 1, of this Agreement, and to the Agreement secretariat after it has been established.

2. As soon as this Agreement enters into force, a certified copy thereof shall be transmitted by the Depositary to the Secretariat of the United Nations for registration and publication in accordance with Article 102 of the Charter of the United Nations.

3. The Depositary shall inform all States and regional economic integration organizations that have signed or acceded to the Agreement, and the Agreement secretariat, of:

    a)   any signature;

    b)   any deposit of an instrument of ratification, acceptance, approval or accession;

    c)   the date of entry into force of this Agreement and of any additional annex as well as of any amendment to the Agreement or to its annexes;

    d)   any reservation with respect to an additional annex or an amendment to an annex;

    e)   any notification of withdrawal of a reservation; and

    f)   any notification of denunciation of this Agreement.

The Depositary shall transmit to all States and regional economic integration organizations that have signed or acceded to this Agreement, and to the Agreement secretariat, the text of any reservation, any additional annex and any amendment to the Agreement or to its annexes.

In witness whereof the undersigned, being duly authorized to that effect, have signed this Agreement.

Done at Monaco on the twenty-fourth day of November 1996.

ANNEX 1

INDICATIVE LIST OF CETACEANS OF THE BLACK SEA TO WHICH THIS AGREEMENT APPLIES

PHOCOENIDAE

Phocoena phocoena Harbour porpoise

DELPHINIDAE
Tursiops truncatus Bottlenose dolphin
Delphinus delphis Common dolphin

INDICATIVE LIST OF CETACEANS OF THE MEDITERRANEAN SEA AND THE CONTIGUOUS ATLANTIC AREA TO WHICH THIS AGREEMENT APPLIES

PHOCOENIDAE
Phocoena phocoena Harbour porpoise

DELPHINIDAE
Steno bredanensis Rough-toothed dolphin

Grampus griseus Risso's dolphin
Tursiops truncatus Bottlenose dolphin
Stenella coeruleoalba Striped dolphin
Delphinus delphis Short-beaked common dolphin

Pseudorca crassidens False killer whale
Orcinus orca Killer whale
Globicephala melas Long-finned pilot whale

ZIPHIIDAE
Mesoplodon densirostris Blainville's beaked whale
Ziphius cavirostris Cuvier's beaked whale

PHYSETERIDAE
Physeter macrocephalus Sperm whale

KOGIIDAE
Kogia simus Dwarf sperm whale

BALAENIDAE
Eubalaena glacialis Northern right whale

BALAENOPTERIDAE
Balaenoptera acutorostrata Minke whale
Balaenoptera borealis Sei whale
Balaenoptera physalus Fin whale
Megaptera novaeangliae Humpback whale

The present Agreement shall also apply to any other cetaceans not already listed in this annex, but which may frequent the Agreement area accidentally or occasionally.

ANNEX 2

CONSERVATION PLAN

The Parties shall undertake, to the maximum extent of their economic, technical, and scientific capacities, the following measures for the conservation of cetaceans, giving priority to conserving those species or populations identified by the Scientific Committee as having the least favourable conservation status, and to undertaking research in areas or for species for which there is a paucity of data.

1. Adoption and enforcement of national legislation

Parties to this Agreement shall adopt the necessary legislative, regulatory or administrative measures to give full protection to cetaceans in waters under their sovereignty and/or jurisdiction and outside these waters in respect of any vessel under their flag or registered within their territory engaged in activities which may affect the conservation of cetaceans. To this end, Parties shall:

(a) develop and implement measures to minimize adverse effects of fisheries on the conservation status of cetaceans. In particular, no vessel shall be allowed to keep on board, or use for fishing, one or more drift nets whose individual or total length is more than 2.5 kilometres;

(b) introduce or amend regulations with a view to preventing fishing gear from being discarded or left adrift at sea, and to require the immediate release of cetaceans caught incidentally in fishing gear in conditions that assure their survival;

(c) require impact assessments to be carried out in order to provide a basis for either allowing or prohibiting the continuation or the future development of activities that may affect cetaceans or their habitat in the Agreement area, including fisheries, offshore exploration and exploitation, nautical sports, tourism and cetacean-watching, as well as establishing the conditions under which such activities may be conducted;

(d) regulate the discharge at sea of, and adopt within the framework of other appropriate legal instruments stricter standards for, pollutants believed to have adverse effects on cetaceans; and

(e) endeavour to strengthen or create national institutions with a view to furthering implementation of the Agreement.

2. Assessment and management of human-cetacean interactions

Parties shall, in co-operation with relevant international organizations, collect and analyse data on direct and indirect interactions between humans and cetaceans in relation to *inter alia* fishing, industrial and touristic activities, and land-based and maritime pollution. When necessary, Parties shall take appropriate remedial measures and shall develop guidelines and/or codes of conduct to regulate or manage such activities.

3. Habitat protection

Parties shall endeavour to establish and manage specially protected areas for cetaceans corresponding to the areas which serve as habitats of cetaceans and/or which provide important food resources for them. Such specially protected areas should be established within the framework of the Convention for the Protection of the Mediterranean Sea against Pollution, 1976, and its relevant protocol, or within the framework of other appropriate instruments.

4. Research and monitoring

Parties shall undertake co-ordinated, concerted research on cetaceans and facilitate the development of new techniques to enhance their conservation. Parties shall, in particular:

(a) monitor the status and trends of species covered by this Agreement, especially those in poorly known areas, or species for which little data are available, in order to facilitate the elaboration of conservation measures;

(b) co-operate to determine the migration routes and the breeding and feeding areas of the species covered by the Agreement in order to define areas where human activities may need to be regulated as a consequence;

(c) evaluate the feeding requirements of the species covered by the Agreement and adapt fishing regulations and techniques accordingly;

(d) develop systematic research programmes on dead, stranded, wounded or sick animals to determine the main interactions with human activities and to identify present and potential threats; and

(e) facilitate the development of passive acoustic techniques to monitor cetacean populations.

5. Capacity building, collection and dissemination of information, training and education

Taking into account the differing needs and the developmental stages of the Range States, Parties shall give priority to capacity building in order to develop the necessary expertise for the implementation of the Agreement. Parties shall co-operate to develop common tools for the collection and dissemination of information about cetaceans and to organize training courses and education programmes. Such actions shall be conducted in concert at the subregional and Agreement level, supported by the Agreement secretariat, the Co-ordination units and the Scientific Committee and carried out in collaboration with competent international institutions or organizations. The results shall be made available to all Parties. In particular, Parties shall co-operate to:

(a) develop the systems for collecting data on observations, incidental catches, strandings, epizootics and other phenomena related to cetaceans;

(b) prepare lists of national authorities, research and rescue centres, scientists and non-governmental organizations concerned with cetaceans;

(c) prepare a directory of existing protected or managed areas which could benefit the conservation of cetaceans and of marine areas of potential importance for the conservation of cetaceans;

(d) prepare a directory of national and international legislation concerning cetaceans;

(e) establish, as appropriate, a subregional or regional data bank for the storage of information collected under paragraphs a) to d) above;

(f) prepare a subregional or regional information bulletin on cetacean conservation activities or contribute to an existing publication serving the same purpose;

(g) prepare information, awareness and identification guides for distribution to users of the sea;

(h) prepare, on the basis of regional knowledge, a synthesis of veterinary recommendations for the rescue of cetaceans; and

(i) develop and implement training programmes on conservation techniques, in particular, on observation, release, transport and first aid techniques, and responses to emergency situations.

6. Responses to emergency situations

Parties shall, in co-operation with each other, and whenever possible and necessary, develop and implement emergency measures for cetaceans covered by this Agreement when exceptionally unfavourable or endangering conditions occur. In particular, Parties shall:

(a) prepare, in collaboration with competent bodies, emergency plans to be implemented in case of threats to cetaceans in the Agreement area, such as major pollution events, important strandings or epizootics; and

(b) evaluate capacities necessary for rescue operations for wounded or sick cetaceans; and

(c) prepare a code of conduct governing the function of centres or laboratories involved in this work.

In the event of an emergency situation requiring the adoption of immediate measures to avoid deterioration of the conservation status of one or more cetacean populations, a Party may request the relevant Co-ordination unit to advise the other Parties concerned, with a view to establishing a mechanism to give rapid protection to the population identified as being subject to a particularly adverse threat.

AGREEMENT on Cooperation in Research, Conservation and Management of Marine Mammals in the North Atlantic (1992)

The Parties,

IN PURSUANCE of the objectives laid down in the Memorandum of Understanding, signed at Tromsø on 19 April 1990, on cooperation between countries bordering the North Atlantic Ocean in research, conservation and management of marine mammals;

HAVING REGARD to their common concerns for the rational management, conservation and optimum utilization of the living resources of the sea in accordance with generally accepted principles of international law as reflected in the 1982 United Nations Convention on the Law of the Sea;

DESIRING to enhance their cooperation in research on marine mammals and their role in the ecosystem, including, where appropriate, multi-species approaches, and on the effects of marine pollution and other human activities;

BEARING IN MIND the need to develop management procedures which take into account the relationship between marine mammals and other marine living resources;

RECALLING the general principles of conservation and sustainable use of natural resources as reflected in the report of the World Commission on Environment and Development;

CONVINCED that regional bodies in the North Atlantic can ensure effective conservation, sustainable marine resource utilization and development with due regard to the needs of coastal communities and indigenous people;

have agreed as follows:

Article 1

There is hereby established an international organization that shall be known as the North Atlantic Marine Mammal Commission (NAMMCO).

Article 2

The objective of the Commission shall be to contribute through regional consultation and cooperation to the conservation, rational management and study of marine mammals in the North Atlantic.

Article 3

The Commission shall consist of:
  (a) a Council;
  (b) Management Committees;
  (c) a Scientific Committee;
  (d) a Secretariat.

Article 4

1. Each Party shall be a member of the Council.

2. The functions of the Council shall be:
  (a) to provide a forum for the study, analysis and exchange of infor-mation among the Parties on matters concerning marine mam-mals in the North Atlantic;
  (b) to establish appropriate Management Committees and coordi-nate their activities;
  (c) to establish guidelines and objectives for the work of the Management Committees;
  (d) to establish working arrangements with the International Council for the Exploration of the Sea and other appropriate organiza-tions;
  (e) to coordinate requests for scientific advice;
  (f) to establish cooperation with States not Parties to this Agreement in order to further the objective set out in Article 2.

3. Decisions of the Council shall be taken by the unanimous vote of those members present and casting an affirmative vote.

Article 5

1. Management Committees shall with respect to stocks of marine mammals within their respective mandates:
    (a) propose to their members measures for conservation and management;
    (b) make recommendations to the Council concerning scientific research.

2. Decisions of Management Committees shall be taken by the unanimous vote of those members present and casting an affirmative vote.

Article 6

1. The Scientific Committee shall consist of experts appointed by the Parties.

2. Subject to the approval of the Council, the Scientific Committee may invite other experts to participate in the conduct of its work.

3. The Scientific Committee shall provide scientific advice in response to requests from the Council, utilizing, to the extent possible, existing scientific information.

Article 7

1. The Council shall establish a Secretariat.

2. The Secretariat shall perform such functions as the Council may determine.

Article 8

The Council may agree to admit observers to meetings of the Commission when such admission is consistent with the objective set out in Article 2.

Article 9

This Agreement is without prejudice to obligations of the Parties under other international agreements.

Article 10

1. This Agreement shall be open for signature on 9 April 1992 by the Faroe Islands, Greenland, Iceland and Norway, and shall enter into force 90 days after signature.

2. It shall remain open for signature by other Parties with the consent of the existing Signatories.

3. Any Party may withdraw from this Agreement upon giving six months' notice.

Done at Nuuk on 9 April 1992.

Index